MODERN
ELEMENTARY
STATISTICS

MODERN
ELEMENTARY
STATISTICS

fourth edition

JOHN E. FREUND

Professor of Mathematics
Arizona State University

PRENTICE-HALL, INC.

Englewood Cliffs, New Jersey

Library of Congress Cataloging in Publication Data

FREUND, JOHN E.
  Modern elementary statistics.

  Includes bibliographies.
  1. Statistics. I. Title.
HA29.F685 1973   519.5   72-13874
ISBN 0-13-593475-3

© 1973, 1967, 1960, 1952
by Prentice-Hall, Inc.
Englewood Cliffs, New Jersey

10   9   8   7   6   5   4   3

Printed in the United States of America

PRENTICE-HALL INTERNATIONAL, INC., *London*
PRENTICE-HALL OF AUSTRALIA, PTY. LTD., *Sydney*
PRENTICE-HALL OF CANADA, LTD., *Toronto*
PRENTICE-HALL OF INDIA PRIVATE LIMITED, *New Delhi*
PRENTICE-HALL OF JAPAN, INC., *Tokyo*

Contents

Preface   xi

1 Introduction   1

   1.1   Modern Statistics   1
   1.2   Descriptive Statistics   2
   1.3   Statistical Inference   3

PART 1   Descriptive Methods

2 Frequency Distributions   9

   2.1   Introduction   9
   2.2   Frequency Distributions   11
   2.3   Graphical Presentations   16

3 Measures of Location   31

   3.1   Introduction   31
   3.2   The Mean   33
   3.3   The Weighted Mean   39
   3.4   The Median and Other Fractiles   47
   3.5   Further Measures of Central Location 53
   3.6   Technical Note (Subscripts and Summations) 58

4 Measures of Variation and Further Descriptions   63

   4.1   Introduction   63
   4.2   The Standard Deviation   66

4.3   Measures of Relative Variation   74
4.4   Some Further Descriptions   79

PART 2   Basic Theory

5 Possibilities and Probabilities   87

5.1   Introduction   87
5.2   Counting   87
5.3   Permutations and Combinations   95
5.4   Probabilities and Odds   102

6 Rules of Probability   112

6.1   Introduction   112
6.2   Sample Spaces and Events   113
6.3   Rules of Probability   124
6.4   Conditional Probability   134
6.5   Bayes' Rule   140

7 Expectations and Decisions   150

7.1   Introduction   150
7.2   Mathematical Expectation   150
7.3   Decision Making   157
7.4   Statistical Decision Problems   165

8 Probability Distributions   170

8.1   Introduction   170
8.2   Probability Functions   172
8.3   The Binomial Distribution   173
8.4   The Hypergeometric Distribution   177
8.5   The Poisson Distribution   180
8.6   The Mean of a Probability Distribution   187
8.7   The Variance of a Probability Distribution   189
8.8   Continuous Distributions   195
8.9   The Normal Distribution   198
8.10 Applications of the Normal Distribution   210
8.11 The Normal Approximation of the
      Binomial Distribution   213

9 Sampling and Sampling Distributions   219

9.1   Random Sampling   219
9.2   Sampling Distributions   225
9.3   The Central Limit Theorem   232
9.4   Technical Note (Simulation)   238

PART 3   Statistical Inference

10 Inferences Concerning Means   247

10.1   Introduction   247
10.2   Problems of Estimation   248
10.3   The Estimation of Means   249
10.4   A Bayesian Estimate   256
10.5   Hypothesis Testing: Two Kinds of Errors   262
10.6   Hypothesis Testing: Null Hypotheses and
        Significance Tests   268
10.7   Tests Concerning Means   275
10.8   Differences Between Means   279

11 Inferences Concerning Standard Deviations   288

11.1   The Estimation of σ 288
11.2   Tests Concerning Standard Deviations   293

12 Inferences Concerning Proportions   299

12.1   Introduction   299
12.2   The Estimation of Proportions   299
12.3   A Bayesian Estimate   306
12.4   Tests Concerning Proportions   312
12.5   Differences Between Proportions   317
12.6   Differences Among k Proportions   320
12.7   Contingency Tables   325
12.8   Tests of Goodness of Fit   329

13 Analysis of Variance   337

13.1   Differences Among k Means   337
13.2   One-Way Analysis of Variance   340
13.3   Two-Way Analysis of Variance   348
13.4   Two-Way Analysis Without Interaction   349

13.5   Two-Way Analysis With Interaction   352
13.6   Some Further Considerations   355

14   Nonparametric Methods   362

14.1   Introduction   362
14.2   The One-Sample Sign Test   363
14.3   The Two-Sample Sign Test   365
14.4   The Two-Sample Median Test   367
14.5   The k-Sample Median Test   368
14.6   Rank-Sum Tests: The Mann-Whitney Test   373
14.7   Rank-Sum Tests: The Kruskal-Wallis Test   376
14.8   Tests of Randomness: Runs   379
14.9   Tests of Randomness: Runs Above and
       Below the Median   382
14.10 Some Further Considerations   383

15   Regression   386

15.1   Introduction   386
15.2   Linear Regression   389
15.3   Non-Linear Regression   400
15.4   Multiple Regression   406
15.5   Regression Analysis   412

16   Correlation   420

16.1   The Coefficient of Correlation   420
16.2   The Interpretation of r 425
16.3   Correlation Analysis   426
16.4   Rank Correlation   433
16.5   Multiple and Partial Correlation   435
16.6   Technical Note (The Calculation of r for
       Grouped Data)   440

17   Planning Surveys and Experiments   448

17.1   Introduction   448
17.2   Sample Designs   448
17.3   Systematic Sampling   449
17.4   Stratified Sampling   449

17.5   Cluster Sampling   453
17.6   Quota Sampling   454
17.7   The Design of Experiments   457
17.8   Latin Squares   459
17.9   Some Further Considerations   464

Statistical Tables   471

Answers to Odd-Numbered Exercises   507

Index   525

Preface

The objectives of this book are like those of the previous editions: *to acquaint beginning students in the biological, social, and physical sciences with the fundamentals of modern statistics.* Although the basic organization has remained unchanged, there are substantial changes, consisting mainly of additions to some of the later chapters. The great majority of the illustrations and exercises are new, and, as in previous editions, they are distributed over a great variety of applications.

The material on probability has been divided into two chapters, consisting of an informal treatment in Chapter 5 and a more formal treatment in Chapter 6 (which may be covered lightly or thoroughly depending on one's interests). The other major change in Part 2 is the more detailed treatment of the binomial, hypergeometric, and Poisson distributions, and the inclusion of binomial tables.

So far as Part 3 is concerned, the material on Bayesian estimation has been reorganized and made somewhat easier, small-sample tests concerning proportions are based directly on binomial tables, the treatment of analysis of variance has been expanded to include the two-way analysis with interaction, several new nonparametric tests have been added, there are new sections on non-linear and multiple regression, also a more detailed treatment of regression and correlation analysis, and a new chapter, Chapter 17, dealing partly with sample designs and partly with the design of experiments.

Mathematical proofs and derivations are still keyed to the lowest level at which, in the opinion of the author, modern statistics can effectively be taught. Since the mathematical training assumed of the reader is a knowledge of arithmetic and some algebra, many theorems (for example, those relating to sampling distributions) are given without proof but with suitable references in annotated bibliographies.

The author would like to express his appreciation to his many colleagues and students whose helpful suggestions and criticisms contributed greatly to previous editions of this text as well as to this fourth edition. In particular, he would like to thank his nephew Ray for checking the calculations in the text and working out the answers to the exercises, and his son Doug for helping with the proofreading. The author would also like to express his appreciation to the staff of Prentice-Hall, Inc., for their courteous cooperation in the production of this book and, above all, to his wife for her cheerful encouragement in spite of the demands made by this project on her husband's time.

Finally, the author is indebted to the Literary Executor of the late Sir Ronald Fisher, F. R. S., Cambridge, and to Oliver and Boyd Ltd., for permission to reprint part of Table IV from their book *Statistical Methods for Research Workers;* and to Professor E. S. Pearson and the *Biometrika* trustees for permission to reproduce parts of Tables 8, 18, and 41 from their *Biometrika Tables for Statisticians.*

Scottsdale, Arizona   JOHN E. FREUND

1

Introduction

1.1 Modern Statistics

Students in the biological, physical, and social sciences often face the study of statistics with mixed emotions; they realize that advanced work in their fields requires some understanding of statistics, but they may also remember some of the difficulties they have experienced previously in the study of mathematics. They are correct on both counts: statistics *is* a branch of applied mathematics, and the claim that an understanding of statistics is needed in the study of science is not an exaggeration. In fact, the growth of statistics has been such in recent years that it has made itself felt in almost every phase of human activity. Statistics no longer consists merely of the collection and presentation of data in tables and charts; it encompasses—indeed, it constitutes—the science of decision making in the face of uncertainty, and we meet uncertainties when we toss a coin, when we experiment with a new drug, when we try to predict an election or the outcome of a game, when we dig for oil, when we decide which of two production processes is more efficient, and so on. It would be presumptuous to say that statistics, in its present state of development, can handle *all* situations involving uncertainties and risks, but new methods are constantly being developed and it can be said that modern statistics, at least, provides the framework for looking at these situations in a logical and systematic way.

There can be no doubt that it is virtually impossible to understand a good part of the work done in the various sciences without having at least a speaking acquaintance with the subject of statistics. Numerical data derived from surveys and experiments constitute the raw material on which interpretations, analyses, and decisions are based, and it is essential to know how to squeeze usable information from such data. This, in fact, is the major objective of statistics.

1

Numerous textbooks have been written on educational statistics, business statistics, psychological statistics, medical statistics, and so forth, and it is true, of course, that these various fields of scientific inquiry demand somewhat different and specialized techniques. Nevertheless, the fundamental principles underlying these techniques are the same regardless of the field of application, and it is hoped that this will become apparent to the reader once he realizes that *statistical concepts and statistical methods are really nothing but a refinement of everyday thinking.*

The approach we shall use in this elementary text is, indeed, keynoted by the above statement. It is our goal to introduce the reader to the ideas and concepts that are basic to modern statistics, and we hope that this, in turn, will enable him to gain a better understanding of scientific principles in general and a clearer picture of the scope and limitations of scientific knowledge.

As we said earlier, the study of statistics may be directed toward applications in various fields of inquiry; furthermore, the subject may be presented in many different degrees of mathematical difficulty and in almost any balance between theory and application. As it is more important, in our opinion, to understand the meaning and implications of basic ideas than it is to memorize an impressive list of formulas, we shall sacrifice some of the details that are sometimes included in introductory courses in statistics. This may be unfortunate in some respects, but it will prevent us from getting lost in an excessive amount of detail which could easily obscure the more important issues. It is hoped that this will avoid some of the unfortunate consequences which often result from the indiscriminate application of so-called standard techniques without a thorough understanding of the basic ideas that are involved.

It cannot be denied that a limited amount of mathematics is a prerequisite for any course in statistics, and that a thorough study of the theoretical principles of statistics would require a knowledge of mathematical subjects taught ordinarily only on the graduate level. Since this book is designed for students with relatively little background in mathematics, our aims, and, therefore, also our prerequisites are considerably more modest. Actually, the mathematical background needed for this study of statistics is amply covered in a course in college algebra; in fact, even a good knowledge of high school algebra provides a sufficient foundation.

1.2 Descriptive Statistics

Everything dealing even remotely with the collection, processing, analysis, interpretation, and presentation of numerical data belongs to the domain of statistics. This includes such diversified tasks as calculating a baseball player's batting average, the collection and presentation of data on marriages and divorces, the evaluation of the reliability of missile components, and even the study of laws governing the behavior of neutrons and electrons. The word "statistics" itself is used in a variety of ways. In the plural it denotes a collection of numerical data such as those found in the financial pages of newspapers or, say, the *Statistical Abstract of the United States*, published each year by the Department of Commerce. A second meaning of "statistics," also in the plural, is that of the totality of methods employed in the collection, processing, analyzing, etc., of any kind of data; in this sense, statistics is a branch of applied mathematics, and it is this field of mathematics which is the subject matter of this book. To complete our linguistic analysis of "statistics," let us also mention that the word "statistic," in the singular, is used to denote a numerical description (such as an average, an index number, or a correlation coefficient) calculated on the basis of sample data.

The origin of statistics may be traced to two areas of interest which are very dissimilar: *games of chance* and what we now call *political science*. Mideighteenth century studies in probability (motivated in part by interest in games of chance) led to the mathematical treatment of errors of measurement and the theory which now forms the foundation of statistics. In the same century, interest in the description and analysis of political units led to the development of methods which nowadays come under the heading of *descriptive statistics*. This includes any treatment designed to summarize, or describe, important features of a set of data *without going any further;* that is, without attempting to infer anything that pertains to *more* than the data themselves. Thus, if someone compiles the necessary data and reports that 2,592 students were enrolled in a certain private university for the academic year 1969–70 and that they spent a total of $2,915,618 on tuition and fees, his work belongs to the domain of descriptive statistics. This is also the case if he determines the average amount one of these students spent that year on tuition and fees, namely, $1,124.85, but *not* if he uses the data to predict the university's future enrollment, its future income from tuition and fees, or to make other kinds of predictions.

1.3 Statistical Inference

Although descriptive statistics is an important branch of statistics and it continues to be widely used, statistical information usually arises from samples (from observations made on only part of a large set of items), and this means that its analysis will require generalizations which go beyond the data. As a result, the most important feature of the recent growth of statistics has been a shift in emphasis from methods which merely describe to methods which serve to make generalizations; that is, a shift in emphasis from descriptive statistics to the methods of *statistical inference*, or *inductive statistics*. To mention but a few examples, the methods of statistical inference are required to predict the operating life span of a record player's needle (on the basis of the performance of similar needles); to estimate the 1980 assessed value of all property in Pima County, Arizona (on the basis of business trends, population projections, and so forth); to compare the effectiveness of two or more "teaching machines" (on the basis of the performance of samples of students); to determine the most effective dosage of a new antibiotic (on the basis of experiments conducted with volunteer patients from selected hospitals); or to predict the flow of traffic on a freeway which has not yet been built. In each of these examples, there are uncertainties because there is only partial, incomplete, or indirect information, and it is the job of statistics to judge the merits of all possible alternatives and, perhaps, suggest a "most profitable" choice, a "most promising" prediction, or a "most reasonable" course of action.

All this would be impossible unless statistics also concerned itself with the question of how data are obtained and how experiments are conducted. As elsewhere, we get "nothing for free," and unless great care is exercised in all phases of an investigation, it may be impossible to reach any valid (or useful) conclusions whatsoever. Generally speaking, *no amount of fancy mathematics or statistical manipulation can salvage poorly planned studies, surveys, or experiments*. A classical example of this is the following "experiment" conducted to judge the merits of a new remedy for seasickness: All the passengers of one ocean liner were given the new remedy while all the passengers on another ocean liner were given a placebo containing nothing but sugar (since there might conceivably be a psychological effect from the mere knowledge of having taken a new kind of pill, or even the knowledge of being part of an "experiment"). Things turned out beautifully—hardly any of the passengers on the first ship got seasick, most of the passengers of the other one did, but, alas, the second ship had run into a severe storm while the first one's sailing had been as smooth as silk.

To avoid situations like this, statistics must encompass all questions concerning the collection of data including the design (or planning) of all phases of a survey or an experiment. Only then will it be possible to distinguish between "good" statistics and "bad" statistics, namely, between statistical techniques that are correctly (and usually profitably) applied, and those that are unintentionally or intentionally perverted.

Some examples of the *biases* that can arise in the collection of data are given in Exercise 3 below.

EXERCISES

1. In three chemistry tests Jim received grades of 84, 27, and 90, while Paul received grades of 65, 67, and 81. Which of the following conclusions can be obtained from these figures by means of purely descriptive methods and which require a statistical inference, namely, a generalization? Explain your answers.
    (a) Jim's grades average 67 while Paul's grades average 71.
    (b) Paul is a better student than Jim.
    (c) Jim was probably ill on the day he took the second test.
    (d) If the instructor discards each student's lowest grade, Jim's average is higher than that of Paul.
    (e) Paul's grades improved from each test to the next.
    (f) Paul studied harder for each successive test.
    (g) Paul's grades on the three tests are spread over a narrower interval of values than Jim's.
    (h) In the next test Paul will probably do better than Jim.

2. A technician working for a consumers' rating service found that four Brand A mattresses lasted, respectively, 763, 655, 702, and 528 hours of continuous torture tests, while four Brand B mattresses lasted, respectively, 752, 887, 713, and 256 hours. Which of the following conclusions can be obtained from these figures by purely descriptive methods and which require a statistical inference, namely, a generalization? Explain your answers.
    (a) The mattress which lasted the longest was of Brand B.
    (b) The four mattresses of Brand A lasted on the average 662 hours while those of Brand B lasted on the average 652 hours.
    (c) The difference between the two averages is less than 15 hours.
    (d) The difference between the two averages is so small that it is impossible to judge whether Brand A is really better than Brand B.
    (e) Probably, the fourth figure for Brand B was recorded incorrectly and should have been 756 instead of 256.
    (f) If the fourth figure for Brand B has been 756 instead of 256, the four mattresses of Brand B would have lasted on the average 777 hours.
    (g) Brand B mattresses have a better chance than Brand A mattresses of surviving the torture test for 700 hours.

3. In all studies based on samples, great care must be exercised to ensure that the samples will lend themselves to valid generalizations, and every precaution must be taken to avoid *biases* of one kind or another. This includes the unfortunate tendency of samples not to be representative

of whatever they are supposed to represent and, thus, lead away from, rather than toward, the truth. Explain why each of the following samples will not yield the desired information:

- (a) In order to predict a municipal election, a public opinion poll telephones persons selected haphazardly from the city's telephone directory.
- (b) To determine the proportion of improperly sealed cans of coffee, a quality control inspector examines every 50th can coming off an assembly line.
- (c) To estimate the average annual income of Princeton graduates ten years after graduation, questionnaires were sent in 1972 to all members of the class of '62 and the estimate was based on the ones returned.
- (d) To ascertain facts about tooth-brushing habits, a sample of the residents of a community are asked how many times they brush their teeth each day.
- (e) To study executives' reaction to its copying machines, the Xerox corporation hires a research organization to ask executives the question, "How do you like using Xerox copies?"
- (f) A house-to-house survey is made to study consumer reaction to a new pudding mix, with no provisions for return visits in case no one is at home.

BIBLIOGRAPHY

A brief and informal discussion of *what statistics is* and *what statisticians do* may be found in a pamphlet titled *Careers in Statistics*, published by the American Statistical Association. It can be obtained by writing to this organization at 806 15th Street, N.W., Washington, D.C., 20005.

PART 1
Descriptive Methods

2

Frequency Distributions

2.1 Introduction

Grouping, classifying, and thus describing measurements and observations is as basic in statistics as it is in science and in many activities of everyday life. To illustrate its importance in statistics, let us consider the problem of an economist who wants to study the size of farms in the United States. Not even giving a thought to the possibility of conducting a survey of his own, since the expense would be staggering, he immediately turns to one of the many organizations that specialize in the gathering of statistical data, namely, the U.S. Department of Commerce. This department not only provides government agencies with statistical data needed for over-all planning and day-by-day operations, but it also makes this information available to businessmen and research workers in various fields. Like other organizations engaged in gathering statistical data, it thus faces the problem of *how to present* the results of its surveys in the most effective and the most usable form. With reference to the information needed by the above-mentioned economist, the Department of Commerce *could* print sheets containing millions of numbers, the actual sizes of all farms in the United States; it is needless to say, however, that this would not be very effective and, without some treatment, not very "usable."

When dealing with large sets of numbers, a good over-all picture and sufficient information can often be conveyed by grouping the data into a number of classes, and the Department of Commerce could, and in fact does, publish its data on the size of farms in tables like the following:

9

Size of Farms in 1964 (acres)	Number of Farms (thousands)
Under 10	183
10–49	637
50–99	542
100–179	633
180–259	355
260–499	451
500–999	210
1,000 and over	145
Total	3,156

This kind of table is called a *frequency distribution* (or simply a *distribution*): It shows the frequencies with which the farm sizes are distributed among the chosen classes. Tables of this sort, in which the data are grouped according to numerical size, are called *numerical* or *quantitative* distributions. In contrast, tables like the one given below, in which the data are sorted according to certain categories, are called *categorical* or *qualitative* distributions:

	1967 Motor Vehicle Registration (thousands)
United States	96,945
Other North and Central America	8,900
South America	5,490
Europe	65,969
Africa	3,822
Asia	13,937
Oceania	5,519

Although frequency distributions present data in a relatively compact form, give a good over-all picture, and contain information which is adequate for many purposes, there are evidently some things which can be obtained from the original data that cannot be obtained from a distribution. For instance, referring to the first of the above tables, we cannot find the exact size of the smallest and largest farms, nor can we find the exact average size of the 542,000 farms in the 50–99 acre group. Nevertheless, frequency distributions present *raw* (unprocessed) data in a more usable form, and the price which we must pay, the loss of certain information, is usually a fair exchange.

Data are sometimes grouped solely to facilitate the calculation of further statistical descriptions. We shall go into this briefly in Chapters 3 and 4, but it is worth noting that this function of frequency distributions is diminishing in importance in view of the ever-increasing availability of high-speed electronic computers.

2.2 Frequency Distributions

The construction of a numerical distribution consists essentially of three steps: (1) we must choose the classes into which the data are to be grouped, (2) we must sort (or tally) the data into the appropriate classes, and (3) we must count the number of items in each class. Since the last two of these steps are purely mechanical, we shall concentrate on the first, namely, the problem of choosing suitable classifications. Note that if the data are recorded on punch-cards or tape, methods that are nowadays widely used, the sorting and counting can be done automatically in a single step.

The two things we shall have to consider in the first step are those of determining the *number of classes* into which the data are to be grouped and the *range of values* each class is to cover, that is, "from where to where" each class is to go. Both of these choices are largely arbitrary, but they depend to some extent on the nature of the data and on the ultimate purpose the distribution is to serve. The following are some rules which are generally observed:

(a) *We seldom use fewer than 6 or more than 15 classes.* This rule reflects sound practice based on experience; in any given example, the actual choice will have to depend on the number of observations we want to group (we would hardly group 5 observations into 12 classes), and on their range.

(b) *We always choose classes which will accommodate all the data.* To this end we must make sure that the smallest and largest values fall within the classification, and that none of the values can fall into possible gaps between successive classes.

(c) *We always make sure that each item goes into only one class.* In other words, we must avoid successive classes which overlap, that is, successive classes having one or more values in common.

(d) *Whenever possible, we make the class intervals of equal length, that is, we make them cover equal ranges of values.* It is generally desirable to make these ranges (intervals) multiples of 5, 10, 100, etc., or other numbers that are easy to work with, to facilitate the tally (perhaps, mechanically) and the ultimate use of the table.

Note that the first three, but not the fourth, of these rules were observed in the construction of the farm-size distribution on page 10,
assuming that the figures were rounded to the nearest acre. (Had these
figures been rounded to the nearest tenth of an acre, a farm of, say, 49.6
acres could not have been accommodated, as it would have fallen between
the second class and the third.) The fourth rule was violated in two
ways: First, the intervals from 10 to 49 acres, 100 to 179 acres, and 260
to 499 acres, among others, cover unequal ranges of values. Second, the
first and last classes are *open*—for all we know, the last class might include
farms of a million acres or more, and if we had grouped profits and losses
instead of acreages, the first class might even have included negative
values. If a set of data contains a few values that are much greater (or
much smaller) than the rest, open classes can help to simplify the over-all
picture by reducing the number of required classes; otherwise, open
classes should be avoided as they can make it impossible (or at least
difficult) to give further descriptions of the data.

As we have pointed out in the preceding paragraph, the appropriate-
ness of a classification may depend on whether the data are rounded to
the nearest acre or to the nearest tenth of an acre. Similarly, it may depend on whether data are rounded to the nearest dollar or the nearest
cent, whether they are given to the nearest inch, the nearest tenth of an
inch, or the nearest hundredth of an inch, and so on. Thus, if we wanted
to group the amounts of the sales made by a saleslady in a department
store, we might use the classification

Size of Sale
(dollars)

0.00– 4.99
5.00– 9.99
10.00–14.99
15.00–19.99
20.00–24.99
etc.

and if we wanted to group the heights of children measured to the nearest
tenth of an inch, we might use the classification

Height
(inches)

20.0–29.9
30.0–39.9
40.0–49.9
50.0–59.9
etc.

Similarly, for the number of empty seats on a bus from Phoenix to El Paso, we might use the classification

Number of
Empty Seats

0– 4
5– 9
10–14
15–19
etc.

Note that in each of these examples the nature of the data is such that a value can fall into one and only one class.

To give a concrete illustration of the construction of a frequency distribution, let us consider the following data representing the scores which 150 applicants for secretarial positions in a large company obtained in an achievement test:

27	79	69	40	51	88	55	48	36	61
53	44	94	51	65	42	58	55	69	63
70	48	61	55	60	25	47	78	61	54
57	76	73	62	36	67	40	51	59	68
27	46	62	43	54	83	59	13	72	57
82	45	54	52	71	53	82	69	60	35
41	65	62	75	60	42	55	34	49	45
49	64	40	61	73	44	59	46	71	86
43	69	54	31	56	51	75	44	66	53
80	71	53	56	91	60	41	29	56	57
35	54	43	39	56	27	62	44	85	61
59	89	60	51	71	53	58	26	77	68
62	57	48	69	76	52	49	45	54	41
33	61	80	57	42	45	59	44	68	73
55	70	39	58	69	51	85	46	55	67

Since the smallest of these scores is 13 and the largest is 94, it would seem reasonable (for most practical purposes) to choose the *nine* classes going from 10 to 19, from 20 to 29, ..., and from 90 to 99. Performing the actual tally and counting the number of values falling into each class, we obtain the results shown in the table at the top of page 14. The numbers shown in the right-hand column of this table are called *class frequencies;* they give the number of items falling into each class. Also, the smallest and the largest values that can go into any given class are referred to as its *class limits;* thus, the class limits of the above table are 10 and 19, 20 and 29, 30 and 39, and so on. More specifically, 10, 20, 30, ..., and 90 are referred to as the *lower class limits*, while 19, 29,

Scores	Tally	Frequency
10–19	/	1
20–29	ℳ /	6
30–39	ℳ ////	9
40–49	ℳ ℳ ℳ ℳ ℳ ℳ /	31
50–59	ℳ ℳ ℳ ℳ ℳ ℳ ℳ ℳ //	42
60–69	ℳ ℳ ℳ ℳ ℳ ℳ //	32
70–79	ℳ ℳ ℳ //	17
80–89	ℳ ℳ	10
90–99	//	2
	Total	150

39, ..., and 99 are referred to as the *upper class limits* of the respective classes.

If we are dealing with figures rounded to the nearest whole number, as in the size-of-farms distribution on page 10, the class which has the limits 10 and 49 actually contains all values between 9.5 and 49.5. Similarly, if we are dealing with measurements rounded to the nearest tenth of an inch, as in the height distribution on page 12, the class which has the limits 30.0 and 39.9 actually contains all values between 29.95 and 39.95, and the class which has the limits 40.0 and 49.9 actually contains all values between 39.95 and 49.95. It is customary to refer to these dividing lines between successive classes as the *class boundaries*, although they are sometimes referred to instead as the *"real" class limits*. In order to make this concept apply also to the classes which are at the two extremes of a distribution, we simply act as if the table were continued in both directions. Thus, the first class of the above distribution of the 150 scores has the lower boundary 9.5, while the last class has the upper boundary 99.5.

It is important to remember that class boundaries should always be "impossible" values, namely, numbers which cannot occur among the values we want to group. We make sure of this by accounting for the extent to which the numbers are rounded when we choose appropriate classifications. For instance, the class boundaries of the size-of-sales distribution on page 12 are -0.005, 4.995, 9.995, 14.995, and so on. Similarly, for the distribution of the scores, the class boundaries are 9.5, 19.5, 29.5, ..., and 99.5, while the figures themselves are, of course, whole numbers. Had there been scores less than 10 in this example, we would have begun the table with the class 0–9, whose boundaries are -0.5 and 9.5.

Two other terms used in connection with frequency distributions are "class mark" and "class interval." A *class mark* is simply the mid-point of a class, and it is obtained by averaging the class limits (or boundaries), that is, by dividing their sum by 2. Thus, the class marks of the distribution of the scores are 14.5, 24.5, 34.5, ..., and 94.5, while those of the size-of-sales distribution on page 12 are 2.495, 7.495, 12.495, and so on. A *class interval* is merely the length of a class (the range of values it can contain), and it is given by the difference between its class boundaries. If the classes of a distribution are all equal in length, their common class interval (which we refer to as the *class interval of the distribution*) is also given by the difference between any two successive class marks. Since $19.5 - 9.5 = 10$, $29.5 - 19.5 = 10$, ..., and $99.5 - 89.5 = 10$, the distribution of the scores has class intervals of length 10, and we say that this is the class interval of the distribution. Note that the class interval is *not* given by the difference between the respective upper and lower class limits, which in our example would equal 9, and not 10.

Suppose now that in connection with the scores of the 150 applicants for secretarial positions, it is of interest to know how many fell below various levels. To provide this information, we have only to convert the distribution on page 14 into what is called a *cumulative frequency distribution* or simply a *cumulative distribution*. Successively adding the frequencies in the table, we thus obtain the following *"less than" cumulative distribution:*

Scores	Cumulative Frequencies
Less than 10	0
Less than 20	1
Less than 30	7
Less than 40	16
Less than 50	47
Less than 60	89
Less than 70	121
Less than 80	138
Less than 90	148
Less than 100	150

Note that in this table we could just as well have written "9 or less" instead of "less than 10," "19 or less" instead of "less than 20," ..., and "99 or less" instead of "less than 100."

If we successively add the frequencies starting at the other end of the distribution, we similarly get a *cumulative "or more" distribution* (or a *cumulative "more than" distribution*), which shows how many of the scores are "10 or more" (or "more than 9"), how many are "20 or more" (or "more than 19"), and so on.

Sometimes it is preferable to show what *percentage* of the items falls into each class, or what *percentage* of the items falls above or below various values. To convert a frequency distribution (or a cumulative distribution) into a corresponding *percentage distribution,* we have only to divide each class frequency (or each cumulative frequency) by the total number of items grouped and multiply by 100. For instance, for the size-of-farm distribution on page 10, it may be more informative to indicate that $\frac{183}{3,156}\cdot 100 = 5.8$ per cent of the farms are under 10 acres, that $\frac{637}{3,156}\cdot 100 = 20.2$ per cent of the farms are from 10 to 49 acres, and so on. Generally speaking, *percentage distributions are useful, especially when we want to compare two or more sets of data.* For instance, it may well be more informative to say that the percentages of farms under 10 acres in two counties are, respectively, 5 per cent and 6 per cent, than to report that in one county 16 of 321 farms and in the other county 43 of 717 farms are under 10 acres.

So far we have discussed only numerical distributions, but the general problem of constructing categorical (or qualitative) distributions is very much the same. Again we must decide how many classes (categories) to use and what kind of items each category is to contain, making sure that all of the items are accommodated and that there are no ambiguities. Since the categories must often be selected before any data are actually obtained, sound practice is to include a category labeled "others" or "miscellaneous."

When dealing with categorical distributions we do not have to worry about such mathematical details as class limits, class boundaries, class marks, etc.; on the other hand, we now have a more serious problem with ambiguities, and we must be careful and explicit in defining what each category is to contain. For instance, if we tried to classify items sold at a supermarket into "meats," "frozen foods," "baked goods," and so on, it would be difficult to decide where to put, for example, frozen beef pies. Similarly, if we wanted to classify occupations, it would be difficult to decide where to put a farm manager, if our table contained (without qualification) the two categories "farmers" and "managers." For this reason, it is often advisable to use standard categories developed by the Bureau of the Census and other government agencies. (For references to such lists see the book by P. M. Hauser and W. R. Leonard in the Bibliography at the end of this chapter.)

2.3 Graphical Presentations

When frequency distributions are constructed primarily to condense large sets of data and display them in an "easy to digest" form, it is usually advisable to present them graphically, that is, in a form that appeals to the human power of visualization. The most common among all graphical presentations of statistical data is the *histogram*, an example of which is shown in Figure 2.1. A histogram is constructed by representing measure-

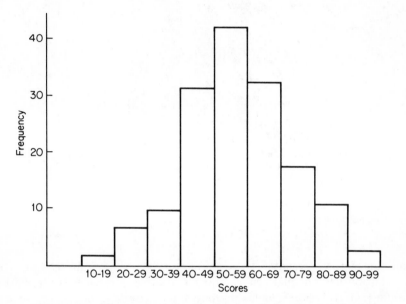

Figure 2.1. Histogram of the distribution of the scores of the 150 applicants.

ments or observations that are grouped (in Figure 2.1 the scores) on a horizontal scale, the class frequencies on a vertical scale, and drawing rectangles whose bases equal the class interval and whose heights are determined by the corresponding class frequencies. The markings on the horizontal scale can be the class limits as in Figure 2.1, the class boundaries, the class marks, or arbitrary key values. For easy readability it is generally preferable to indicate the class limits, although the bases of the rectangles actually go from one class boundary to the next. Similar to histograms are *bar charts*, like the one of Figure 2.2, where the lengths of the bars are proportional to the class frequencies, but there is no pretense of having a continuous (horizontal) scale.

There are several points that must be watched in the construction of histograms. First, it must be remembered that this kind of figure cannot be used for distributions with *open* classes. Second, it should be noted

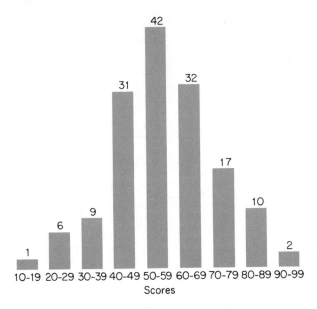

Figure 2.2. Bar chart of the distribution of the scores of the 150 applicants.

that the picture presented by a histogram can be very misleading if a distribution has unequal classes and no suitable adjustments are made. To illustrate this point, let us regroup the distribution of the 150 scores by combining all those from 60 to 79 into one class. Thus, the distribution becomes

Scores	*Frequency*
10–19	1
20–29	6
30–39	9
40–49	31
50–59	42
60–79	49
80–89	10
90–99	2

and its histogram (with the class frequencies represented by the heights of the rectangles) is shown in Figure 2.3. This figure gives the impression that just about *half* the scores fall on the interval from 60 to 79, whereas

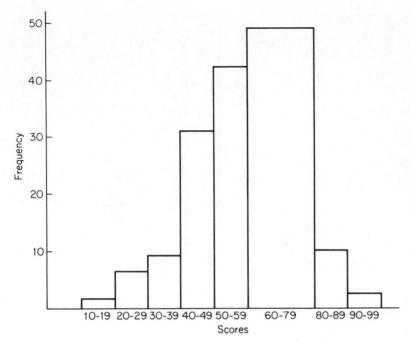

Figure 2.3. Incorrectly modified histogram of the distribution of the scores.

the correct proportion is close to $\frac{1}{3}$, $\frac{49}{150}$ to be exact. This error is due to the fact that when we compare the size of rectangles, triangles, and other plane figures, we instinctively compare their *areas* and not their sides. In order to correct for this, we simply draw the rectangles of the histogram so that the class frequencies are represented by their areas, and not by their heights. In Figure 2.4 we accomplished this by reducing the height of the rectangle representing the class 60–79 to *half* of what it was in Figure 2.3.

The practice of representing class frequencies by means of areas is especially important if histograms are to be approximated with smooth curves. For instance, if we wanted to approximate the histogram of Figure 2.1 with a smooth curve, we could say that the number of scores exceeding 69 is given by the shaded area of Figure 2.5. Clearly, this area is approximately equal to the sum of the areas of the corresponding three rectangles.

An alternate, though less widely used, form of graphical presentation is the *frequency polygon* (see Figure 2.6). Here the class frequencies are plotted at the class marks and the successive points are connected by means of straight lines. Note that we added classes with zero frequencies

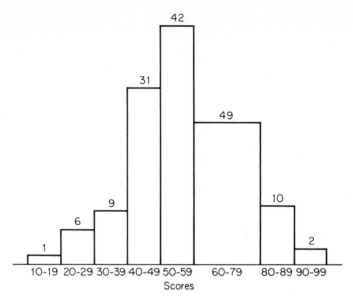

Figure 2.4. Correctly modified histogram of the distribution of the scores.

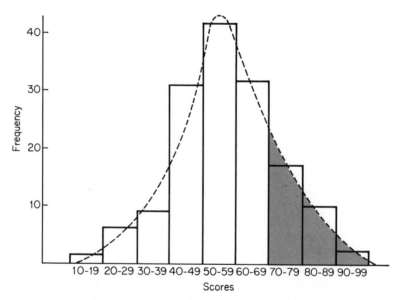

Figure 2.5. Histogram of the distribution of the scores approximated with a smooth curve.

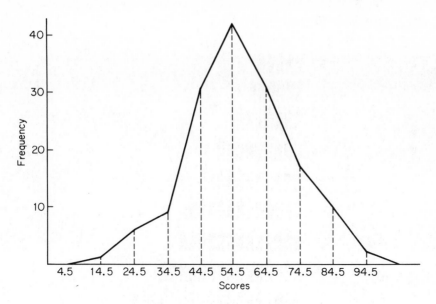

Figure 2.6. Frequency polygon of the distribution of the 150 scores.

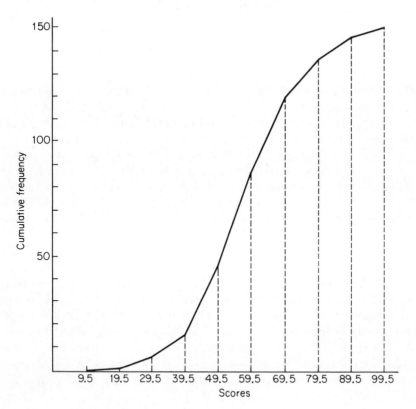

Figure 2.7. Ogive of the distribution of the 150 scores.

21

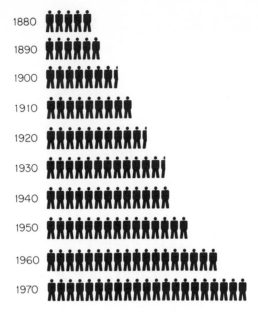

Each symbol = 10 million people

Figure 2.8. Pictogram of the population of the United States.

at both ends of the distribution in order to "tie down" the graph to the horizontal scale.

If we apply the same technique to a cumulative distribution, we obtain what is called an *ogive*. Note, however, that now the cumulative frequencies are *not* plotted at the class marks—it stands to reason that the cumulative frequency corresponding, say, to "less than 20" in our example should be plotted at 20, or preferably at the class boundary of 19.5, since "less than 20" actually includes everything up to 19.5. Figure 2.7 shows an ogive representing the cumulative "less than" distribution of the scores of the 150 applicants.

Although the visual appeal of histograms, frequency polygons, and ogives exceeds that of frequency tables, there are ways in which distributions can be presented even more dramatically and probably also more effectively. We are referring here to the various kinds of pictorial presentations (see, for example, Figures 2.8 and 2.9) with which the reader must surely be familiar through newspapers, magazines, advertising, and other sources. The number of ways in which distributions (and other statistical data) can be displayed pictorially is almost unlimited, depending only on the imagination and artistic talent of the individual preparing the presentations.

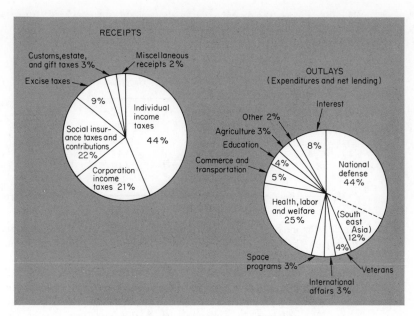

Figure 2.9. Pie charts of the Annual Federal Budget: 1966–69.

EXERCISES

1. Decide for each of the following quantities whether it can be determined on the basis of the distribution of the 150 scores on page 14; if possible, give a numerical answer:
   (a) The number of scores which were at least 50.
   (b) The number of scores which were greater than 50.
   (c) The number of scores which were 80 or less.
   (d) The number of scores which were less than 80.
   (e) The number of scores which were more than 90.
   (f) The number of scores which were greater than 39 but at most 69.

2. If the amounts paid for the repairs of cars damaged in accidents are grouped into a frequency table with the classes $0.00–$99.99, $100.00–$199.99, $200.00–$299.99, $300.00–$399.99, $400.00–$499.99, and $500.00 or more, decide for each of the following quantities whether it can be determined on the basis of this distribution:
   (a) How many of the amounts were less than $200.00.
   (b) How many of the amounts were at least $200.00.
   (c) How many of the amounts were more than $200.00.
   (d) How many of the amounts were $200.00 or more.

3. The following is the distribution of the weekly earnings of 1,216 secretaries in the Phoenix, Arizona, metropolitan area in March, 1969:

Weekly Earnings	Number of Secretaries
Under $80	21
$80– $99	296
$100–$119	494
$120–$139	247
$140–$159	119
$160 and over	39

Decide for each of the following quantities whether it can be determined on the basis of this distribution; if possible give a numerical answer:

(a) The number of secretaries with weekly earnings of at least $120.

(b) The number of secretaries with weekly earnings of more than $120.

(c) The number of secretaries with weekly earnings of more than $180.

(d) The number of secretaries with weekly earnings of less than $100.

(e) The number of secretaries with weekly earnings of at most $100.

(f) The number of secretaries with weekly earnings of at least $60.

4. The number of students absent from school each day are grouped into a distribution having the classes 3–10, 11–18, 19–26, 27–34, and 35–42. Find (a) the limits of each class, (b) the class boundaries, (c) the class marks, and (d) the class interval.

5. The following is the distribution of the actual weight (in ounces) of 50 "one-pound" bags of coffee, which a grocery clerk filled from bulk stock:

Weight	Number of Bags
15.5–15.6	3
15.7–15.8	9
15.9–16.0	17
16.1–16.2	14
16.3–16.4	6
16.5–16.6	1

Find (a) the limits of each class, (b) the class marks, (c) the class boundaries, and (d) the class interval of the distribution.

6. The weights of certain laboratory animals, given to the nearest tenth of an ounce, are grouped into a table having the class boundaries 11.45, 13.45, 15.45, 17.45, and 19.45 ounces. What are the limits of the four classes of this distribution?

7. The class marks of a distribution of temperature readings, given to the nearest degree Fahrenheit, are 113, 128, 143, 158, and 173. Find the class boundaries of this distribution, and also the class limits.

8. Class limits and class boundaries have to be interpreted very carefully when we are dealing with ages, for the age group from 5 through 9, for example, includes all those who have passed their fifth birthday but not yet reached their tenth. Taking this into account, what are the boundaries and the class marks of the following age groups: 10–19, 20–29, 30–39, and 40–49.

9. A study of air pollution in a city yielded the following daily readings of the concentration of sulfur dioxide (in parts per million):

```
.04  .11  .05  .01  .15  .12  .19  .06  .13  .03
.18  .01  .08  .11  .08  .14  .02  .14  .08  .10
.17  .09  .14  .07  .13  .11  .09  .05  .15  .08
.06  .05  .12  .10  .27  .12  .16  .10  .09  .15
.07  .10  .17  .13  .20  .18  .11  .17  .14  .04
.22  .11  .09  .02  .12  .16  .15  .12  .13  .07
.05  .14  .04  .16  .19  .10  .06  .03  .16  .13
.18  .13  .11  .09  .06  .23  .11  .12  .07  .11
```

(a) Group these data into a table having the classes .00–.04, .05–.09, .10–.14, .15–.19, .20–.24, and .25–.29.

(b) Convert the distribution obtained in (a) into a cumulative "less then" distribution.

(c) Construct a histogram of the distribution obtained in (a).

(d) Draw an ogive of the cumulative "less than" distribution obtained in (b) and use it to read off (roughly) the value below which we should find the lowest half of the data.

10. The following are the number of customers a restaurant served for lunch on 120 weekdays:

```
50  64  55  51  60  41  71  53  63  64  46  59
66  45  61  57  65  62  58  65  55  61  50  55
53  57  58  66  53  56  64  46  59  49  64  60
58  64  42  47  59  62  56  63  61  68  57  51
61  51  60  59  67  52  52  58  64  43  60  62
48  62  56  63  55  73  60  69  53  66  54  52
56  59  65  60  61  59  63  56  62  56  62  57
57  52  63  48  58  64  59  43  67  52  58  47
63  53  54  67  57  61  65  78  60  66  63  58
60  55  61  59  74  62  49  63  65  55  61  54
```

(a) Group these figures into a table having the classes 40–44, 45–49, 50–54, 55–59, 60–64, 65–69, 70–74, and 75–79.

(b) Draw a histogram of the distribution obtained in (a).

(c) Convert the distribution obtained in (a) into a percentage distribution and draw its frequency polygon.

(d) Convert the percentage distribution obtained in (c) into a cumulative "less than" percentage distribution and draw its ogive.

11. The following are the amounts (in dollars) which 60 students attending various 4-year colleges paid for room and board during the academic year 1969–70:

950	851	875	945	910	965	948	824	946	1043
920	900	903	854	964	773	916	920	983	926
912	812	943	940	876	942	1010	873	830	995
985	962	927	922	974	925	940	928	880	970
945	945	859	928	860	895	837	985	910	948
893	920	1063	880	914	892	933	850	930	882

(a) Group these amounts into a distribution having the classes 750–799, 800–849, 850–899, 900–949, 950–999, 1000–1049, and 1050–1099.

(b) Draw a bar chart of the distribution obtained in (a).

(c) Convert the distribution obtained in (a) into a cumulative "or more" distribution and draw its ogive.

12. In a two-week study of the productivity of workers, the following data were obtained on the average number of acceptable pieces which 100 workers produced per hour:

3.6	6.7	4.3	2.8	5.6	7.9	8.4	4.9	3.6	6.5
8.2	2.2	6.2	5.5	7.2	6.8	4.0	3.7	7.8	4.3
6.5	7.3	5.7	3.9	4.6	5.7	5.6	6.0	5.0	8.8
4.5	5.6	7.5	4.0	5.1	7.0	7.4	7.6	4.8	5.9
3.4	7.4	5.3	6.4	8.0	3.2	6.3	5.2	6.2	3.5
5.1	3.5	4.4	4.5	5.4	5.1	5.5	4.8	6.0	7.6
6.8	8.5	6.0	7.7	6.1	3.3	4.5	6.1	5.3	2.1
4.7	5.2	6.8	5.2	6.9	4.2	6.7	3.4	5.3	4.5
5.4	4.1	5.9	5.3	5.0	7.3	6.1	5.5	6.5	6.2
7.0	3.8	5.0	4.7	3.5	2.6	5.8	8.2	7.4	4.1

(a) Group these figures (which are rounded to the nearest tenth) into a distribution having the classes 2.0–2.9, 3.0–3.9, 4.0–4.9, 5.0–5.9, 6.0–6.9, 7.0–7.9, and 8.0–8.9.

(b) Draw a histogram and a frequency polygon of the distribution obtained in (a).

(c) Convert the distribution obtained in (a) into a cumulative "less than" distribution and draw its ogive.

(d) Use the ogive obtained in (c) to read off roughly how many acceptable pieces, at most, the lowest 25 per cent of the workers can be expected to average.

13. The following are measurements of the breaking strength (in ounces) of a sample of 60 linen threads:

32.5	15.2	35.4	21.3	28.4	26.9	34.6	29.3	24.5	31.0
21.2	28.3	27.1	25.0	32.7	29.5	30.2	23.9	23.0	26.4
27.3	33.7	29.4	21.9	29.3	17.3	29.0	36.8	29.2	23.5
20.6	29.5	21.8	37.5	33.5	29.6	26.8	28.7	34.8	18.6
25.4	34.1	27.5	29.6	22.2	22.7	31.3	33.2	37.0	28.3
36.9	24.6	28.9	24.8	28.1	25.4	34.5	23.6	38.4	24.0

(a) Group these measurements into a distribution having the classes 15.0–19.9, 20.0–24.9, 25.0–29.9, 30.0–34.9, and 35.0–39.9.

(b) Construct a histogram and a frequency polygon of the distribution obtained in (a).

(c) Convert the distribution obtained in (a) into a cumulative "or more" percentage distribution and draw its ogive.

14. The following are the scores which 200 students obtained in a 12th grade achievement test in American history:

53	96	40	56	79	60	74	66	73	86	42	69	53	95	80	37
45	54	12	65	60	43	52	72	29	68	30	32	70	56	41	51
62	83	64	47	71	65	62	46	43	51	64	72	98	59	34	42
20	67	85	56	57	75	42	54	63	49	58	67	18	76	62	64
72	41	73	70	27	63	66	36	82	75	53	57	62	47	25	70
71	32	51	59	53	86	94	69	40	67	26	75	33	84	67	72
55	57	51	42	68	69	17	75	61	77	65	71	50	61	40	87
46	88	45	63	97	38	45	53	35	77	92	45	83	77	67	48
61	63	67	55	54	64	66	81	64	54	52	65	60	59	90	63
37	25	84	53	56	49	23	85	65	62	62	24	54	46	78	76
93	50	76	69	51	71	56	72	52	50	81	70	56	63	66	39
58	74	30	87	62	82	61	95	57	60	76	82	48	56	44	53
59	66	52	33	64	40	76	58								

(a) Group these scores into a distribution having the classes 10–19, 20–29, 30–39, . . ., and 90–99, and convert it into a percentage distribution.

(b) Draw a bar chart of the original distribution obtained in (a).

(c) Convert the distribution obtained in (a) into a cumulative "less than" percentage distribution and draw its ogive.

(d) Use the ogive obtained in (c) to read off roughly the score below which are the lowest 20 per cent of the scores and the score above which are the highest 20 per cent of the scores.

15. Choose a local television station and construct a table showing how many of the half-hour periods between 2 P.M. and 6 P.M. it broadcasts during one week are situation comedies, game shows, serious drama, educational programs, sports, news, and so forth.

16. Take that part of the classified ads of a large daily newspaper where individuals (not builders) advertise houses for sale, and construct a table showing how many of these have a fireplace and air conditioning, a fireplace but no air conditioning, air conditioning but no fireplace, neither a fireplace nor air conditioning.

17. Use a daily newspaper listing prices on the New York Stock Exchange and construct a table showing how many of the D, E, F, and G stocks traded on a certain day showed a net increase, a net decrease, or no change in price.

18. Categorical distributions are often presented graphically as *pie charts* like those of Figure 2.9, where a circle is divided into sectors which are proportional in size to the frequencies (or percentages) of the corresponding categories. Making use of the fact that 1 per cent of the data is thus represented by a central angle of 3.6 degrees, draw a pie chart of the following categorical distribution:

Freight Carried in 1967 on
Inland Waterways
(millions of ton-miles)

Atlantic coast rivers	28,760
Gulf coast rivers	25,002
Pacific coast rivers	6,242
Mississippi river system	114,579
Great lakes system	106,809

19. Draw a pie chart (see Exercise 18) of the distribution of motor vehicle registrations on page 10.

20. Draw a pie chart (see Exercise 18) of the following distribution of the gifts received by a liberal arts college in 1970:

Total of Gifts

Alumni	$43,900
Parents of students	$12,500
Corporations	$68,200
Foundations	$81,600
Friends	$ 6,800
Trustees	$40,500

21. The pictogram of Figure 2.10 is intended to illustrate the fact that the total value of corporation stocks held by U.S. life insurance companies has tripled from 1960 to 1970. Explain why this pictogram does not convey the correct impression of the change, and indicate how it should be modified.

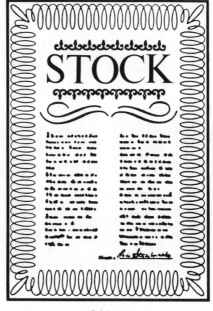

5 billion
dollars in 1960

15 billion
dollars in 1970

Figure 2.10. Value of corporation stock held by U.S. life insurance companies.

BIBLIOGRAPHY

Stimulating discussions of *what not to do* in the presentation of statistical data may be found in

HUFF, D., *How to Lie with Statistics.* New York: W. W. Norton & Company, Inc., 1954.

REICHMAN, W. J., *Use and Abuse of Statistics.* Baltimore: Penguin Books, Inc., 1971.

Useful references to lists of standard categories may be found in

HAUSER, P. M., and LEONARD, W. R., *Government Statistics for Business Use,* 2nd ed. New York: John Wiley & Sons, Inc., 1956.

3

Measures of Location

3.1 Introduction

Descriptions of statistical data can be quite brief or quite elaborate, depending partly on the nature of the data themselves, and partly on the purpose for which they are to be used. Sometimes, we even describe the same set of data in several different ways. To draw an analogy, a large motel might describe itself to the public as having luxurious facilities, a heated swimming pool, and TV in every room; on the other hand, it might describe itself to the fire department by giving the floor space of each unit, the number of sprinklers, and the number of employees. Both of these descriptions may serve the purpose for which they are designed, but they would hardly satisfy the State Corporation Commission in passing on the owner's application for issuing stock. This would require detailed information on the management of the motel, various kinds of financial statements, and so on.

Whether we describe things statistically or whether we simply describe them verbally, it is always desirable to say neither too little nor too much. Thus, it may sometimes be satisfactory to present data simply as they are and let them "speak for themselves"; in other instances it may be satisfactory to group, classify, and present them using the methods of Chapter 2. However, most of the time it is necessary to summarize them further by means of one or more well-chosen descriptions. In this chapter and in Chapter 4 we shall concentrate mainly on two kinds of descriptions, called *measures of location* and *measures of variation;* some others are mentioned briefly in Section 4.4.

The measures of location we shall study in this chapter are also referred to at times as "measures of central tendencies," "measures of central values," and "measures of position." Except for some of the measures discussed in Section 3.4, they may also be referred to crudely as "averages" in the sense that they provide numbers that are indicative of the "center," "middle," or the "most typical" of a set of data.

When we said that the choice of a statistical description depends partly on the nature of the data themselves, we were referring among other things to the following distinction: *if a set of data consists of all conceivably possible (or hypothetically possible) observations of a certain phenomenon, we refer to it as a population; if it contains only part of these observations, we refer to it as a sample.* The qualification "hypothetically possible" was added to take care of such clearly hypothetical situations where, say, twelve flips of a coin are looked upon as a sample from the population of all possible flips of the coin, or where we shall want to look upon the weights of eight 30-day-old calves as a sample of the weights of all (past, present, and future) 30-day-old calves. In fact, we often look upon the results obtained in an experiment as a sample of what we might obtain if the experiment were repeated over and over again.

In actual practice, whether a set of data is looked upon as a sample or as a population depends to some extent on what we intend to do with the data. Suppose, for example, that we are offered a lot of 5,000 ceramic tiles which we may or may not be interested in buying, depending on their strength. If we measure the breaking strength of ten of these tiles to estimate the average breaking strength of all the tiles, these ten measurements constitute a sample from the population which consists of the breaking strengths of all the tiles. We thus have a sample of *size* 10 from a population of *size* 5,000. In a different context, however, we might look upon the 5,000 tiles and their breaking strengths as only a sample of all the millions of tiles which the tile manufacturer produces throughout the years. To consider another illustration of this important distinction, suppose that we are interested in the monthly mileages of police cars, and that we have at our disposal complete figures on the mileages that police cars were driven in Orange County, California during June 1972. If we do not generalize about corresponding mileages for other counties or other years (including the future), we are justified in saying that the data constitute a population; they contain all the information that is relevant to the phenomenon with which we are concerned. On the other hand, if we want to make generalizations about corresponding mileages for the whole year 1972, for the whole State of California, or for the entire United States, then the June 1972 figures for Orange County are only a sample.

As we have defined it here, the word "sample" is used in very much the same way as it is used in everyday language. An employer considers the opinions of 25 of his 600 employees a sample of all their opinions on a given matter, and a consumer considers a box of Blum's candy a sample of the firm's product. (Later, in Chapters 9 through 16, we shall interpret the term "sample" in a somewhat narrower fashion, limiting it to data that can reasonably serve for making generalizations about the population from which they are obtained. Thus, the above-mentioned sample of data pertaining to Orange County may not be acceptable as a sample if generalizations are to be made about police-car mileages in the entire United States.) The fact that the word "universe" is sometimes used instead of "population" makes it evident that neither term is used here in its colloquial sense. In statistics, both terms refer to the actual or hypothetical totality of measurements or observations with which we are concerned, and not necessarily (or directly) to human beings or animals.

In this chapter and the next we shall limit ourselves to methods of description without making generalizations, but it is important even here to distinguish between samples and populations. As we have said before, the kind of description we may want to use will depend on what we intend to do later on, whether we merely want to present facts about populations or whether we want to generalize from samples. We shall, thus, begin in this chapter with the practice of using different symbols depending on whether we are describing samples or populations; in Chapter 4 we shall carry this distinction one step further by even using different formulas.

3.2 The Mean

There are many problems in which we have to represent data by means of a single number which, in its way, is descriptive of the entire set. The most popular measure used for this purpose is what the layman calls an "average" and what, in statistics, is called an *arithmetic mean*, or simply a *mean*. We gave the word "average" in quotes because it generally has a loose connotation and different meanings—for example, when we speak of a batting average, an average housewife, a person with average taste, and so on. (Since there also exist *geometric means* and *harmonic means*, see Exercises 22 and 24 on pages 45 and 46, it must be kept in mind that when we speak of *the mean*, we refer to the arithmetic mean and not to the others.)

The arithmetic mean of a set of n numbers is defined simply as their sum divided by n. For instance, given that the total attendance at major league baseball games in the years 1965, 1966, 1967, and 1968 was, respectively, 22.4, 25.2, 23.8, and 23.0 million, we find that the *mean*, namely, the "average" annual attendance for these four years was

$$\frac{22.4 + 25.2 + 23.8 + 23.0}{4} = 23.6 \text{ million}$$

In order to develop a simple formula for the mean that is applicable to any set of data, it will be necessary to represent the figures (measurements or observations) to which the formula is to be applied with some general symbols such as x, y, or z. In the above example, we could have represented the annual attendance figures with the letter x and referred to the four values as x_1 (x *sub-one*), x_2 (x *sub-two*), x_3, and x_4. More generally, if we have n measurements which we designate x_1, x_2, x_3, $\ldots$, and x_n, we can write

$$\text{mean} = \frac{x_1 + x_2 + \ldots + x_n}{n}$$

This formula is perfectly general and it will take care of any set of data, but it is still somewhat cumbersome. To make it more compact, we introduce the symbol Σ (capital *sigma*, the Greek letter for S), which is simply a mathematical shorthand notation indicating the process of summation or addition. If we write $\Sigma\, x$, this represents the "sum of the x's," and we now have

$$\text{mean} = \frac{\Sigma\, x}{n}$$

Using the sigma notation in this form, the number of terms to be added is not stated explicitly; it is tacitly understood, however, to refer to all the x's with which we happen to be concerned. If we wanted to use a more explicit notation, we could write the sum $x_1 + x_2 + \ldots + x_n$ as $\sum\limits_{i=1}^{n} x_i$ instead of $\Sigma\, x$; this would indicate explicitly that we are adding the x's with subscripts from 1 to n. For a further discussion of the use of subscripts and the Σ notation, see Section 3.6.

To go one step further, we shall finish simplifying our notation by assigning a special symbol to the mean itself. If we look upon the x's as a sample, we write their mean as $\bar{x}$ (*x-bar*); if we look upon them as a population, we write their mean as μ (*mu*, the Greek letter for m). If we refer to sample data as y's or z's, we correspondingly write their means as $\bar{y}$ or $\bar{z}$. To further emphasize the distinction between samples and populations, we denote the number of values in a sample, the *sample size*, with the letter

n and the number of values in a population, the *population size*, with the letter N. We thus have the formulas*

$$\blacktriangle \qquad \bar{x} = \frac{\Sigma\, x}{n} \quad \text{or} \quad \mu = \frac{\Sigma\, x}{N} \qquad \blacktriangle$$

depending on whether we are dealing with a sample or a population. In order to distinguish between descriptions of samples and descriptions of populations, statisticians not only use different symbols, but they refer to the first as *statistics* and the second as *parameters*. Hence, we say that $\bar{x}$ is a statistic and that μ is a parameter.

To illustrate this terminology and notation, let us consider the problem of a consumer-testing service which wants to determine the "true" average lifetime of a manufacturer's 75-watt light bulbs, namely, the average number of hours they can be expected to last in continuous use. As it is obviously impossible to examine *all* of the manufacturer's 75-watt bulbs for he would have none left to sell, suppose that a technician working for the testing service selects *five* of the bulbs and that they last, respectively, 967, 889, 940, 922, and 952 hours. These measurements constitute a sample with the mean

$$\bar{x} = \frac{967 + 889 + 940 + 922 + 952}{5} = 934 \text{ hours}$$

and this figure, 934 hours, can be used as an *estimate* of the parameter μ (the actual average lifetime of *all* the manufacturer's 75-watt bulbs), provided that sufficient care was taken in obtaining the sample. Note how this notation eliminates such confusing language as "we use a mean to estimate a mean"—the statement "we use $\bar{x}$ as an estimate of μ" makes it clear that we are using a statistic to estimate a parameter, namely, a sample mean to estimate the mean of a population.

The popularity of the mean as a measure describing the "middle" or "center" of a set of data is not just accidental. Anytime we use a single number to describe a set of data, there are certain desirable properties we must keep in mind. Thus, some of the noteworthy properties of the mean are: (1) it is familiar to most persons, although they may not call it by this name; (2) it always exists, that is, it can be calculated for any kind of numerical data; (3) it is always unique, or in other words, a set of data has one and only one mean; (4) it takes into account each individual item;

---

* Formulas marked ▲ are actually used for practical computations. This will make it easier for the reader to distinguish between formulas needed for calculations and those given primarily as part of definitions or derivations.

(5) it lends itself to further statistical manipulation (as we shall see on page 41, it is possible to combine the means of several sets of data into an over-all mean without having to refer back to the original raw data); and (6) it is relatively *reliable* in the sense that it does not vary too much when repeated samples are taken from one and the same population, at least not as much as some other kinds of statistical descriptions. This question of reliability is of fundamental importance when it comes to problems of estimation, hypothesis testing, and making predictions, and we shall have a good deal more to say about it later in this book.

Whether the fourth property of the mean listed above is actually desirable is open to some doubt; a single extreme (very large or very small) value can affect the mean to such an extent that it is debatable whether it is really "representative" or "typical" of the data. To illustrate, suppose that someone told us that the average age of those attending a birthday party was 21. This figure certainly would be misleading if it so happened that the party was attended by eight teenagers aged, respectively, 16, 16, 18, 17, 19, 16, 17, and 18, and one of the teenagers' mothers who is 52.

To give another illustration, suppose that in the example concerning the 75-watt light bulbs, the last figure had been recorded *incorrectly* as 592 instead of 952. The technician would thus have obtained a mean of

$$\frac{967 + 889 + 940 + 922 + 592}{5} = 862 \text{ hours}$$

instead of the correct value of $\bar{x} = 934$ hours. This shows how one careless mistake can have a pronounced effect on the mean. There could have been serious consequences if the testing service had accepted the erroneous figure and estimated μ, the actual mean lifetime of the manufacturer's 75-watt bulbs, as 862 hours instead of 934. Note that in Section 3.4 we shall meet another kind of "average," called the *median*, which has the important feature that it is not so readily affected by a very large or a very small value.

Since the computation of means is quite easy, involving only addition and one division, there is usually no need to look for short-cuts or simplifications. However, if the numbers are unwieldy, that is, if each number has many digits, or if the sample (or population) size is very large, it may be advantageous to group the data first and then compute the mean from the resulting distribution. Another reason why we shall investigate the problem of obtaining means from grouped data is that published data are very often available only in the form of distributions.

Earlier, on page 10, we observed that the grouping of data entails some loss of information. Each item, so to speak, loses its identity (we know only how many items fall into each class), and the *actual* mean of the data can no longer be calculated. However, a good approximation can be obtained by assigning the value of the class mark to each item falling into a given class, and this is how we *define* the mean of a distribution. Thus, the six values which fall into the second class of the distribution of the 150 scores on page 14 are treated as if they all equaled 24.5, the mid-point of the class going from 20 to 29. Similarly, the value which falls into the first class is treated as if it equaled 14.5, the nine values which fall into the third class are treated as if they all equaled 34.5, and so forth. This procedure is generally very satisfactory, since the errors which are thus introduced will more or less "average out."

To obtain a formula for the mean of a distribution, let us write the successive class marks as $x_1, x_2, \ldots, x_k$ (assuming that there are k classes) and the corresponding class frequencies as $f_1, f_2, \ldots, f_k$. The total that goes into the numerator of the formula for the mean is thus obtained by adding f_1 times the value x_1, f_2 times the value x_2, $\ldots$, and f_k times the value x_k; in other words, it is equal to $x_1 f_1 + x_2 f_2 + \ldots + x_k f_k$. Using the Σ notation introduced on page 34, we can now write the formula for the mean of a distribution as

$$\bar{x} = \frac{\Sigma \, x \cdot f}{n}$$

where $\Sigma \, x \cdot f$ represents, in words, the sum of the products obtained by multiplying each class mark by the corresponding class frequency, and n equals $f_1 + f_2 + \ldots + f_k$, the sum of the class frequencies, or Σf. (When dealing with a population instead of a sample, we have only to substitute μ for $\bar{x}$ in this formula and N for n.)

To illustrate the calculation of the mean of a distribution, let us refer again to the distribution of the scores of the 150 applicants on page 14. Writing the class marks in the second column, we get

Scores	Class Marks x	Frequencies f	Products $x \cdot f$
10–19	14.5	1	14.5
20–29	24.5	6	147.0
30–39	34.5	9	310.5
40–49	44.5	31	1379.5
50–59	54.5	42	2289.0
60–69	64.5	32	2064.0
70–79	74.5	17	1266.5
80–89	84.5	10	845.0
90–99	94.5	2	189.0
	Total	150	8505.0

and it follows that the mean of the distribution is

$$\bar{x} = \frac{8505.0}{150} = 56.7$$

It is of interest to note that the mean of the original *raw* data on page 13 is $\frac{8500}{150} = 56.67$, so that the difference between the two means is extremely small.

The calculation of the mean of the distribution of the 150 scores was fairly easy because the frequencies were all small. Even so, the calculations can be simplified by performing a *change of scale;* that is, we replace the class marks with numbers that are easier to handle. This is also referred to as "coding," and in our example, we might replace the class marks of the distribution of the scores with the consecutive integers -4, -3, -2, -1, 0, 1, 2, 3, and 4. Of course, when we do something like this, we also have to account for it in the formula we use to calculate the mean. Referring to the new (coded) class marks as u's, it can easily be shown (see Exercise 7 on page 61) that the formula for the mean of a distribution becomes

▲
$$\bar{x} = x_0 + \frac{\Sigma\, u{\cdot}f}{n}{\cdot}c$$
▲

where x_0 is the class mark (in the original scale) to which we assign 0 in the new scale, c is the class interval, n is the number of items grouped, and $\Sigma\, u{\cdot}f$ is the sum of the products obtained by multiplying each of the coded class marks by the corresponding frequency.

Illustrating this short-cut technique by recalculating the mean of the distribution of the scores of the 150 applicants, we obtain

Class Marks *x*	*u*	*f*	*u·f*
14.5	−4	1	−4
24.5	−3	6	−18
34.5	−2	9	−18
44.5	−1	31	−31
54.5	0	42	0
64.5	1	32	32
74.5	2	17	34
84.5	3	10	30
94.5	4	2	8
	Total	150	33

and

$$\bar{x} = 54.5 + \frac{33}{150}\cdot 10 = 56.7$$

It should be noted that this agrees with the result obtained earlier; the short-cut formula does *not* entail any further approximation, and it should always yield the same result as the formula on page 37.

Unless one can use an automatic computer, the short-cut method will generally save a good deal of time; about the only time that the short-cut method will not provide appreciable savings in time and energy is when the original class marks are already easy-to-use numbers. In order to reduce the work to a minimum, it is generally advisable to put the zero of the u-scale near the middle of the distribution, preferably at a class mark having one of the highest frequencies.

A fact worth noting is that this short-cut method cannot be used for distributions with *unequal* classes, although there exists a modification which makes it applicable also in that case. Neither the short-cut formula nor the formula on page 37 is applicable to distributions with *open classes;* the means of such distributions cannot be found without going back to the raw data or making special assumptions about the values which fall into an open class.

3.3 The Weighted Mean

There are many situations in which it would be very misleading to average quantities without accounting in some way for their relative importance in the over-all picture we are trying to describe. For instance, if we are given the information that in 1969 the average salary of elementary school teachers in California, Oregon, and Washington was $9,100, $7,789, and $7,950, respectively, we *cannot* conclude that the average salary paid elementary school teachers in 1969 in these three Pacific states was

$$\frac{9,100 + 7,789 + 7,950}{3} = \$8,280$$

To get a meaningful figure, we would have to know how many elementary school teachers were employed that year in each of the states. Similarly, it would be pointless to calculate the mean of the prices of various food items without accounting in some way for the respective roles which they play in the average family's budget, or in the average store's volume of sales. To give one more example, a student cannot calculate his average grade in a course for the whole semester, unless he knows how much importance his instructor assigns to each quiz and the mid-term and final examinations.

Returning to the first example, let us now add the information that in 1969 there were 106,000, 11,707, and 17,500 elementary school teachers, respectively, in California, Oregon, and Washington. Then, simple calculations show that the total earnings of these $106,000 + 11,707 + 17,500 = 135,207$ teachers were

$$106{,}000 \cdot 9{,}100 + 11{,}707 \cdot 7{,}789 + 17{,}500 \cdot 7{,}950 = \$1{,}194{,}910{,}823$$

and, hence, that they earned on the average

$$\frac{106{,}000 \cdot 9{,}100 + 11{,}707 \cdot 7{,}789 + 17{,}500 \cdot 7{,}950}{106{,}000 + 11{,}707 + 17{,}500} = \frac{1{,}194{,}910{,}823}{135{,}207} = \$8{,}838$$

to the nearest dollar. The average we have calculated here is called a *weighted mean*—we averaged the mean salaries for the three states giving due weight to their relative importance.

In general, the *weighted mean* of a set of n numbers $x_1, x_2, \ldots,$ and x_n, whose relative importance is expressed numerically by a corresponding set of numbers $w_1, w_2, \ldots,$ and w_n, called the *weights*, is given by the formula

▲
$$\bar{x}_w = \frac{\Sigma \, w \cdot x}{\Sigma \, w}$$
▲

Here $\Sigma \, w \cdot x$, written more fully as $\displaystyle\sum_{i=1}^{n} w_i \cdot x_i$, represents the sum of the products obtained by multiplying each of the x's by the corresponding weight, while $\Sigma \, w$, or $\displaystyle\sum_{i=1}^{n} w_i$, represents the sum of the weights. Note that when the weights are all equal, the formula for the weighted mean reduces to that of the ordinary (arithmetic) mean.

To give another example, suppose that someone invests \$2,000 at 4 per cent in a savings account, \$5,000 at 6 per cent in a certificate of deposit, and \$20,000 at 8 per cent in second mortgages. To average these three percentage rates, we must weight them with the respective amounts which the person invests, and hence, substitution into the formula yields

$$\frac{2{,}000 \cdot 4 + 5{,}000 \cdot 6 + 20{,}000 \cdot 8}{2{,}000 + 5{,}000 + 20{,}000} = \frac{198{,}000}{27{,}000} = 7.33 \text{ per cent}$$

A special form of the weighted mean arises when we want to determine the over-all mean of several sets of data on the basis of their individual means and the number of items in each. Given n_1 numbers whose mean is $\bar{x}_1$, n_2 numbers whose mean is $\bar{x}_2$, $\ldots$, and n_k numbers whose mean is $\bar{x}_k$, the over-all mean of all these numbers is given by

▲
$$\frac{\Sigma\, n\cdot\bar{x}}{\Sigma\, n}$$
▲

Note that the numerator, written more fully as $\sum_{i=1}^{k} n_i\bar{x}_i$, represents the sum of the products obtained by multiplying each individual mean by the size of the corresponding set of data and, hence, it represents *the actual total of all the data;* the denominator, $\sum_{i=1}^{k} n_i$, stands for the total number of items in the combined data.

Actually, this was the case in the first example of this section, the one dealing with the teachers' salaries in the three states, but to give another example, suppose that in the four sections of freshman English offered at a two-year college there are, respectively, 36, 45, 28, and 41 students. If in the final examination they averaged, respectively, 62.5, 70.2, 59.5, and 67.0, substitution into the above formula yields

$$\frac{36(62.5) + 45(70.2) + 28(59.5) + 41(67.0)}{36 + 45 + 28 + 41} = 65.48$$

for the average final examination grade of *all* of the students. This illustrates what we meant on page 36 when we said that the mean lends itself readily to "further statistical manipulation."

The choice of the weights did not pose any problems in the examples of this section, but there are situations in which their selection is not quite so obvious. For instance, if we wanted to compare figures representing the cost of living in different locations or at different times, we would face the difficult task of having to account for the relative importance in the average person's budget of such items as food, rent, entertainment, medical care, and so on.

EXERCISES

1. Suppose we are given the high temperature recorded each day of the year 1972 in Atlanta, Georgia. Give one illustration each of a situation where these data would be looked upon (a) as a population, and (b) as a sample.

2. Suppose that the final election returns from a given county show that the two candidates for a certain office received, respectively, 16,283 and 13,559 votes. What office might these candidates be running for so that we can look upon these figures (a) as a sample and (b) as a population?

3. The dean of a college has in his files a complete record of how many A's, B's, C's, etc., each instructor gave to his students during the academic year 1971–72. Give one illustration each of a problem (situation) in which the dean would look upon this information (a) as a sample and (b) as a population.

4. The following are the speeds (in miles per hour) at which 25 cars were timed on the San Bernardino Freeway in early-morning traffic: 52, 56, 54, 78, 71, 66, 69, 60, 70, 53, 55, 62, 67, 60, 56, 72, 73, 61, 68, 59, 67, 66, 67, 73, and 65. Find the mean of these speeds and comment on the (misleading?) argument that "on the average cars do not exceed the speed limit of 65 miles per hour on this freeway in early-morning traffic."

5. The following are the monthly water bills which a resident of Scottsdale, Arizona, received in 1971: $10.26, $9.29, $11.24, $12.22, $19.07, $21.03, $22.50, $26.41, $18.09, $23.96, $16.18, and $15.60. Find the mean, namely, the average water bill this person paid per month in 1971.

6. The following are the number of seconds which 16 insects survived after being sprayed with a certain insecticide: 121, 115, 79, 52, 102, 126, 81, 65, 109, 119, 115, 121, 103, 75, 59, and 110.
   (a) Calculate the mean of these 16 measurements.
   (b) Recalculate the mean of these 16 measurements by first subtracting 100 from each value, finding the mean of the numbers thus obtained, and then adding 100 to the result. (What general simplification does this suggest for the calculation of means?)

7. Twenty-four cans of a floor wax, randomly selected from a large production lot, have the following net weights (in ounces): 12.0, 11.9, 12.2, 12.0, 11.9, 12.0, 12.0, 12.1, 11.8, 12.0, 12.0, 12.1, 11.9, 11.9, 12.2, 12.1, 12.0, 11.9, 11.9, 12.1, 12.0, 12.0, 11.9, and 12.0.
   (a) Calculate the mean of these 24 weights.
   (b) Recalculate the mean of these 24 weights by first subtracting 12.0 from each value, finding the mean of the numbers thus obtained, and then adding 12.0 to the result. (What general simplification does this suggest for the calculation of means?)

8. The following are the number of twists that were required to break 20 forged alloy bars: 37, 29, 34, 21, 54, 38, 30, 26, 48, 37, 24, 33, 39, 51, 44, 38, 35, 29, 46, and 31. Find the mean of these values.

9. A student took six readings on the direction from which the wind was blowing at a certain place, obtaining 9, 349, 350, 4, 18, and 350 degrees. (These angles are measured clockwise with 0° being due North.) Averaging these figures he obtains a mean of 180°, which means that the wind blew from the South. Find the fallacy of this argument and a more appropriate way of averaging these readings.

10. A bridge is designed to carry a maximum load of 120,000 pounds. If at a given moment it is loaded with 32 vehicles having an average (mean) weight of 3,400 pounds, is there any danger that it might collapse?

11. Find the mean of the distribution of Exercise 5 on page 24.

12. Find the mean of the sulfur dioxide concentrations of Exercise 9 on page 25 (a) on the basis of the raw (ungrouped) data, and (b) on the basis of the distribution obtained in that exercise *without coding*.

13. Referring to Exercise 10 on page 25, calculate the mean of the number of customers which the restaurant served for lunch on the 120 days
    (a) on the basis of the raw (ungrouped) data;
    (b) on the basis of the distribution obtained in that exercise, but without coding;
    (c) on the basis of the distribution obtained in that exercise, with coding.

14. Use the distribution obtained in Exercise 11 on page 26 to find the mean of the students' expenses for room and board
    (a) without coding;
    (b) with coding.

15. Referring to the data of Exercise 12 on page 26, find the mean
    (a) on the basis of the raw (ungrouped) data;
    (b) by suitably coding the class marks of the distribution obtained in that exercise.

16. Suitably coding the class marks of the distribution obtained in Exercise 13 on page 27, find the mean of the 60 breaking strengths.

17. Suitably coding the class marks of the distribution obtained in Exercise 14 on page 27, find the mean of the 200 scores.

18. In business and economics, there are many problems in which we are interested in *index numbers*, that is, in measures of the changes that have taken place in the prices (quantities, or values) of various commodities. In general, the year or period we want to compare by means of an index number is called the *given year* or *given period*, while the year or period relative to which the comparison is made is called the *base year* or *base period*. Furthermore, given-year prices are de-

noted p_n, base-year prices are denoted p_0, and the ratio p_n/p_0 for a given commodity is called the corresponding *price relative*. A very simple kind of index number is given by the *mean of the price relatives* of the commodities with which we are concerned, multiplied by 100 to express the index as a percentage.

(a) Find the mean of the price relatives comparing the 1969 prices of the given processed fruits and vegetables (in cents) with those of 1965:

	1965	1969
Fruit cocktail, No. 303 can	26.1	27.9
Pears, No. $2\frac{1}{2}$ can	47.0	50.9
Frozen orange juice, 6 oz.	23.7	24.3
Peas, No. 303 can	23.7	24.6
Tomatoes, No. 303 can	16.1	19.6
Frozen broccoli, 10 oz.	26.4	27.6

(b) Find the mean of the price relatives comparing the following 1967 prices with those of 1960, where all prices are in cents per pound:

	1960	1967
Copper	32.4	38.6
Lead	11.9	14.0
Zinc	12.9	13.8

19. If we substitute q's for p's in the index number of Exercise 18, where given-year quantities (produced, sold, or consumed) are denoted q_n and base-year quantities are denoted q_0, we obtain a corresponding *quantity index*. Given the following data in thousands of short tons, find the mean of the quantity relatives comparing the 1967 production figures with those of 1960:

	1960	1967
Copper	1080	954
Lead	247	317
Zinc	435	549

20. Another way of obtaining an index comparing given-year prices with a corresponding set of base-year prices (see Exercise 18) is to average the two sets of prices separately, take the ratio of the two means, and then multiply by 100 to express the index as a percentage. Canceling denominators, the formula for such a *simple aggregative index* is thus

given by $\dfrac{\Sigma\,p_n}{\Sigma\,p_0}\!\cdot\!100$, where $\Sigma\,p_n$ is the sum of the given-year prices and $\Sigma\,p_0$ is the sum of the base-year prices.

   (a) Referring to part (a) of Exercise 18, calculate a simple aggregative index comparing the 1969 prices of the processes foods with those of 1965.

   (b) Referring to part (b) of Exercise 18, calculate a simple aggregative index comparing the 1967 prices of the three metals with those of 1960.

21. In actual practice, simple aggregative indexes are seldom used because they fail the so-called *units test*. This means that the value of the index depends on the units for which the different prices are quoted. To illustrate this, change the prices of tomatoes in part (a) of Exercise 18 to 16,100.0 and 19,600.0 cents per *thousand* No. 303 cans (while leaving the other prices unchanged), and compare the resulting simple aggregative index with the one obtained in part (a) of Exercise 20.

22. The *geometric mean* of a set of n numbers is given by the nth root of their product, namely, by $\sqrt[n]{x_1\!\cdot\!x_2\!\cdot\ldots\cdot x_n}$, and it is used mainly to average ratios, rates of change, index numbers, and the like.

   (a) Find the geometric mean of 7 and 28.

   (b) Find the geometric mean of 6, 50, and 90.

   (c) Find the geometric mean of 1, 1, 1, 4, and 8.

   (d) In his first season as a professional, a hockey player scored 6 goals, in his second season he scored 9 goals, and in his third season he scored 24 goals. Thus, from the first season to the second his "output" was multiplied by $\dfrac{9}{6}=\dfrac{3}{2}$, and from the second season to the third his "output" was multiplied by $\dfrac{24}{9}=\dfrac{8}{3}$. Find the geometric mean of these two "growth rates," and apply it to the number of goals he scored in his third season to predict the number of goals he will score in his fourth season.

23. In actual practice, the *geometric mean* is often obtained by making use of the fact that the logarithm of the geometric mean of a set of numbers equals the (arithmetic) mean of the logarithms of the numbers.

   (a) Using logarithms (Table X at the end of this book), calculate the geometric mean of 113, 106, 124, 128, and 114, which are the 1970 values of the index of exports of manufactured goods of five large European countries.

   (b) Referring to part (b) of Exercise 18, use logarithms to find the geometric mean of the price relatives comparing the 1967 prices of the three metals with those of 1960. Multiply by 100 to express the resulting index as a percentage and compare with the mean of the price relatives obtained in that exercise.

24. The *harmonic mean* of n numbers is given by n divided by the sum of the reciprocals of the n numbers, namely, by $\dfrac{n}{\Sigma\, 1/x}$, and it is used only in very special situations. For instance, if \$12 is spent on pills costing 40 cents a dozen and another \$12 is spent on pills costing 60 cents a dozen, the average price is *not* $\dfrac{40 + 60}{2} = 50$ cents a dozen; since a total of \$24 is spent on a total of 50 dozen pills, the actual cost per dozen is $\dfrac{2400}{50} = 48$ cents.

   (a) Verify that 48 is, in fact, the harmonic mean of 40 and 60.
   (b) If a plane flies the first 240 miles of a trip at 300 miles per hour, the next 240 miles at 450 miles per hour, and the last 240 miles at 360 miles per hour, how long does it take the plane to travel these 720 miles and what is its average speed? Verify that the harmonic mean of 300, 450, and 360 gives the correct answer for the average speed.

25. If an instructor counts the final examination in a course four times as much as each one-hour examination, what is the weighted average grade of a student who received grades of 69, 75, 56, and 72 in four one-hour examinations and a final examination grade of 78?

26. If the mean weight of the 4 offensive starting backs of a professional football team is 213 pounds and the mean weight of the 7 offensive starting linemen is 238 pounds, what is the mean weight of this starting eleven?

27. If an investor bought 40 shares of General Motors' stock at \$78 a share, 80 shares at \$112 a share, and another 80 shares at \$94 a share, what was his average cost per share?

28. A butcher sells three grades of beef, respectively, for \$1.59, \$1.79, and \$2.19 a pound. If in a certain week he sells 600 pounds of the cheapest grade, 400 pounds of the medium-priced grade, and 200 pounds of the most expensive grade, what is the average price he gets per pound?

29. In 1965, the average annual wage payments to workers in the manufacturing, mining, and construction industries in Arizona were \$6,551, \$7,878, and \$7,136, respectively. Find the average annual wages of these workers, if 63,700 of them were employed in manufacturing, 15,900 in mining, and 22,500 in construction.

30. In 1971, a university paid its 64 instructors an average salary of \$7,258, its 185 assistant professors an average salary of \$9,772, its 132 associate professors an average salary of \$11,156, and its 69 full professors an average salary of \$14,582. What was the mean salary paid to the 450 members of this faculty?

31. Aggregative index numbers (see Exercise 20) can be improved by weighting the various prices by the corresponding quantities (produced, sold, or consumed) in the given year, the base year, or some other fixed period of time. Using the given-year quantities as weights we obtain the *weighted aggregative index* $\dfrac{\Sigma \, q_n p_n}{\Sigma \, q_n p_0} \cdot 100$, and using the base-year quantities as weights we obtain the *weighted aggregative index* $\dfrac{\Sigma \, q_0 p_n}{\Sigma \, q_0 p_0} \cdot 100$, where the notation is the same as in Exercises 18 and 19. Note that these index numbers are essentially ratios of weighted means with the sum of the weights canceled.

    (a) Using the prices of part (b) of Exercise 18 and the 1967 quantities of Exercise 19 as weights, calculate a weighted aggregative index comparing the 1967 prices of the three metals with those of 1960.

    (b) Repeat part (a) using as weights the 1960 quantities instead of the 1967 quantities.

    (c) The following are data on the prices of five major crops in dollars per bushel and their production in millions of bushels:

	1960 Prices	*1968 Prices*	*1968 Quantities*
Wheat	1.74	1.22	1,570
Corn	1.00	1.05	4,375
Oats	0.60	0.59	930
Barley	0.84	0.88	418
Soybeans	2.13	2.42	1,080

    Using the 1968 quantities as weights, calculate a weighted aggregative index comparing the 1968 prices of these crops with those of 1960.

32. When averaging price relatives, we often use as weights the given-year *values* of the corresponding commodities, namely, the products $p_n q_n$, where the notation is the same as in Exercises 18 and 19.

    (a) Use the data of part (b) of Exercise 18 and Exercise 19 to calculate such a weighted mean of price relatives comparing the 1967 prices of the three metals with those of 1960.

    (b) Use the data of part (c) of Exercise 31 to calculate such a weighted mean of price relatives comparing the 1968 prices of the crops with those of 1960.

3.4 The Median and Other Fractiles

To avoid the difficulty met on page 36, where we showed that an extreme value (perhaps, a gross error) can have a pronounced effect on the mean, we sometimes describe the "middle" or "center" of a set of data with other kinds of statistical descriptions. One of these is the *median*, which is defined simply as *the value of the middle item (or the mean of the values of the two middle items) when the data are arranged in an increasing or decreasing order of magnitude*.

If we have an *odd* number of items, there is always a middle item whose value is the median. For example, the median of the five numbers 5, 10, 2, 7, and 8 is 7, as can easily be verified by first arranging these numbers according to size, and the median of the nine numbers 3, 5, 6, 9, 9, 10, 10, 12, and 13 is 9. Note that there are two 9's in this last example and that we do not refer to either of them as *the* median. The median is a number and not an item, namely, the value of the middle item. Generally speaking, if there are n items and n is *odd*, the median is the value of the $\dfrac{n+1}{2}$ th largest item. Thus, the median of 25 numbers is given by the value of the $\dfrac{25+1}{2} = $ 13th largest, the median of 49 numbers is given by the value of the $\dfrac{49+1}{2} = $ 25th largest, and the median of 81 numbers is given by the value of the $\dfrac{81+1}{2} = $ 41st largest.

If we have an *even* number of items, there is never a middle item, and the median is defined as the mean of the values of the two middle items. For instance, the median of the six numbers 3, 6, 8, 10, 13, and 15 (which are already ordered according to size) is $\dfrac{8+10}{2} = 9$. It is halfway between the two middle values (here the 3rd and the 4th) and, if we interpret it correctly, the formula $\dfrac{n+1}{2}$ again gives the *position* of the median.

For the six given numbers the median is, thus, the value of the $\dfrac{6+1}{2} = $ 3.5th largest, and we interpret this as "halfway between the values of the third and the fourth." Similarly, the median of 100 numbers is given by the value of the $\dfrac{100+1}{2} = $ 50.5th largest item, or halfway between the values of the 50th and the 51st.

It is important to remember that the formula $\dfrac{n+1}{2}$ is *not* a formula for the median, itself; it merely tells us the position of the median, namely, the number of items we have to count until we reach the item whose value is the median (or the two items whose values have to be averaged to obtain the median).

It should not be surprising that the median and the mean of a set of data do not always coincide. Both of these measures describe the "middle" of a set of data, but they describe it in different ways. The median is central in the sense that it divides the data so that the values of half the items are less than or equal to the median while the values of the other half are greater than or equal to the median. The mean is central in a different sense, namely, in the sense of a *center of gravity*. (If we were to cut the histogram of a distribution out of cardboard, its center of gravity would lie on a vertical line through the point on the horizontal scale which corresponds to the mean.) Sometimes the median and the mean do coincide, though, and we interpret this as indicative of the *symmetry* of the data, a further property we shall discuss in Chapter 4.

The median has certain desirable properties, some of which it shares with the mean. Like the mean it always exists—that is, it can be calculated for any kind of numerical data—and it is always unique. Once the data are arranged according to size, the median is simple enough to find, but (unless the work is done with automatic equipment) ordering large sets of data can be a very tedious job. Unlike the mean, the median is *not* easily affected by extreme values, as is illustrated in Exercise 7 on page 56. Also unlike the mean, the median can be used to define the middle of a number of objects, properties, or qualities, which do not permit a quantitative description. It is possible, for instance, to rank a number of tasks according to their difficulty and then describe the middle one as being of "average" difficulty; also, we might rank samples of chocolate sauce according to their consistency and then describe the middle one as having "average" consistency. Perhaps the most important distinction between the median and the mean is that in problems of inference (estimation, prediction, and so on) the mean is usually *more reliable* than the median. In other words, the median is usually subject to greater chance fluctuations than the mean, as is illustrated in Exercise 8 on page 56 and in Exercise 6 on page 236.

So far as symbolism is concerned, we shall write the median of a set of x's (looked upon as a sample) as $\tilde{x}$. Some statisticians correspondingly represent population medians with the symbol $\tilde{\mu}$, but there is no real need to introduce this notation. In most problems of estimation we assume that the population mean and median coincide; in fact, we shall be interested in the median mainly in connection with problems where *sample medians* are used to estimate *population means*.

If we want to determine the median of a set of data that has already been grouped, we find ourselves in a position similar to the one in which we found ourselves on page 37; we can no longer determine the *actual* value of the median, although we can find the class into which the median must fall. What we do then is most easily understood with the aid of a diagram like that of Figure 3.1, showing again the histogram of the distribution of the scores of the 150 applicants. The median of a distribution is defined as a number, a point, which is such that *half the total area of the rectangles of the histogram lies to its left and half lies to its right.* This means that the sum of the areas of the rectangles to the left of the dashed line of Figure 3.1 equals the sum of the areas of the rectangles to its right. Note that

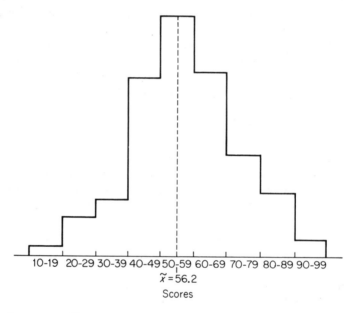

$$\tilde{x} = 56.2$$

Scores

Figure 3.1. Histogram showing the median of the distribution of the scores of the 150 applicants.

this definition is equivalent to the assumption that, within each class, the items are distributed (or spread out) evenly throughout the class interval.

In contrast to the problem of finding the median of ungrouped data, where we looked for the middle item (or items) and counted off $\dfrac{n+1}{2}$, we are now looking for a number, a dividing line, that divides the total area of the histogram into two equal parts, each representing a frequency of $\dfrac{n}{2}$. *Hence, to find the median of a distribution with a total frequency of n, we must, so to speak, count $\dfrac{n}{2}$ items starting at either end.*

To illustrate this procedure, let us refer again to the distribution of the 150 scores on page 14. Since $n = 150$ in this example, we will have to count $\frac{n}{2} = 75$ items from either end. Beginning at the bottom of the distribution, we find that 47 of the values are less than 50 while 89 are less than 60, so that the median must fall into the class whose limits are 50–59. Since 47 of the values fall below this class, we must count another $75 - 47 = 28$ of its 42 values, and we accomplish this by adding $\frac{28}{42}$ of the class interval of 10 to 49.5, the lower boundary of the class. (We add $\frac{28}{42}$ of the class interval because we want to count 28 of the 42 values contained in this class.) We thus get

$$\tilde{x} = 49.5 + \frac{28}{42} \cdot 10 = 56.2$$

rounded to one decimal.

Generally speaking, if L is the lower boundary of the class containing the median, f is its frequency, c the class interval, and j the number of items we still lack when reaching L, then the *median of the distribution* is given by the formula

$$\tilde{x} = L + \frac{j}{f} \cdot c$$

It is possible, of course, to arrive at the median of a distribution by starting at the other end and *subtracting* an appropriate fraction of the class interval from the upper boundary U of the class into which the median must fall. For the distribution of the scores we thus obtain

$$\tilde{x} = 59.5 - \frac{14}{42} \cdot 10 = 56.2$$

and the two answers are identical, as they should be. A general formula for the case where we start counting from the top of a distribution is given by

$$\tilde{x} = U - \frac{j'}{f} \cdot c$$

where j' is the number of items we still lack when reaching U.

The median belongs to a general class of statistical descriptions called *fractiles*, and we shall study them briefly at this time because the method we have just described can be used to determine any fractile of a distribution. Generally speaking, *a fractile is a value below which lies a given fraction of a set of data;* for the median this fraction is $\frac{1}{2}$. Similarly, the three quartiles Q_1, Q_2, and Q_3 are such that 25 per cent of the data fall below Q_1, 25 per cent fall between Q_1 and Q_2, 25 per cent fall between Q_2 and Q_3, and 25 per cent fall above Q_3. To find the first and third quartiles of a distribution (Q_2 actually *is* the median), we can use either of the two formulas given above. For Q_1 we count 25 per cent of the items starting at the bottom of the distribution, and for Q_3 we count 75 per cent of the items starting at the bottom or 25 per cent starting at the top. Referring again to the distribution of the scores, we count $\frac{150}{4} = 37.5$ values from both ends and get

$$Q_1 = 39.5 + \frac{21.5}{31} \cdot 10 = 46.4$$

and

$$Q_3 = 69.5 - \frac{8.5}{32} \cdot 10 = 66.8$$

The *deciles* D_1, D_2, ..., and D_9 are such that 10 per cent of the data fall below D_1, 10 per cent fall between D_1 and D_2, 10 per cent fall between D_2 and D_3, ..., and 10 per cent fall above D_9. For example, counting 40 per cent of the items starting at the bottom of the distribution of the scores we find that

$$D_4 = 49.5 + \frac{13}{42} \cdot 10 = 52.6$$

and counting 10 per cent of the items starting at the other end, we find that

$$D_9 = 79.5 - \frac{3}{17} \cdot 10 = 77.7$$

The *percentiles* of a distribution, P_1, P_2, ..., and P_{99} are such that 1 per cent of the data falls below P_1, 1 per cent falls between P_1 and P_2, ..., and 1 per cent falls above P_{99}. (Note that the fifth decile and the fiftieth percentile are both equal to the median, and that the twenty-fifth and seventy-fifth percentiles equal Q_1 and Q_3, respectively.)

Directly, fractiles serve to indicate above or below what values we can find certain percentages of a set of data; indirectly, they can also provide short-cut estimates of other kinds of statistical descriptions. For

Here 3 and 8 both occur with the highest frequency of three. (The fact that there is more than one mode is sometimes an indication that the data are not homogeneous, that is, they constitute a combination of several sets of data.) Another disadvantage of the mode is that if no two values are alike, the mode does not exist.

So far as the mode of grouped data is concerned, we have already pointed out that the class with the highest frequency is called the *modal class*. If one wants to be more specific, one can define the mode of a distribution as the class mark of the modal class or in one of the more elaborate ways which are discussed in some of the older texts.

The question of what particular "average" should be used in a given problem is not always easily answered. As we saw in Exercises 22 and 24 on pages 45 and 46, there are problems in which the nature of the data dictates the use of such averages as the geometric and harmonic means. The nature of the data may also dictate the use of a weighted mean, perhaps even a *weighted* geometric mean. We also saw that when dealing with qualitative data, we may have no choice but to use the mode or in some special instances the median.

A very interesting distinction between the mean, median, and mode will be brought out in Section 7.4, where we shall discuss their use in a problem of *decision making*. At that time we shall see that the choice between these three measures of central location (and others) may well depend on the risks, penalties, and rewards that are involved.

Generally speaking, the mean is by far the most widely used measure of central location, and we shall use it almost exclusively in the third part of this book. In problems of inference the mean has generally such definite advantages that other measures—for instance, the median—are used mainly because of their computational ease.

The fact that there is a certain amount of arbitrariness in the selection of statistical descriptions has led some persons to believe that they can take a set of data, apply the magic of statistics, and prove almost anything they want. In fact, a nineteenth century British statesman once said that there are three kinds of lies: lies, damned lies, and statistics. To give an example where such a criticism might be justified, suppose that a paint manufacturer asks his research department to "prove" that on the average a gallon of his paint covers more square feet than that of his two principal competitors. Suppose, furthermore, that the research department tests five cans of each brand, getting the following results (in square feet per gallon can):

Brand A: 488, 513, 515, 523, 526
Brand B: 521, 496, 522, 496, 520
Brand C: 495, 500, 494, 530, 506

instance, either the median or the *mid-quartile*, $\frac{1}{2} \cdot (Q_1 + $

to estimate the mean of a population, and in Chapter 11
a certain measure based on fractiles can be used to estim
standard deviation (which measures the variability of a p
this purpose it may be necessary to determine fractiles
ungrouped data, and how this can be done is explained in
page 58. Note also that fractiles can be read directly off
distribution without much loss of accuracy; in fact, we al
method in Exercises 9, 12, and 14 on pages 25 through 27

3.5 Further Measures of Central Location

Besides the mean and the median, there are numerous
describing the "middle," "center," or "average" of a set of
these, the *geometric* and *harmonic means* were introduced ea
cises 22 and 24 on pages 45 and 46, and the *mid-quartile* w
in the last paragraph. Two other measures of central locatio
tioning are the *mid-range* and the *mode*: The *mid-range* of a
given by the mean of the smallest value and the largest, and
serves as a quick estimate of the mean of the population fro
data were obtained; the *mode* is simply the value which occ
highest frequency. Thus, if more applicants for a job are
than any other age, we say that 26 is their *modal age* (that is,
26); similarly, in the distribution of the scores the class wh
limits 50–59 is the *modal class* (it has the highest frequency.)
advantage of the mode is that it requires no calculations, and i
value lies in the fact that it can be used with qualitative as we
titative data. For instance, if we wanted to study consumers'
for different kinds of food, different kinds of packaging, or diffe
of advertising, we could in each case determine the modal choic
problems we might even compare the modal preferences exp
several different groups. In situations like these we could not
the median or the mean; although the median can sometimes b
describe the "middle" of qualitative data, it is not applicable
objects can be *ordered* in some way.

 When dealing with quantitative data, the disadvantages of
outweigh its desirable features, and it is, in fact, rarely used. A
disadvantage of the mode is that it need not be unique, as is ill
by the following data:

3 5 8 3 6 3 4 9 8 5 8

If the manufacturer's own brand is Brand A, the person who is analyzing the data will find to his delight that the means of the three samples are, respectively, 513, 511, and 505. Hence, he can claim that in actual tests a can of his employer's product covered on the average more square feet than those of his competitors.

Now suppose, for the sake of argument, that the manufacturer's own brand is Brand B. Clearly, the person doing the analysis can no longer base the comparison on the sample means—this would not prove his point. Trying instead the sample medians, he finds that they are, respectively, 515, 520, and 500, and this provides him with exactly the kind of "proof" he wants. The median is a perfectly respectable measure of the "average" or "middle" of a set of data, and using the medians he can claim that his employer's product came out best in the test.

Finally, suppose the manufacturer's own brand is Brand C. After trying various measures of central location, the person doing the analysis finally comes upon one that does the trick: the *mid-range*, which we defined on page 53. Since the mid-ranges of the three samples are, respectively, 507, 509, and 512, he can claim that "on the average" his employer's product scored highest in the test. The moral of this example is that *if data are to be compared, the method of comparison should be decided on beforehand, or at least without looking at the actual data.* All this is aside from the fact that comparisons based on samples are often far from conclusive. It is quite possible that whatever differences there may be among the three means (or other descriptions) can be attributed entirely to chance (see Exercise 5 on page 346).

EXERCISES

1. During a crackdown on speeding, the members of a police department gave the following number of citations for speeding on thirteen consecutive days: 57, 61, 68, 57, 62, 50, 55, 62, 53, 54, 51, 59, and 52. Calculate the mean, the median, and the mid-range of these data.

2. In Belgium, 61.7, 59.9, 61.9, 59.4, 59.8, 62.7, 63.9, and 68.4 thousand metric tons of fish were caught in the years 1961 through 1968. Find the median.

3. Find the median of the 25 speeds of Exercise 4 on page 42.

4. Find the median of the water bills of Exercise 5 on page 42.

5. Find the median of the 16 measurements of Exercise 6 on page 42.

6. With reference to Exercise 8 on page 42 find
   (a) the median of the 20 values;

(b) the median of the 20 values after correcting the first one, which should have been 39 instead of 37;

(c) the median of the 17 values which remain after discarding the three largest values, which were presumably recorded incorrectly.

7. Each of 15 insurance salesmen was assigned a quota (number of policies) which he should write during the first three months of the year, and the following are the percentages of their respective quotas which they actually attained: 104, 97, 80, 111, 94, 92, 88, 104, 385, 76, 93, 105, 108, 82, and 90. Calculate the median and the mean of these percentages, and indicate which of the two measures is a better indication of these insurance salesmen's "average" performance.

8. To verify the claim that the mean is generally *more reliable* than the median (namely, that it is subject to smaller chance fluctuations), a student conducted an experiment consisting of 12 tosses of three dice. The following are his results: 2, 4, and 6; 5, 3, and 5; 4, 5, and 3; 5, 2, and 3; 6, 1, and 5; 3, 2, and 1; 3, 1, and 4; 5, 5, and 2; 3, 3, and 4; 1, 6, and 2; 3, 3, and 3; 4, 5, and 3.

(a) Calculate the twelve medians and the twelve means.

(b) Group the medians and the means obtained in (a) into separate distributions having the classes 1.5–2.5, 2.5–3.5, 3.5–4.5, and 4.5–5.5. (Note that there will be no ambiguities since the medians of three whole numbers and the means of three whole numbers cannot equal 2.5, 3.5, or 4.5.)

(c) Draw histograms of the two distributions obtained in (b) and explain how they illustrate the claim that the mean is generally more reliable than the median.

9. Repeat Exercise 8 with your own data by repeatedly rolling three dice (or one die three times) and construct corresponding distributions for the twelve medians and the twelve means. (If no dice are available, *simulate* the experiment mentally or by drawing numbered slips of paper out of a hat.)

10. Referring to the weight distribution of Exercise 5 on page 24, find

(a) the median;

(b) Q_1 and Q_3;

(c) the deciles D_1 and D_9.

11. Find the median of the air pollution data of Exercise 9 on page 25

(a) without grouping the data;

(b) using the distribution obtained in that exercise.

Also compare the two values obtained with the one read off the ogive in part (d) of the exercise.

12. Referring to the distribution of customers served by a restaurant, which the reader was asked to obtain in Exercise 10 on page 25, find

(a) the median;

(b) Q_1, Q_3, and the mid-quartile;

(c) the deciles D_2 and D_8;

(d) the percentiles P_5 and P_{95}.

13. Find the median of the amounts of Exercise 11 on page 26
    (a) without grouping the data;
    (b) using the distribution obtained in that exercise.

14. Use the distribution obtained in Exercise 12 on page 26 to find
    (a) the median;
    (b) Q_1 and Q_3.

    Also compare the value of Q_1 with that read off the ogive in part (d) of that exercise.

15. Use the distribution obtained in Exercise 12 on page 26 to find the nine deciles D_1, D_2, D_3, ..., D_8 and D_9.

16. Find the median of the breaking strengths of Exercise 13 on page 27
    (a) without grouping the data;
    (b) using the distribution obtained in that exercise.

17. Use the distribution obtained in Exercise 13 on page 27 to find
    (a) Q_1 and Q_3;
    (b) the percentiles P_{15} and P_{85}.

18. Use the distribution obtained in Exercise 14 on page 27 to find
    (a) the median;   (c) the deciles D_2 and D_8;
    (b) Q_1 and Q_3;   (d) the percentiles P_1 and P_{99}.

    Also compare the results of part (c) with the values read off the ogive in part (d) of Exercise 14 on page 28.

19. Find the mode and the mid-range of the raw (ungrouped) scores on page 13.

20. Find the mode (if it exists) of each of the following sets of golf scores:
    (a) 73, 77, 74, 75, 77, 75, 76, 76, 78, 76, 76, 75, and 74;
    (b) 79, 79, 76, 77, 79, 76, 71, 76, 73, 75, 79, 76, and 72;
    (c) 73, 99, 86, 78, 82, 74, 108, 90, 105, 81, and 97.

21. Referring to Exercise 4 on page 42, find the mode and the mid-range of the 25 speeds.

22. Referring to the ungrouped data of Exercise 9 on page 25, find the mode of the 80 air pollution readings.

23. After a visit to the Phoenix Zoo, fifty elementary children were asked what animals they liked best, and their replies were: lions, zebras, bears, monkeys, tigers, lions, giraffes, monkeys, elephants, coyotes, bears, camels, monkeys, giraffes, giraffes, elephants, monkeys, tigers, monkeys, lions, coyotes, bears, bears, bears, lions, monkeys, tigers, monkeys, giraffes, bears, otters, monkeys, monkeys, lions, bears, lions, alligators, giraffes, elephants, tigers, elephants, elephants, monkeys, zebras, lions, coyotes, giraffes, bears, elephants, and monkeys. What is their modal choice?

24. In general, we define the jth k-tile, $F_{j/k}$, to be such that j/k is the fraction of the data falling below $F_{j/k}$. For grouped data, we calculate $F_{j/k}$ by the method indicated in the text; for ungrouped data, we arrange the n observations according to size and let $F_{j/k}$ be given by the value of the $\dfrac{j(n+1)}{k}$ th largest observation.

(a) Is this rule consistent with the definition of the median for ungrouped data?

(b) Find $F_{1/3}$ and $F_{2/3}$ for the distribution of the 200 grades obtained in Exercise 14 on page 27.

(c) Find the two quartiles $F_{1/4}$ and $F_{3/4}$ for the 25 speeds of Exercise 4 on page 42. Note that if $\dfrac{j(n+1)}{k}$ is not a whole number, we have to *interpolate* between successive observations. For instance, to find the $5\frac{1}{4}$th largest value in an ordered array of numbers, we must go $\frac{1}{4}$ of the way from the 5th to the 6th. Thus, if the 5th largest number is 13 and the 6th largest number is 15, then the "$5\frac{1}{4}$th largest" is 13.5.

(d) Find $F_{1/5}$ and $F_{4/5}$, namely, the two deciles D_2 and D_8, for the twenty figures of Exercise 8 on page 20. [See comment to part (c).]

3.6 Technical Note (Subscripts and Summations)

In order to make the formulas of statistics applicable to any kind of data, it is customary to represent measurements or observations by means of letters such as x, y, and z. Furthermore, to make it easy to refer to individual measurements within a set of data we use *subscripts*, writing, for example, x_1, x_2, and x_3 for the weights of three guinea pigs. Thus, n measurements may be referred to as x_1, x_2, ..., and x_n, and if we want to refer to one of these measurements in general, we write x_i, where i can assume any one of the values from 1 to n.

Instead of writing the subscript as i we could just as well have used j, k, or some other letter, and instead of x we could have used y, z, or some other letter. Generally speaking, *it is customary to use different letters for different kinds of measurements and the same letter with different subscripts for different measurements of the same kind.* Thus, we might write x_{31} and x_{57} for the heights of the thirty-first and fifty-seventh persons on a list, and we might write x_{24}, y_{24}, and z_{24} for the I.Q., age, and income of the twenty-fourth person in a certain sample.

In connection with this notation we must be careful not to confuse, say, the subscript 24 appearing in x_{24} with the double subscript "2 and 4" used, for example, to indicate that x_{24} is the morning temperature of the 2nd patient on the 4th day after an operation. In advanced work in statistics we frequently use two or more subscripts to indicate, for example, that x_{ijk} is the cost of operating the ith washing machine with the jth kind of detergent and the kth kind of load. Except for Chapter 13 and Section 17.8, we shall use only single subscripts in the remainder of this book.

When we introduced the Σ notation on page 34, we referred to it as a kind of mathematical shorthand; in fact, we pointed out that it is an abbreviated sort of shorthand, as it does not tell us explicitly what x's, or how many of them, we are supposed to add. This is taken care of in the more explicit notation

$$\sum_{i=1}^{n} x_i = x_1 + x_2 + \ldots + x_n$$

where it is made clear that we are adding the x's whose subscript i is 1, 2, ..., and n. Generally, we shall not use this explicit notation in the text, in order to simplify the over-all appearance of the formulas, assuming that it is clear in each case what x's we are referring to and how many there are.

Using the Σ notation, we shall also have the occasion to write such expressions as Σx^2, Σxy, $\Sigma x^2 f$, ..., which (more explicitly) represent the sums

$$\sum_{i=1}^{n} x_i^2 = x_1^2 + x_2^2 + x_3^2 + \ldots + x_n^2$$

$$\sum_{j=1}^{m} x_j y_j = x_1 y_1 + x_2 y_2 + \ldots + x_m y_m$$

$$\sum_{i=1}^{n} x_i^2 f_i = x_1^2 f_1 + x_2^2 f_2 + \ldots + x_n^2 f_n$$

Working with several subscripts, we may have to evaluate a *double summation* such as

$$\sum_{j=1}^{3} \sum_{i=1}^{4} x_{ij} = \sum_{j=1}^{3} (x_{1j} + x_{2j} + x_{3j} + x_{4j})$$
$$= x_{11} + x_{21} + x_{31} + x_{41} + x_{12} + x_{22}$$
$$+ x_{32} + x_{42} + x_{13} + x_{23} + x_{33} + x_{43}$$

To verify some of the formulas involving summations that are stated, but not proved, in the text, the reader will find it convenient to use the following rules:

$$Rule\ A:\quad \sum_{i=1}^{n} (x_i \pm y_i) = \sum_{i=1}^{n} x_i \pm \sum_{i=1}^{n} y_i$$

$$Rule\ B:\quad \sum_{i=1}^{n} k \cdot x_i = k \cdot \sum_{i=1}^{n} x_i$$

$$Rule\ C:\quad \sum_{i=1}^{n} k = k \cdot n$$

The first of these rules states that the summation of the sum (or difference) of two terms equals the sum (or difference) of the individual summations, and it can be extended to the sum or difference of more than two terms. The second rule states that we can, so to speak, factor a constant out of a summation, and the third rule states that the summation of a constant is simply n times that constant. All of these rules can be proved by actually writing out in full what each of the summations represents.

EXERCISES

1. Write each of the following in full, that is, without summation signs:

   (a) $\sum_{i=1}^{8} x_i$;   (c) $\sum_{i=1}^{4} x_i y_i$;   (e) $\sum_{i=2}^{6} x_i^2$;

   (b) $\sum_{i=1}^{6} y_i$;   (d) $\sum_{j=1}^{4} x_j f_j$;   (f) $\sum_{j=1}^{5} (x_j - y_j)$.

2. Write each of the following as summations:

   (a) $z_1 + z_2 + z_3 + z_4$;

   (b) $x_3 + x_4 + x_5 + x_6 + x_7 + x_8 + x_9 + x_{10}$;

   (c) $x_1 f_1 + x_2 f_2 + x_3 f_3 + x_4 f_4 + x_5 f_5$;

   (d) $y_1^2 + y_2^2 + y_3^2 + y_4^2$;

   (e) $3x_1 + 3x_2 + 3x_3 + 3x_4 + 3x_5$;

   (f) $(x_2 - y_2) + (x_3 - y_3) + (x_4 - y_4) + (x_5 - y_5) + (x_6 - y_6)$;

   (g) $(z_1 - 2) + (z_2 - 2) + (z_3 - 2)$;

   (h) $x_1^2 f_1 + x_2^2 f_2 + x_3^2 f_3 + x_4^2 f_4 + x_5^2 f_5 + x_6^2 f_6$.

3. Given $x_1 = 3$, $x_2 = 2$, $x_3 = -1$, $x_4 = 1$, $x_5 = -2$, and $x_6 = 4$, find

   (a) $\sum_{i=1}^{6} x_i$; (b) $\sum_{i=2}^{5} x_i$; and (c) $\sum_{i=1}^{6} x_i^2$.

4. Given $x_1 = 2$, $x_2 = 5$, $x_3 = 4$, $x_4 = 3$, $f_1 = 2$, $f_2 = 5$, $f_3 = 1$, and $f_4 = 2$, find

   (a) $\sum_{i=1}^{4} x_i$;   (c) $\sum_{i=1}^{4} x_i f_i$;

   (b) $\sum_{i=1}^{4} f_i$;   (d) $\sum_{i=1}^{4} x_i^2 f_i$.

5. Given $x_1 = 1$, $x_2 = 2$, $x_3 = 2$, $y_1 = -1$, $y_2 = 3$, $y_3 = 1$, $z_1 = 3$, $z_2 = -1$, and $z_3 = 2$, find

  (a) $\sum_{i=1}^{3} x_i y_i$;   (c) $\sum_{i=1}^{3} y_i z_i$;   (e) $\sum_{i=1}^{3} (x_i + y_i + z_i)$;

  (b) $\sum_{i=1}^{3} x_i z_i$;   (d) $\sum_{i=1}^{3} x_i y_i z_i$;   (f) $\sum_{i=1}^{3} (x_i + 2y_i - z_i)$.

6. Prove that $\sum_{i=1}^{n} (x_i - \bar{x}) = 0$, where $\bar{x} = \frac{1}{n} \cdot \sum_{i=1}^{n} x_i$.

7. If the class marks of a distribution are coded so that $x_i = c \cdot u_i + x_0$, where the x_i are the class marks in the original scale, the u_i are the class marks in the coded scale, c is the class interval, and x_0 is the class mark (in the original scale) which is the origin of the coded scale, prove that $\bar{x} = c \cdot \bar{u} + x_0$, where $\bar{x}$ is the mean of the distribution in the original scale and $\bar{u}$ is its mean in the coded scale.

8. Given $x_{11} = 3$, $x_{12} = 1$, $x_{13} = -2$, $x_{14} = 2$, $x_{21} = 1$, $x_{22} = 4$, $x_{23} = -2$, $x_{24} = 5$, $x_{31} = 3$, $x_{32} = -1$, $x_{33} = 2$, and $x_{34} = 3$, find

  (a) $\sum_{i=1}^{3} x_{ij}$ separately for $j = 1, 2, 3$, and 4;

  (b) $\sum_{j=1}^{4} x_{ij}$ separately for $i = 1, 2$, and 3.

9. With reference to the data of Exercise 8, evaluate the double summation $\sum_{i=1}^{3} \sum_{j=1}^{4} x_{ij}$ using

  (a) the results obtained in part (a) of Exercise 8;
  (b) the results obtained in part (b) of Exercise 8.

10. Using the values of Exercise 8, find

  (a) $\sum_{i=1}^{3} \sum_{j=1}^{4} x_{ij}^2$ (b) $\sum_{i=1}^{3} \left[\sum_{j=1}^{4} x_{ij} \right]^2$ (c) $\sum_{j=1}^{4} \left[\sum_{i=1}^{3} x_{ij} \right]^2$

11. Is it true in general that $\left[\sum_{i=1}^{n} x_i \right]^2 = \sum_{i=1}^{n} x_i^2$? (*Hint:* check whether the equation holds for $n = 2$).

12. Prove that $\sum_{i=1}^{n} (x_i - k)^2 = \sum_{i=1}^{n} x_i^2 - 2k \cdot \sum_{i=1}^{n} x_i + nk^2$.

13. Given $\sum_{i=1}^{4} x_i = 7$ and $\sum_{i=1}^{4} x_i^2 = 15$, use the formula of Exercise 12 to evaluate $\sum_{i=1}^{4} (x_i + 3)^2$.

BIBLIOGRAPHY

More detailed treatments of index numbers may be found in most textbooks on business statistics, for instance, in

FREUND, J. E., and WILLIAMS, F. J., *Elementary Business Statistics: The Modern Approach*, 2nd ed. Englewood Cliffs, N.J.: Prentice-Hall, Inc., 1972.

An informal discussion of the ethics involved in choosing measures of location is given in the book by D. Huff listed on page 30.

4

Measures of Variation

and Further Descriptions

4.1 Introduction

An important characteristic of most sets of data is that the values are generally *not all alike;* indeed, the extent to which they are unalike, or vary among themselves, is of fundamental importance in statistics. Since the measures discussed in Chapter 3 do not tell us anything about this characteristic, we shall now investigate other measures, so-called *measures of variation*, which tell us something about the extent to which data are dispersed, spread out, or bunched.

First, let us give a few examples to illustrate the importance of measuring variability. Suppose, for instance, that a none-too-honest land developer claims that the average temperature at his development is a "comfortable" 75 degrees. This figure may be correct, but it certainly does not convey the right impression if a good part of the time the temperature is a frigid 35 degrees, while more than half the time it is close to 100 degrees. What is needed here is not only an average, but also a measure of the size of the fluctuations, namely, a measure of the variability of the temperature.

To consider another example, suppose that the owner of a garage has to choose between two mechanics applying for a job and that, to be objective, he asks each of them to remove, clean, and replace the carburators of five cars. If it turns out that they both average about three hours per car, it may seem that they are equally good, but this would be misleading if the first mechanic took, respectively, 168, 195, 212, 149, and 176 minutes to work on the five cars, while the second mechanic took, respectively, 174, 183, 165, 186, and 192 minutes. Thus, it would seem that the work of the second mechanic is fairly *consistent*, while that of the first mechanic is rather *erratic*—he couldn't have done a very good job in only 149 minutes, and he must have been thinking about something else when he took as much as 212 minutes. This, of course, should influence the garage owner's decision, and it illustrates the importance of considering, and in some way measuring, the variability of data, even if it is only by inspection.

The concept of variability is of special importance in statistical inference, because it is here that we have to cope with questions of *chance variation*. To illustrate the meaning of this term, suppose that a balanced coin is flipped 100 times. Although we may expect 50 heads and 50 tails, we would certainly not be surprised if we got, say, 54 heads and 46 tails, 49 heads and 51 tails, or 53 heads and 47 tails. Most likely, we would ascribe the occurrence of a few extra heads or a few extra tails to chance. In order to study this phenomenon called chance, suppose we repeatedly flip a balanced coin 100 times and that in 10 such "experiments" we obtain 48, 56, 50, 53, 49, 46, 51, 48, 44, and 56 heads. This gives us some idea about the magnitude of the fluctuations (variations) produced by chance in an experiment of this kind, and knowledge like this would be important, for example, if we had to decide whether there was anything "phony" about a coin (or the person by whom it is tossed), if 100 tosses yield 28 heads and 72 tails. Judging by the above data, it would seem that most of the time we should get anywhere from 40 to 60 heads (in the experiment the number of heads ranged from 44 to 56); hence, we may well conclude that there is something wrong with the coin or the way in which it was flipped. We have presented this argument here on an intuitive basis to demonstrate the importance of measuring variability; it will be treated more formally in Chapter 8.

As a final example, let us consider a problem in which the *variability of the population* from which the sample is drawn plays an important role. Suppose we want to estimate the average I.Q. of two groups of individuals. The first group consists of the high school seniors in a fairly large community, and the second group consists of all the graduate students enrolled in a state university. So far as intelligence is concerned, the first group is not very homogeneous, and the I.Q.'s of the high school students may well vary anywhere from 85 to 140. The graduate students, on the other hand, must all have pretty high I.Q.'s, most of which will probably fall on the interval from 115 to 135. Thus, it stands to reason that if we take the same number of measurements for each group (say, we measure the I.Q.'s of 10 of the high school students and 10 of the graduate students), the information gained from the graduate students is apt to be *more reliable.* In other words, the sample mean obtained for the graduate students is apt to be closer than that of the high school seniors to the respective *true* average I.Q.'s. This illustrates the fact that *in order to evaluate the "closeness" of an estimate or the "goodness" of a generalization one must know something about the variability of the population from which the sample is obtained.* There is not much variability among the I.Q.'s of the graduate students, and a sample mean cannot help but be fairly close; so far as the high school seniors are concerned, their I.Q.'s vary considerably and a sample mean may well be off by a large amount.

We have given all these examples to show that the concept of variability plays a fundamental role in the analysis of statistical data; in what follows, we shall learn how to measure it statistically. To introduce one way of measuring variability, let us refer back to the second example, and let us observe that the times of the first mechanic ranged from 149 minutes to 212 minutes, while those of the second mechanic ranged from 165 to 192 minutes. These *extreme* (smallest and largest) *values* give us some indication of the variability of the respective sets of data, and more or less the same would be accomplished by giving the *differences* between the respective extreme values, which we refer to as the *ranges* of the two samples. Thus, the times of the first mechanic have a *range* of $212 - 149 = 63$ minutes, while those of the second mechanic have a *range* of $192 - 165 = 27$ minutes.

In spite of the obvious advantage of the range that it is *easy to calculate* and *easy to understand,* it is generally not looked upon as a very useful measure of variation. As it is based only on the two extreme values, its main shortcoming is that it does not tell us anything about the actual *dispersion* of the data which fall in between. Each of the following three sets of data

Set 1:	8	20	20	20	20	20	20	20	20	20
Set 2:	8	8	8	8	8	20	20	20	20	20
Set 3:	8	9	11	12	13	15	15	18	19	20

has a range of $20 - 8 = 12$, but the dispersions of the data are by no means the same. As a result, the range is used mainly in situations where it is desired to get a quick, though not necessarily very accurate, picture of the variability of a set of data. When the sample size is very small, the range is quite adequate, and it is, thus, used widely in *quality control,* where it is important to keep a continuous check on the variability of raw materials, machines, or manufactured products (see also Exercise 8 on page 292).

4.2 The Standard Deviation

Since the variation of a set of numbers is *small* if they are bunched closely about their mean and it is *large* if they are spread over considerable distances away from their mean, it would seem reasonable to define variation in terms of the distances (deviations) by which numbers depart from their mean. If we have a set of numbers $x_1, x_2, \ldots,$ and x_n, whose mean is $\bar{x}$, we can write the amounts by which they differ from their mean as $x_1 - \bar{x}$, $x_2 - \bar{x}, \ldots,$ and $x_n - \bar{x}$. These quantities are called the *deviations from the mean* and it suggests itself that we might use their average, namely, their *mean*, as a measure of the variation of the n numbers. This would not be a bad idea, if it were not for the fact that we would always get 0 for an answer, no matter how widely dispersed the data might be. As the reader was asked to show in Exercise 6 on page 61, the quantity $\Sigma (x - \bar{x})$ is *always* equal to zero—some of the deviations are positive, some are negative, but they "average out," that is, their sum as well as their mean are always equal to zero.

Since we are really interested in the *magnitude* of the deviations and not in their signs, we might simply "ignore" the signs and, thus, define a measure of variation in terms of the *absolute values* of the deviations from the mean. Indeed, if we added the values of the deviations from the mean as if they were all positive and divided by n, we would obtain a measure of variation called the *mean deviation* (see Exercises 5 and 6 on page 75). Unfortunately, this measure of variation has the drawback that, owing to the absolute values, it is difficult to subject it to any sort of theoretical treatment; for instance, it is difficult to study mathematically how *in problems of sampling*, mean deviations are affected by chance. However, there exists another way of eliminating the signs of the deviations from the mean, which is preferable on theoretical grounds: The *squares* of the deviations from the mean cannot be negative; in fact, they are positive unless a value happens to coincide with the mean, in which case both $x - \bar{x}$ and $(x - \bar{x})^2$ are equal to zero. Thus, if we average the squared deviations from the mean and then take the square root to compensate for the fact that the deviations were squared, we obtain the formula

$$\sqrt{\frac{\Sigma (x - \bar{x})^2}{n}}$$

and this is how, traditionally, the *standard deviation* has been defined. Expressing literally what we have done here mathematically, it has also been called the *root-mean-square deviation*.

Nowadays, it has become the custom among most statisticians and research workers to make a slight modification in this definition, which consists of dividing the sum of the squared deviations from the mean by $n - 1$ instead of n. Following this practice, which will be explained later, let us thus formally define s, the *sample standard deviation*, as

$$\blacktriangle \qquad s = \sqrt{\frac{\Sigma\,(x - \bar{x})^2}{n - 1}} \qquad \blacktriangle$$

and its square, the *sample variance*, as

$$\blacktriangle \qquad s^2 = \frac{\Sigma\,(x - \bar{x})^2}{n - 1} \qquad \blacktriangle$$

Note that to facilitate the calculation of standard deviations a table of square roots, Table XII, is given at the end of the book; the use of this table is explained on page 496.

The formulas we have given so far in this section are meant to apply to samples, but if we substitute μ for $\bar{x}$ and N for n, we obtain analogous formulas for the standard deviation and the variance of a *population*. It has become fairly general practice to write population standard deviations as σ (small Greek *sigma*) when dividing by N and as S when dividing by $N - 1$; symbolically,

$$\blacktriangle \qquad \sigma = \sqrt{\frac{\Sigma\,(x - \mu)^2}{N}} \quad \text{and} \quad S = \sqrt{\frac{\Sigma\,(x - \mu)^2}{N - 1}} \qquad \blacktriangle$$

To explain why we divide by $n - 1$ instead of n and $N - 1$ instead of N in the formulas for s and S, let us point out that if we wanted to use sample variances to estimate the respective variances of the populations from which the samples were obtained, division by n instead of $n - 1$ would give us values which *on the average* are too small. We cannot prove the following at the level of this book, but it is shown in most textbooks on mathematical statistics that the values would be too small on the average by the factor $\dfrac{n - 1}{n}$. For instance, for $n = 5$ the estimates would on the average be $\dfrac{5 - 1}{5} = 0.80$ or 80 per cent of what they should be, and hence 20 per cent too small. To compensate for this we divide by $n - 1$ instead of n in the formulas for the sample standard deviation and the sample variance. As the statisticians say, this makes the sample variance s^2 *unbiased;* that is, if we calculate s^2 for several samples taken from the same population, the values we get should average S^2, the variance of the population. Note, however, that this modification is of no significance unless n is small; generally, its effect is negligible when n is large, say 100 or more. The same applies to the difference between σ^2 and S^2, which is negligible unless the size of the population is very small, and in actual practice this is usually not the case.

To illustrate the calculation of a sample standard deviation, let us find s for the following data on the number of burglaries reported in a town during the first six weeks of 1972: 12, 18, 7, 11, 15, and 9. First calculating $\bar{x}$, we get

$$\bar{x} = \frac{12 + 18 + 7 + 11 + 15 + 9}{6} = 12$$

and then the remainder of the calculations are as shown in the following table

x	$x - \bar{x}$	$(x - \bar{x})^2$
12	0	0
18	6	36
7	−5	25
11	−1	1
15	3	9
9	−3	9
	0	80

and

$$s = \sqrt{\frac{80}{6 - 1}} = \sqrt{16} = 4$$

Thus, $\bar{x} = 12$ provides us with an estimate of the average number of burglaries in this town per week, and the value of the standard deviation, $s = 4$, tells us something about the variability of the figures from week to week. How such a value of s is to be interpreted will be discussed on page 72, and how it can be used to judge how close $\bar{x} = 12$ might be to μ, the *true* average number of burglaries in this town per week, will be discussed in Chapter 10. (Note that in the above table we totaled, as a check, the $x - \bar{x}$ column, and as we have indicated on page 66, the sum of its values must always equal zero.)

The calculation of s was very easy in this example, and this was due largely to the fact that the x's, their mean, and hence also the deviations from the mean were all whole numbers. Had this not been the case, it might have been profitable to use the following *short-cut formula for s:*

$$s = \sqrt{\frac{n(\Sigma\, x^2) - (\Sigma\, x)^2}{n(n - 1)}}$$

This formula does not involve any approximations and it can be derived from the other formula for s by using the rules for summations given on page 60 (see Exercise 32 on page 78). The advantage of this short-cut formula is that we do not have to go through the process of actually finding the deviations from the mean; instead we calculate $\Sigma\, x$, the sum of the x's, $\Sigma\, x^2$, the sum of their squares, and substitute directly into the formula. Referring again to the burglary data, we now have

x	x^2
12	144
18	324
7	49
11	121
15	225
9	81
72	944

and

$$s = \sqrt{\frac{6(944) - (72)^2}{6 \cdot 5}} = 4$$

It appears that in this particular example the "short-cut" method is actually more involved; this may be the case, but in actual practice, when we are dealing with realistically complex data, the short-cut formula usually provides considerable simplifications.

To demonstrate the advantages of the short-cut formula, let us determine the sample variance of the numbers 12, 7, 9, 5, 4, 8, 17, 2, 11, 14, 13, and 9, using first the formula on page 67 and then the short-cut formula on page 68. Without using the short-cut formula we get

x	$(x - \bar{x})$	$(x - \bar{x})^2$
12	2.75	7.5625
7	−2.25	5.0625
9	−0.25	0.0625
5	−4.25	18.0625
4	−5.25	27.5625
8	−1.25	1.5625
17	7.75	60.0625
2	−7.25	52.5625
11	1.75	3.0625
14	4.75	22.5625
13	3.75	14.0625
9	−0.25	0.0625
111	0	212.2500

and

$$\bar{x} = \frac{111}{12} = 9.25, \qquad s^2 = \frac{212.2500}{11} = 19.3$$

and working with the short-cut formula we get

x	x^2
12	144
7	49
9	81
5	25
4	16
8	64
17	289
2	4
11	121
14	196
13	169
9	81
111	1,239

and

$$s^2 = \frac{12(1,239) - (111)^2}{12 \cdot 11} = 19.3$$

Here the mean was 9.25, not a whole number, and the short-cut formula provided considerable simplifications.

A further simplification in the calculation of s or s^2 consists of adding an arbitrary positive or negative number to each measurement. It is easy to prove that this would have no effect on the final result, and had we used this trick in the last example, we might have subtracted 10 (added -10) from each number, getting 2, -3, -1, -5, -6, -2, 7, -8, 1, 4, 3, and -1 instead of the original numbers. The sum of these numbers is -9, the sum of their squares is 219, and substitution into the formula for s^2 yields

$$s^2 = \frac{12(219) - (-9)^2}{12 \cdot 11} = 19.3$$

which is exactly what we had before. Since the purpose of this trick is to reduce the size of the numbers with which we have to work, it is usually desirable to subtract a number that is close to the mean. In our example the mean was 9.25, and the calculations might have been even simpler if we had subtracted 9 instead of 10. Although the short-cut formula on page 68 was given for use with samples, we have only to substitute N for n throughout to make the formula applicable to the calculation of S^2 or S. In Exercise 33 on page 79, the reader will be asked to derive a similar short-cut formula for the calculation of σ^2 or σ.

If we want to calculate the standard deviation of data which have already been grouped, we are faced with the same problem as on page 37. Proceeding as we did in connection with the mean, and assigning the value of the class mark to each value falling into a given class, we obtain the formula

$$s = \sqrt{\frac{\Sigma\,(x - \bar{x})^2 \cdot f}{n - 1}}$$

and, if we substitute μ for $\bar{x}$ and N or $N - 1$ for $n - 1$, we obtain analogous formulas for σ and S. Note that in this formula the x's are now the class marks and the f's are the corresponding class frequencies.

The above formula serves to *define* s for grouped data, but it is seldom used in actual practice. Either we use a *computing formula* analogous to the one on page 68, namely,

$$\blacktriangle \qquad s = \sqrt{\frac{n(\Sigma\,x^2 f) - (\Sigma\,x f)^2}{n(n - 1)}} \qquad \blacktriangle$$

where the x's are the class marks and the f's the corresponding class frequencies, or we use the same kind of *coding* as in the calculation of the mean of grouped data. Following the notation on page 38, we obtain

$$\blacktriangle \qquad s = c\sqrt{\frac{n(\Sigma\,u^2 f) - (\Sigma\,u f)^2}{n(n - 1)}} \qquad \blacktriangle$$

This is the *short-cut formula for computing the standard deviation of grouped data*. Note that this formula can be used only when the class intervals are all equal.

Although this short-cut formula may look fairly complicated, it makes the calculation of s very easy. Instead of having to work with the actual class marks and the deviations from the mean, we have only to find the sum of the products obtained by multiplying each u by the corresponding f, the sum of the products obtained by multiplying the square of each u by the corresponding f, and substitute into the formula.

To illustrate the use of this short-cut formula for the calculation of s for grouped data, let us refer again to the distribution of the scores of the

150 applicants on page 14. Using the same u-scale as on page 38, we get

Class Marks x	u	f	$u \cdot f$	$u^2 \cdot f$
14.5	−4	1	−4	16
24.5	−3	6	−18	54
34.5	−2	9	−18	36
44.5	−1	31	−31	31
54.5	0	42	0	0
64.5	1	32	32	32
74.5	2	17	34	68
84.5	3	10	30	90
94.5	4	2	8	32
			33	359

and

$$s = 10 \cdot \sqrt{\frac{150(359) - (33)^2}{150 \cdot 149}} = 15.4$$

The variation of the scores of the 150 applicants is, thus, measured by a standard deviation of 15.4, and we shall indicate below how such a figure might be interpreted.

We have computed the standard deviation of grouped data under the assumption that all measurements falling into a class are located at its class mark. The error introduced by this assumption, which is appropriately called a *grouping error*, can be appreciable when the class interval c is very large. A correction which compensates for this error, called *Sheppard's correction*, is mentioned in the Bibliography on page 84.

In the argument which led to the definition of the standard deviation we observed that the dispersion of a set of data is small if the values are bunched closely about their mean and that it is large if the values are spread over considerable distances away from the mean. Correspondingly, we can now say that *if the standard deviation of a set of data is small, the values are concentrated near the mean, and if the standard deviation is large, the values are scattered widely about the mean.* To present this argument on a less intuitive basis (after all, *what is small and what is large?*), let us mention an important theorem, called *Chebyshev's Theorem* after the Russian mathematician P. L. Chebyshev (1821–1894). According to this theorem, which we shall study further in Chapter 8,

at least $\left(1 - \dfrac{1}{k^2}\right) \cdot 100$ *per cent of any set of data fall within* k *standard deviations of their mean.*

Thus, for $k = 2$ we can say that least $\left(1 - \dfrac{1}{2^2}\right)\cdot 100 = 75$ per cent of any data must fall within 2 standard deviations of their mean, and with reference to the 150 scores on page 13, we find that at least 75 per cent of the scores must fall between $\bar{x} - 2s = 56.7 - 2(15.4) = 25.9$ and $\bar{x} + 2s = 56.7 + 2(15.4) = 87.5$. Checking the data, themselves, we find that actually 144 of the 150 values, or 96 per cent, lie between 25.9 and 87.5, so that "at least 75 per cent" *is* correct.

Chebyshev's Theorem also tells us that for $k = 5$ *at least* $\left(1 - \dfrac{1}{5^2}\right)\cdot 100$ $= 96$ per cent of any data must fall within 5 standard deviations of their mean, and that for $k = 10$ *at least* $\left(1 - \dfrac{1}{10^2}\right)\cdot 100 = 99$ per cent of any data must fall within 10 standard deviations of their mean. Thus if all the 1-pound cans of coffee filled by a food processor have a mean weight of 16 ounces with a standard deviation of 0.02 ounces, we find that *at least* 96 per cent of the cans contain between 15.9 and 16.1 ounces of coffee, and *at least* 99 per cent of the cans contain between 15.8 and 16.2 ounces of coffee. Actually, Chebyshev's Theorem has more significant applications in problems of inference, that is, in connection with problems where we make generalizations, and some of these will be discussed in Chapters 8 and 9. So far as the work of this chapter is concerned, we have mentioned Chebyshev's Theorem only to give the reader an intuitive feeling of *how* the standard deviation is indicative of the "spread" or "dispersion" of a set of data.

In the beginning of this section we demonstrated that there are many ways in which knowledge of the variability of a set of data can be important. Another interesting application arises in the comparison of numbers belonging to two or more *different* sets of data. To illustrate, let us suppose that Tom and George, who attend *different* universities, received grades of 72 and 76, respectively, in final examinations in Freshman Sociology. If their respective classes averaged 54 and 52, we can conclude that they both scored above average, but we could say more if we also knew the corresponding standard deviations. Thus, let us suppose that the Freshman Sociology grades of students in the two universities had standard deviations of 20 and 12 points, respectively. With this information we can now say that Tom was $\dfrac{72 - 54}{20} = 0.9$ standard deviations above average in his class, while George was $\dfrac{76 - 52}{12} = 2.0$ standard deviations above average in his. This indicates that George's performance was, relatively speaking, much better than Tom's, and this was not apparent by looking only at the grades and the corresponding means.

What we have done here consisted of converting the grades into so-called *standard units*. Generally speaking, if x is a measurement belonging to a set of data having the mean $\bar{x}$ (or μ) and the standard deviation s (or σ), then its value in *standard units* is

▲
$$\frac{x - \bar{x}}{s} \quad \text{or} \quad \frac{x - \mu}{\sigma}$$
▲

Standard units, *standard scores*, or *z-scores*, as they are also called, tell us how many standard deviations an item is above or below the mean of the set of data to which it belongs. Their use is particularly important in the comparison of different *kinds* of measurements; for instance, the information that a man's weight is 20 pounds above average, his blood pressure is 12 mm above average, and his pulse rate is 5 beats per minute below average, might contribute *much more* to a doctor's ability to diagnose the man's troubles *if these quantities were all expressed in terms of standard units*.

4.3 Measures of Relative Variation

The standard deviation of a set of measurements is often used as an indication of their inherent precision. If we repeatedly measure the *same* quantity, for example, a person's temperature, the mileage a person gets with a certain gasoline, or the weight of a piece of rock brought down from the moon, we would hardly expect always to get the same result. Consequently, the amount of variation we do find in repeated measurements of the same kind provides us with information about their precision. To give an example, suppose that 5 measurements of the length of a certain object have a standard deviation of 0.20 in. Although this information may be important, it does not allow us to judge the *relative precision* of these measurements; for this purpose we would also have to know something about the actual size of the quantity we are trying to measure. Clearly, a standard deviation of 0.20 in. would indicate that the measurements are very precise if we measured the span of a bridge; on the other hand, they would be far from precise if we measured the diameter of a small ball bearing.

This illustrates the need for measures of *relative variation*, that is, measures which express the magnitude of the variation relative to the size of whatever is being measured. The most widely used measure of relative variation is the *coefficient of variation*, V, which is defined as

▲
$$V = \frac{s}{\bar{x}} \cdot 100$$
▲

This simply expresses the standard deviation of a set of data (or distribution) as a percentage of its mean. When dealing with populations, we analogously define the coefficient of variation as $\frac{\sigma}{\mu}\cdot 100$ or $\frac{S}{\mu}\cdot 100$.

If in the above example the standard deviation $s = 0.20$ in. had referred to 5 measurements of the length of a room and if the mean of the measurements had been 240 in., we would have had

$$V = \frac{0.2}{240}\cdot 100 = \frac{20}{240} = 0.083 \text{ per cent}$$

or approximately *eight hundredths of one per cent*. For most practical purposes this would be regarded as very precise.

By using the coefficient of variation, it is also possible to compare the dispersions of two or more sets of data that are given in different units of measurement. Instead of having to compare, say, the variability of weights in pounds, lengths in inches, ages in years, and prices in dollars, we can instead compare the respective coefficients of variation—they are all percentages. Another measure of relative variation, the *coefficient of quartile variation*, is referred to in Exercise 29 below.

EXERCISES

1. Find the range of the data of Exercise 1 on page 55.

2. Referring to Exercise 2 on page 55, find the range of the quantities of fish caught in Belgium in the years 1961 through 1968.

3. Find the range of the daily pollution data of Exercise 9 on page 25.

4. Find the range of the measurements of breaking strength of Exercise 13 on page 27.

5. Write a formula for the *mean deviation* of n numbers $x_1, x_2, \ldots,$ and x_n, namely, the mean of the *absolute values* of the deviations from their mean $\bar{x}$*. Then use the formula to find the mean deviation of the six weekly burglary figures on page 68.

6. Find the mean deviation (see Exercise 5) of the data of Exercise 1 on page 55.

7. On page 35 we gave the lifetimes of five light bulbs as 967, 889, 940, 922, and 952 hours.
   (a) Calculate s using the formula on page 67.

---

* The symbol $|x|$ denotes the *absolute value* of x. If x is a positive number or zero $|x| = x$, and if x is a negative number $|x| = -x$. For example $|3| = 3$ and $|-3| = 3$.

(b) Calculate s using the short-cut formula on page 68.

(c) Rework part (b) after subtracting 900 from each of the five lifetimes.

(d) Calculate the mean deviation (see Exercise 5), and determine the amount by which it differs from s.

8. Use the short-cut formula to calculate s for the insecticide data of Exercise 6 on page 42.

9. Use the short-cut formula to calculate s for the 24 weights of Exercise 7 on page 42.

10. Use the short-cut formula to find s for the ungrouped scores of the 150 applicants on page 13, and compare the result with that obtained for the corresponding grouped data on page 72.

11. Use the short-cut formula to find s for the ungrouped breaking strengths of Exercise 13 on page 27.

12. Find s^2 for the twelve means and the twelve medians obtained in part (a) of Exercise 8 on page 56. What is illustrated by the difference in the size of these two variances?

13. Find s for the distribution of weights of Exercise 5 on page 24
    (a) without coding;
    (b) with coding.

14. Suppose we are planning to select a sample from among the scores of Exercise 14 on page 27 for further study, so that the data given in that exercise can be looked upon as a population. Calculate S on the basis of the distribution obtained in that exercise.

15. The following is the distribution of the percentage of students belonging to a certain minority group in 40 schools:

Percentage	Frequency
0– 4	14
5– 9	11
10–14	7
15–19	6
20–24	2

Calculate s^2 for this distribution
    (a) without coding;
    (b) with coding.
Also write a formula for the *mean deviation* (see Exercise 5) of a distri-

bution with the class marks x_1, x_2, ..., x_k, the corresponding frequencies f_1, f_2, ..., f_k, and the mean $\bar{x}$. Use it to find

    (c) the mean deviation of the given distribution.

16. Find s for the distribution of the air pollution data obtained in part (a) of Exercise 9 on page 25.

17. Looking upon the data of Exercise 10 on page 25 as a population, calculate S on the basis of the distribution obtained in part (a) of that exercise.

18. Find s for the distribution of the students' expenses for room and board which was obtained in part (a) of Exercise 11 on page 26.

19. Calculate s^2 for the distribution obtained in part (a) of Exercise 12 on page 26 for the productivity of the 100 workers.

20. Having kept records for many years, Mrs. Brown knows that it takes her on the average 44 minutes to cook dinner; the standard deviation is 1.6 minutes. If she always starts cooking dinner at 5, *at least* what percentage of the time does she have dinner ready before 6?

21. In a certain large city, the monthly pay of all full-time city employees averages \$617 with a standard deviation of \$51. At least what fraction of all these employees receive a monthly pay of between \$464 and \$770?

22. Rephrasing Chebyshev's Theorem, we can say that at most $\frac{1}{k^2} \cdot 100$ per cent of any data differs from the mean by k standard deviations or more. Thus, if the final examination grades of the students taking a course in biology average 67 with a standard deviation of 10.5, at most what percentage of the students received a grade of 46 or less?

23. Use the values of $\bar{x}$ and s obtained, respectively, on pages 38 and 72 to calculate the coefficient of variation for the scores of the 150 applicants.

24. Use the result of part (a) of Exercise 7 on page 75 to calculate the coefficient of variation for the lifetimes of the five 75-watt light bulbs.

25. Use the result of Exercise 9 on page 76 to calculate the coefficient of variation of the 24 weights of Exercise 7 on page 42.

26. Use the results of Exercise 11 on page 43 and Exercise 13 on page 76 to find the coefficient of variation for the distribution of the weights of Exercise 5 on page 24.

27. Use the result of Exercise 17 on page 43 and the value of S obtained in Exercise 14 on page 76 to find the coefficient of variation for the 200 scores.

28. In order to compare the precision of two micrometers, a laboratory technician studies recent measurements made with both instruments. The first micrometer was recently used to measure the diameter of a ball bearing and the measurements had a mean of 4.93 mm and a standard deviation of 0.019 mm; the second was recently used to measure the unstretched length of a spring and the measurements had a mean of 2.57 in. with a standard deviation of 0.011 in. Which of the two micrometers is relatively more precise?

29. An alternate measure of variation is given by the *interquartile range* $Q_3 - Q_1$; it represents the length of the interval which contains the middle 50 per cent of a set of data. Some research workers prefer to use the *semi-interquartile range* (also called the *quartile deviation*) which, as its name implies, is given by the formula $(Q_3 - Q_1)/2$. Also, when working with the quartiles, a corresponding measure of *relative* variation is the so-called *coefficient of quartile variation*, which is given by the ratio of the semi-interquartile range to the *mid-quartile* $\dfrac{Q_1 + Q_3}{2}$ multiplied by 100.

    (a) Use the values of Q_1 and Q_3 obtained on page 52 to calculate the interquartile range as well as the coefficient of quartile variation for the scores of the 150 applicants.

    (b) Use the results of part (b) of Exercise 12 on page 56 to calculate the semi-interquartile range and the coefficient of quartile variation for the data on the number of customers which a certain restaurant served for lunch.

    (c) Use the values of Q_1 and Q_3 obtained in Exercise 17 on page 57 to calculate the semi-interquartile range and the coefficient of quartile variation for the distribution of the breaking strengths.

30. Mr. Green belongs to an age group for which the average weight is 143 pounds with a standard deviation of 14 pounds, and Mr. Black belongs to an age group for which the average weight is 165 pounds with a standard deviation of 17 pounds. If Mr. Green weighs 175 pounds and Mr. Black weighs 198 pounds, which of the two is more seriously overweight compared to his age group?

31. In a city in the Southwest, restaurants charge on the average $4.35 for a steak dinner (with a standard deviation of $0.40), $2.35 for a chicken dinner (with a standard deviation of $0.25), and $5.55 for a lobster dinner (with a standard deviation of $0.30). If a restaurant in this city charges $4.95 for a steak dinner, $2.75 for a chicken dinner, and $5.95 for a lobster dinner, which of the three dinners is relatively most overpriced?

32. Use the rules of summation of Section 3.6 to verify that the formulas for s given on pages 67 and 68 are equivalent. [*Hint:* make use of the *binomial* expansion $(a + b)^2 = a^2 + 2ab + b^2$.]

33. Verify the following short-cut formulas for the variance σ^2 of a population

(a) for ungrouped data $\sigma^2 = \dfrac{\Sigma\, x^2}{N} - \left(\dfrac{\Sigma\, x}{N}\right)^2$;

(b) for grouped data $\sigma^2 = \dfrac{\Sigma\, x^2 \cdot f}{N} - \left(\dfrac{\Sigma\, x \cdot f}{N}\right)^2$.

34. Use the short-cut formula of part (a) of Exercise 33 to calculate σ for a population consisting of the numbers 1, 2, 3, 4, 5, and 6.

35. Use the short-cut formula of part (b) of Exercise 33 to calculate σ for the distribution of breaking strengths obtained in part (a) of Exercise 13 on page 27.

4.4 Some Further Descriptions

So far we have discussed statistical descriptions coming under the general heading of "measures of location" and "measures of variation." Actually, there is no limit to the number of ways in which statistical data can be described, and statisticians are continually developing new methods of describing characteristics of numerical data that are of interest in particular problems. In this section we shall briefly study the problem of describing the *over-all shape* of a distribution.

Although frequency distribution can assume almost any shape or form, there are certain standard types which fit most distributions we meet in actual practice. Foremost among these is the aptly described *bell-shaped* distribution, which is illustrated by the histogram of Figure 4.1. One often runs into this kind of distribution when dealing with actual data, and there are certain theoretical reasons why, in many problems, one can actually *expect* to get bell-shaped distributions. Although the distribution of Figure 4.2 is also more or less bell-shaped, it differs from the one of Figure 4.1 inasmuch as the latter is *symmetrical* while the former is *skewed*. Generally speaking, a distribution is said to be *symmetrical* if we can picture the histogram folded (say, along the dotted line of Figure 4.1) so that the two halves will more or less coincide. If a distribution has a more pronounced "tail" on one side, such as the distribution of Figure 4.2, we say that the distribution is *skewed*. Note that most income or wage distributions (for instance, that of Exercise 3 on page 24) are skewed with a tail on the right; we say that such distributions have *positive skewness* or that they are *positively skewed*. Correspondingly, if a distribution has a pronounced tail on the left, we say that the distribution has *negative skewness* or that it is *negatively skewed*.

There are several ways of measuring the extent to which a distribution

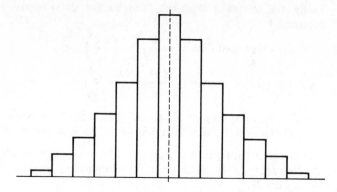

Figure 4.1. Histogram of a bell-shaped distribution.

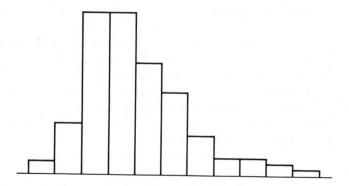

Figure 4.2. Histogram of a skewed distribution.

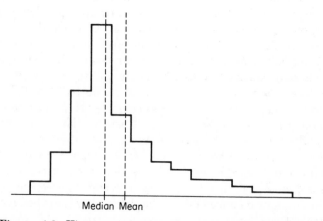

Median Mean

Figure 4.3. Histogram showing the mean and the median of a positively skewed distribution.

is skewed. A relatively easy one is based on the fact illustrated in Figure 4.3, namely, that if a distribution has a "tail" on the right, its median will generally be exceeded by its mean. (If the "tail" is on the left, this order will be reversed, and the median will generally exceed the mean.) Based on this difference, the so-called *Pearsonian coefficient of skewness* measures the skewness of a distribution by means of the formula

▲
$$\frac{3(\text{mean} - \text{median})}{\text{standard deviation}}$$
▲

Here the difference between the mean and the median is divided by the standard deviation, to make this description of the shape of the distribution independent of the units of measurement we may happen to use.

If we substitute into this formula the values obtained for the mean, the median, and the standard deviation of the distribution of the 150 scores on pages 38, 51, and 72, we find that the Pearsonian coefficient of skewness equals

$$\frac{3(56.7 - 56.2)}{15.4} = 0.097$$

The fact that this is close to zero indicates that the distribution is fairly symmetrical.

Besides bell-shaped distributions, two other kinds—*J-shaped* and *U-shaped* distributions—are sometimes, though less frequently, met in actual practice. As is illustrated by the histograms of Figure 4.4, the

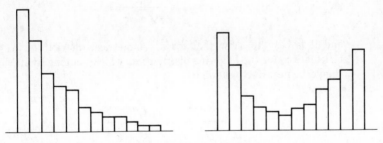

Figure 4.4. Histograms of a J-shaped distribution and a U-shaped distribution.

names of these distributions literally describe their shapes, and we find that the distribution of Exercise 15 on page 76, for example, is a J-shaped distribution. Other examples of J-shaped and U-shaped distributions may be found in Exercises 7 and 8 below.

EXERCISES

1. Use the values of the mean, the median, and the standard deviation of the air pollution data of Exercise 9 on page 25, which were obtained in Exercise 12 on page 43, Exercise 11 on page 56, and Exercise 16 on page 77, to find the Pearsonian coefficient of skewness of their distribution.

2. Using the results of Exercise 14 on page 43, Exercise 13 on page 57, and Exercise 18 on page 77, calculate the Pearsonian coefficient of skewness for the distribution of the room-and-board expenses of the 60 students of Exercise 11 on page 26.

3. Using the results of Exercise 15 on page 43, Exercise 14 on page 57, and Exercise 19 on page 77, calculate the Pearsonian coefficient of skewness for the distribution of the productivity data of Exercise 12 on page 26.

4. An alternate measure of skewness, α_3 (*alpha-three*), is defined in terms of so-called *moments about the mean*. For grouped data the kth moment about the mean is given by

$$m_k = \frac{\Sigma\,(x - \bar{x})^k \cdot f}{n}$$

where the x's are the class marks of the distribution and the f's are the corresponding frequencies.* Note that if the divisor were $n - 1$ instead of n, the second moment about the mean, m_2, would be the variance of the distribution. Using these moments, alpha-three is defined by the ratio

$$\alpha_3 = \frac{m_3}{(\sqrt{m_2})^3}$$

and it is easy to show that for a symmetrical distribution $\alpha_3 = 0$. Calculate α_3 for the following distribution, which, judging by its shape, should be negatively skewed:

Classes	Frequency
1–3	5
4–6	10
7–9	20
10–12	50
13–15	15

* The calculation of moments about the mean can be greatly simplified by using the same kind of coding as was used on page 71 in calculating the standard deviation of a distribution.

5. Using moments about the mean (see Exercise 4) the peakedness (or *kurtosis*) of a distribution is often measured by the statistic α_4 (*alpha-four*), which is defined by the formula

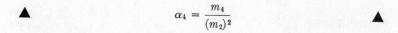

$$\alpha_4 = \frac{m_4}{(m_2)^2}$$

Draw a histogram and calculate α_4 for each of the following distributions, of which the first is relatively peaked while the other is relatively flat:

(a)

x	f
-3	6
-2	9
-1	10
0	50
1	10
2	9
3	6

(b)

x	f
-3	4
-2	11
-1	20
0	30
1	20
2	11
3	4

Also verify that in spite of their differences in shape, the two distributions have the same standard deviation.

6. Give one example each of actual data whose distribution might reasonably be expected to be
    (a) bell-shaped and symmetrical;
    (b) bell-shaped but not symmetrical;
    (c) positively skewed;
    (d) negatively skewed;
    (e) J-shaped;
    (f) U-shaped.

7. Roll a pair of dice 120 times and construct a distribution showing how many times there were 0 sixes, how many times there was 1 six, and how many times there were 2 sixes. Draw a histogram of this distribution, which should be J-shaped, and calculate α_3 (see Exercise 4).

8. If a coin is flipped five times, the result may be represented by means of a sequence of H's and T's (for example, HHTTH), where H stands for *heads* and T for *tails*. Having obtained such a sequence of H's and T's, we can then check after each successive flip whether the number of heads exceeds the number of tails. For example, for the sequence HHTTH, heads is ahead after the first flip, after the second flip, after the third flip, *not* after the fourth flip, but again after the fifth flip; altogether, it is ahead *four times*. Repeating this experiment 50 times,

construct a histogram showing in how many cases heads was ahead altogether 0 times, 1 time, 2 times, ..., and 5 times. Explain why the resulting distribution should be U-shaped.

BIBLIOGRAPHY

Sheppard's correction for the grouping error introduced when calculating the standard deviation of grouped data is discussed in

MILLS, F. C., *Introduction to Statistics.* New York: Holt, Rinehart & Winston, Inc., 1956.

PART 2
Basic Theory

5

Possibilities
and Probabilities

5.1 Introduction

Directly or indirectly, the concept of probability plays an important role in all problems of science, business, and everyday life which in any way involve an element of uncertainty. Hence, if we identify statistics with the art, or science, of making decisions in the face of uncertainty—as we did in Chapter 1—it follows that questions concerning probabilities, their meaning, their determination, and their mathematical manipulation are basic to any treatment of statistics. Thus, the study of statistics requires some knowledge of probability, and we shall take care of this informally in Section 5.4 and then in some detail in Chapter 6. First, though, we shall devote the next two sections to methods which enable us to determine *what is possible* in a given situation—after all, how can anyone be expected to judge what is likely or unlikely, what is probable or improbable, or what is credible or incredible, unless he knows at least what is possible?

5.2 Counting

In contrast to the complexity of most of the methods used in modern science, the simple process of counting still plays an important role. One still has to count 1, 2, 3, 4, ..., for example, to determine the number of persons affected by an epidemic, the size of the response to a questionnaire, the number of items damaged while being shipped, the number of earthquakes or floods, and so on. Sometimes, the process of counting can be simplified by using mechanical devices (for instance, when counting spectators passing through turnstiles), or by performing counts indirectly (for instance, when subtracting the serial numbers of invoices to determine the total number of sales). At other times, the process of counting can be simplified greatly by means of special mathematical techniques; this is what we shall demonstrate below.

In the study of "what is possible," there are essentially two kinds of problems: first there is the problem of *listing everything that can happen in a given situation,* and then there is the problem of *determining how many different things can happen (without actually constructing a complete list).* The second kind of problem is especially important, because there are many problems in which we really do not need a complete list, and hence, can save ourselves a great deal of unnecessary work.

Although the first kind of problem may seem straightforward and easy, this is not always the case. Suppose, for instance, that in a small community, tests for drivers' licenses are given only once a week, and that we are interested in what can happen to three applicants in three consecutive weeks. To be specific, we are interested only in *how many of them pass the test each week.* Clearly, there are many possibilities: all three of the applicants might pass on the first try; one might pass on the first try, another on the third try, while the third applicant fails every time; one applicant might pass on the first try and the other two on the second try; and to mention one more possibility, all three of the applicants might fail every time. Continuing this way carefully, we might come up with the correct answer that there are altogether 20 possibilities.

To handle problems like this systematically, it helps to refer to a diagram like that of Figure 5.1, which is called a *tree diagram.* This diagram shows that for the first try there are four possibilities (four branches) corresponding to 0, 1, 2, or 3 of the applicants passing the test; for the second try there are four branches emanating from the top branch, three from the second branch, two from the third branch, and none from the bottom branch. Clearly, there are still four possibilities (0, 1, 2, or 3) when none of the applicants passes on the first try, but only three possibilities (0, 1, or 2) when one of the applicants passes on the first try, two possibilities (0 or 1) when two of the applicants pass on the first try, and there is no need to go on when all three of the applicants pass on the first try. The same sort of reasoning applies also to the third try, and we thus find that (going from left to right) there are altogether 20 different paths along the "branches" of the tree diagram of Figure 5.1. In other words, 20 different things can happen in the given situation. It can also be seen from this diagram that in *ten* of the cases all three of the applicants pass the test (sooner or later) in the first three tries, in *six* of the cases two of

First try   Second try   Third   Number
                                        try   that
                                                      passed

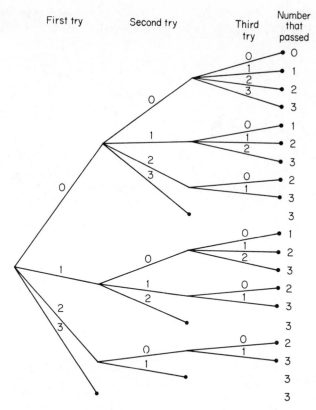

Figure 5.1. Tree diagram for the driver's license example.

the applicants pass the test in the first three tries, in *three* of the cases only one of the applicants passes the test in the first three tries, and in *one* case none of the applicants passes the test in the first three tries.

To consider another example in which a tree diagram can be of some aid (at least, until we have studied other techniques), suppose that a scholarship contest has led to the selection of six finalists, whom we shall refer to as Mr. A, Ms. B, Ms. C, Mr. D, Mr. E, and Mr. F. Now, each of the judges is to give his first choice and his second choice, and the problem is to determine *in how many different ways this can be done*. Drawing a tree diagram like that of Figure 5.2, we find practically by inspection that there are 30 possibilities, corresponding to the 30 different paths along the "branches of the tree." Starting at the top, the first path corresponds to the first choice being Mr. A and the second choice being Ms. B; the second path corresponds to the first choice being Mr. A and the second choice being Ms. C; ...; the thirteenth path corresponds to

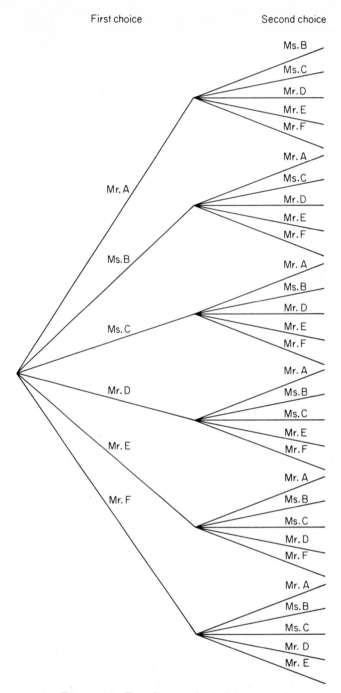

Figure 5.2. Tree diagram for judging example.

the first choice of Ms. C and the second choice of Mr. D; ...; and the thirtieth path (namely, the one along the bottom branches of the tree) corresponds to the first choice being Mr. F and the second choice being Mr. E.

Note that the answer we obtained in this example is the *product* of 6 and 5, namely, the *product* of the number of ways in which each judge can make his first choice and the number of ways in which he can subsequently make his second choice. In fact, our example illustrates the following general rule:

> *If a selection consists of two separate steps, of which the first can be made in m different ways and the second in n different ways, then the whole selection can be made in m·n different ways.*

Thus, if a restaurant offers 8 different desserts, which it serves with coffee, tea, or milk, there are altogether $8 \cdot 3 = 24$ different ways in which one can order a dessert and a drink. Also, if a college schedules 4 lecture sections and 15 laboratory sections for a course in freshman chemistry, there are $4 \cdot 15 = 60$ different ways in which a student can choose one of each.

By using appropriate tree diagrams, it is easy to generalize the above rule so that it will apply to selections involving more than two steps. For k steps, where k is a positive integer, we thus obtain the following rule:

> *If a selection consists of k separate steps, of which the first can be made in n_1 different ways, the second can be made in n_2 different ways, ..., and the kth can be made in n_k different ways, then the whole selection can be made in $n_1 \cdot n_2 \cdot \ldots \cdot n_k$ different ways.*

Thus, we simply multiply the number of ways in which each step can be made, and if each of the judges in the above example had to give also a third choice, the total number of possibilities for each judge would have been $6 \cdot 5 \cdot 4 = 120$. Also, if a new-car buyer has the choice of four body styles, three different engines, and ten colors, there are $4 \cdot 3 \cdot 10 = 120$ ways in which he can select one of these cars, and if he also has the option of ordering the car with or without air conditioning, with or without an automatic transmission, and with or without power brakes, the total number of possibilities goes up to $4 \cdot 3 \cdot 10 \cdot 2 \cdot 2 \cdot 2 = 960$.

To consider another example, suppose that a test consists of 12 multiple choice questions, with each permitting a choice of three alternatives.

Applying the above rule with $n_1 = 3$, $n_2 = 3$, ..., and $n_{12} = 3$, we find that there are

$$3 \cdot 3 \cdot 3 \cdot 3 \cdot 3 \cdot 3 \cdot 3 \cdot 3 \cdot 3 \cdot 3 \cdot 3 \cdot 3 = 531{,}441$$

ways in which a student can check off his answers, and in *only one* of these cases will all his answers be correct. To make matters worse, in

$$2 \cdot 2 \cdot 2 \cdot 2 \cdot 2 \cdot 2 \cdot 2 \cdot 2 \cdot 2 \cdot 2 \cdot 2 \cdot 2 = 4{,}096$$

of the 531,441 possibilities *all of his answers* will be wrong. As can be seen from these examples, in the "multiplication of choices," as the above rules are sometimes called, the number of possibilities will usually increase very rapidly.

EXERCISES

1. A theatrical promoter plans to put $20,000 into *one* Broadway production each year, so long as the number of flops in which he invests does not exceed the number of hits.
   (a) Draw a tree diagram showing the 8 possible situations that can arise during the first four years the plan is in operation. (*Hint:* if the first year's play is a flop, that branch of the tree diagram ends right then and there.)
   (b) In how many of the situations described in part (a) will he continue to invest in the fifth year?
   (c) If the promoter loses his total investment in a flop and doubles his money in a hit, in how many of the situations described in part (a) is he exactly even after four years?

2. In a traffic court, violators are classified according to whether or not they are properly licensed, whether their violations are major or minor, and whether or not they have committed any other violations in the preceding year.
   (a) Construct a tree diagram showing the various ways in which this traffic court classifies violators.
   (b) If there are 20 violators in each of the eight categories of part (a) and the judge gives each violator who is not properly licensed a stern lecture, how many of the violators will receive a stern lecture?
   (c) If, furthermore, the judge gives a $50 fine to everybody who has committed a major violation and/or another violation in the preceding year, how many of the violators will receive a $50 fine?
   (d) How many of the violators will receive a stern lecture from the judge as well as a $50 fine?

3. Among four different routes, A, B, C, and D, Mr. Jones can take only Routes A, B, and C to work because Route D is one-way in the other direction, and he can take only Routes A, B, and D on the way home because Route C is one-way in the other direction. In how many different ways can he go to and from work?

4. An artist has two paintings in a show which lasts two days.
   (a) If we are interested only in how many of his paintings are sold on each of the two days, show by means of a tree diagram that there are altogether six possibilities.
   (b) If we *do* care which painting is sold on what day, show by means of a tree diagram that there are altogether nine possibilities.

5. A student can study either 0, 1, or 2 hours for a French test on any given night. Construct a tree diagram to show that there are 10 different ways in which he can study altogether 6 hours for the test on four consecutive nights.

6. On the faculty of a college there are three professors named Jones: Henry Jones, Frank Jones, and Richard Jones. Draw a tree diagram to show the six ways in which the payroll department can distribute their paychecks so that each of them receives a check made out to himself or one of the other two Joneses. In how many of these possibilities will
   (a) only one of them get his own check;
   (b) none of them get the right check?

7. If the five finalists in the Miss Universe contest are Miss Spain, Miss U.S.A., Miss Argentina, Miss Italy, and Miss Japan, draw a tree diagram to show the various ways in which the judges could choose the winner and the first runner-up.

8. Without drawing a tree diagram, find the number of ways in which the judges could choose the winner, the first runner-up, and the second runner-up from among the five finalists in the Miss Universe contest (see Exercise 7).

9. There are five different trails to the top of a certain mountain. In how many different ways can a person hike up and down the mountain if
   (a) he must take the same trail both ways;
   (b) he can, but need not, take the same trail both ways;
   (c) he does not want to take the same trail both ways?

10. In a primary election, there are three candidates for mayor, four candidates for city treasurer, and two candidates for county attorney.
    (a) In how many different ways can a voter mark his ballot for these three offices?
    (b) In how many different ways can a person mark his ballot if he exercises his option of not voting for any candidate for any or all of these offices? (This includes the possibility of leaving the ballot blank.)

11. A psychologist preparing four-letter nonsense words for use in a memory test chooses the first letter from among the consonants q, r, t, w, and z, the second letter from among the vowels a, e, i, o, and u, the third letter from among the consonants k, x, and v, and the fourth letter from among the vowels a, e, o, and u.
   (a) How many different nonsense words can he construct?
   (b) How many of these nonsense words will begin either with the letter t or the letter w?
   (c) How many of these nonsense words will have the vowel a in both places?

12. A true-false test consists of twenty questions. In how many different ways can a student check off his answers to these questions?

13. In a certain city, a child watching television on a Saturday morning has the choice between two different cartoon shows from 8 to 8:30, three different cartoon shows from 8:30 to 9, three different cartoon shows from 9 to 9:30, and two different cartoon shows from 9:30 to 10. In how many different ways can one of these children plan his Saturday morning entertainment of watching cartoons continuously from 8 to 10?

14. A cafeteria offers 12 different soups or salads, 18 main dishes, and 10 desserts. In how many different ways can one choose a soup or salad, a main dish, and a dessert?

15. A multiple-choice test consists of eight questions, each permitting a choice of 4 alternatives.
   (a) In how many different ways can a student check off his answers to these questions?
   (b) In how many different ways can a student check off his answers to these questions and get them all wrong?

16. In a drive-in restaurant, a customer can order a hamburger rare, medium rare, medium, or well done, and also with or without mustard, with or without onions, and with or without relish. In how many different ways can a person order a hamburger in this restaurant?

17. How many different sums of money can be formed with one or more of the following coins: a dollar, a half-dollar, a quarter, a dime, a nickel, and a penny? (*Hint:* there are two possibilities for each coin depending on whether or not it is included.)

18. An art dealer has 9 large oil paintings and 8 small oil paintings for sale. If his first customer buys one of the paintings and his second customer buys one of each size, how many different choices can the second customer make if
   (a) the first customer buys one of the large paintings;
   (b) the first customer buys one of the small paintings?

5.3 Permutations and Combinations

The rules of the preceding section are often applied when *several selections are made from one and the same set, and the order in which they are made is of significance*. This was the case in the example on page 89, where the judges had to make a first and second choice from among the six finalists in the scholarship contest, and it would also be the case, say, if the 85 members of a union had to elect a president, a vice-president, and a secretary-treasurer.

In general, if r objects are selected from a set of n objects, any particular arrangement of these r objects is referred to as a *permutation*. For instance, 3 5 2 4 1 is one of many possible permutations of the first five positive integers, and so are 1 3 2 4 5 and 2 4 5 1 3; *Maine, Connecticut, and New Hampshire* is a permutation (a particular ordered arrangement) of 3 of the 6 New England states; E C D G, A F G C, and D B C F are three of many possible permutations of 4 of the first 7 letters of the alphabet, and if we were asked to list *all possible* permutations of 2 of the 5 vowels a, e, i, o, and u, our answer would be

ae	*ai*	*ao*	*au*	*ei*	*eo*	*eu*	*io*	*iu*	*ou*
ea	*ia*	*oa*	*ua*	*ie*	*oe*	*ue*	*oi*	*ui*	*uo*

So far as the counting of possible permutations is concerned, direct application of the second rule on page 91 leads to the following result:

> The number of permutations of r objects selected from a set of n objects is $n(n-1)(n-2)\cdot\ldots\cdot(n-r+1)$, which we shall denote $_nP_r$.*

To prove this formula, we have only to observe that the first selection is made from the whole set of n objects, the second selection is made from the $n-1$ objects which remain after the first selection has been made, the third selection is made from the $n-2$ objects which remain after the first two selections have been made, and the rth and final selection is made from the

$$n - (r - 1) = n - r + 1$$

objects which remain after the first $r-1$ selections have been made. Thus, the number of ways in which 4 new cars can be assigned to a company's 8 executives is $8\cdot7\cdot6\cdot5 = 1,680$, and the number of ways in

---

* The following are some alternate symbols used to denote the number of permutations of r objects selected from a set of n objects: $P(n, r)$, P_r^n, $P_{n,r}$, and $(n)_r$.

which a recent college graduate can make a first, second, and third choice among 21 potential employers is $21 \cdot 20 \cdot 19 = 7,980$. Incidentally, it has been assumed in our discussion that the objects with which we are dealing are all *distinguishable* in some way; otherwise, there are complications, which we shall discuss briefly in Exercise 8 on page 99.

Since products of consecutive integers arise in many problems involving permutations and other kinds of special arrangements, it generally simplifies matters if we use the *factorial notation,* in which $1! = 1$, $2! = 2 \cdot 1 = 2$, $3! = 3 \cdot 2 \cdot 1 = 6$, $4! = 4 \cdot 3 \cdot 2 \cdot 1 = 24$, $5! = 5 \cdot 4 \cdot 3 \cdot 2 \cdot 1 = 120$, ..., and in general, $n!$, which reads "*n* factorial," denotes the product

$$n(n - 1)(n - 2) \cdot \ldots \cdot 3 \cdot 2 \cdot 1$$

for any positive integer n. In this notation it is customary, also, to let $0! = 1$ *by definition,* because this makes it easier to write certain general formulas relating to permutations and other kinds of special arrangements.

Using the factorial notation, we can now say that

> *The number of permutations of n objects taken all together is* *n!.*

This is simply a special case of the rule for $_nP_r$ with $r = n$, where the last factor becomes $n - r + 1 = n - n + 1 = 1$, and $_nP_n = n(n - 1)(n - 2) \cdot \ldots \cdot 1 = n!$. Thus, the number of ways in which six instructors can be assigned to teach six sections of a history course is $6! = 720$, and the number of ways in which the starting eleven of a football team can be introduced before a game is $11! = 39,916,800$.

There are many problems in which we are interested in the number of ways in which r objects can be selected from a set of n objects, but where we do not care about the order in which the selection is made. We may thus want to know in how many ways a committee of 6 can be selected from among the 85 union members referred to on page 95, or the number of ways in which a census taker can select 5 of the 64 families in an apartment building, say, to fill in a more detailed questionnaire than the other 59 families. To obtain answers to these questions, let us first examine the following 24 permutations of three of the first four letters of the alphabet:

abc	*acb*	*bac*	*bca*	*cab*	*cba*
abd	*adb*	*bad*	*bda*	*dab*	*dba*
acd	*adc*	*cad*	*cda*	*dac*	*dca*
bcd	*bdc*	*cbd*	*cdb*	*dbc*	*dcb*

If we are interested only in the number of ways in which three of these four letters can be selected, and not in the order in which they are selected, it should be noted that each row in the above table contains $3 \cdot 2 \cdot 1 = 6$ permutations of the same three letters selected from among the first four letters of the alphabet. In fact, there are only four ways in which three letters can be selected from among the first four letters of the alphabet if we are not interested in the order in which the selection is made: they are a, b, and c; a, b, and d; a, c, and d; and b, c, and d. In general, there are $r!$ permutations of r specific objects selected from a set of n objects and, hence, the $n(n - 1) \cdot \ldots \cdot (n - r + 1)$ permutations of r objects selected from a set of n objects contain each set of r objects $r!$ times. (In our example, the 24 permutations of 3 letters selected from among the first 4 letters of the alphabet contained each set of 3 letters $3! = 6$ times.) Dividing $n(n - 1) \cdot \ldots \cdot (n - r + 1)$ by $r!$, we thus arrive at the following rule:

The number of ways in which r objects can be selected from a set of n objects is

$$\frac{n(n - 1) \cdot \ldots \cdot (n - r + 1)}{r!}$$

Symbolically, we write the number of ways in which r objects can be selected from a set of n objects as $\binom{n}{r}$ and we refer to it as the number of *combinations* of n objects taken r at a time.* In Exercise 9 on page 100, the reader will be asked to verify that in the factorial notation

$$\binom{n}{r} = \frac{n!}{(n - r)! \cdot r!}$$

for $r = 0, 1, 2$, and n, and this is one reason why we let $0! = 1$—clearly, there is only *one way* in which we can take all n of the objects, and the formula yields

$$\frac{n!}{(n - n)! \cdot n!} = \frac{n!}{0! \cdot n!}$$

which equals 1 when $0! = 1$.

An easy way of calculating the quantities $\binom{n}{r}$, also called *binomial*

---

* The following are some alternate symbols used to denote the number of combinations of r objects selected from a set of n objects: $C(n, r)$, C_r^n, $_nC_r$, and $C_{n,r}$.

coefficients, is indicated in Exercise 20 on page 101; otherwise, their values up to $n = 20$ can be obtained from Table VIII at the end of the book. The reason why these quantities are called "binomial coefficients" is explained in Exercise 19 on page 101.

Applying these results to the two examples on page 96, we find that the committee of 6 can be selected from among the 85 union members in

$$\frac{85 \cdot 84 \cdot 83 \cdot 82 \cdot 81 \cdot 80}{6!} = 437,353,560 \text{ ways}$$

and that the 5 families can be chosen from among the 64 families in the apartment building in

$$\frac{64 \cdot 63 \cdot 62 \cdot 61 \cdot 60}{5!} = 7,624,512 \text{ ways}$$

Also, from Table VIII we find that a person can invite 5 of his 9 friends to a party in $\binom{9}{5} = 126$ ways, and that a light fixture with 12 bulbs can have 3 bulbs burn out in $\binom{12}{3} = 220$ ways.

EXERCISES

1. In how many different ways can a medical research worker inject 3 of his 18 laboratory mice with 3 different dosages of a serum.

2. In how many different ways can a student arrange his 5 textbooks on a shelf?

3. In how many different ways can 3 persons each buy one of the 25 cars in a used-car lot?

4. In an English class, the students are given the choice of 12 different essay topics. In how many different ways can four students each choose a topic if
   (a) no two students can choose the same topic;
   (b) there is no restriction on the choice of topics?

5. The price of a European tour includes 5 stop-overs to be selected from among 10 cities. In how many ways can one plan such a tour if
   (a) the order of the stop-overs matters;
   (b) the order of the stop-overs does not matter?

6. In how many ways can a television director schedule a sponsor's 6 different commercials during the 6 time slots allocated to commercials during the telecast of the first quarter of a football game?

7. Three married couples have bought six seats in a row for a performance of the Ice Follies.
   (a) In how many different ways can they be seated?
   (b) In how many ways can they be seated if all the men are to sit together and all the women are to sit together?
   (c) In how many ways can they be seated if each couple is to sit together with the husband to the left of his wife?

8. If some of the n objects from which we make a selection are indistinguishable, the formulas for the number of permutations and combinations no longer apply. Suppose, for instance, that we want to determine the number of ways in which we can arrange the letters in the word "book." If we distinguish for the moment between the two o's by referring to them as o_1 and o_2, there are indeed $_4P_4 = 4! = 24$ different permutations of the symbols b, o_1, o_2, and k. However, if we drop the subscripts, then bo_1ko_2 and bo_2ko_1, for example, represent the *same* permutation *boko*. Since each permutation *without* subscripts will then correspond to two permutations *with* subscripts, the answer to our original question, namely, the total number of permutations of the letters in the word "book." is $\dfrac{24}{2} = 12$. Similarly, if we refer to the e's in "receive" as e_1, e_2, and e_3, there are $_7P_7 = 5{,}040$ permutations of the symbols r, e_1, c, e_2, i, v, and e_3, but since the three e's can be arranged among themselves in $3! = 6$ different ways, each permutation *without* subscripts corresponds to 6 permutations *with* subscripts, and, hence, there are only $\dfrac{7!}{3!} = \dfrac{5{,}040}{6} = 840$ different permutations of the letters in "receive." In general, this argument leads to the result that if r_1 of the n objects are alike, r_2 others are also alike, $\ldots$, and r_k others are alike, then the total number of permutations of the n objects is given by

$$\frac{n!}{r_1! \cdot r_2! \cdot \ldots \cdot r_k!}$$

For instance, the number of permutations of the letters in "minimum" is

$$\frac{7!}{3!2!} = \frac{5{,}040}{6 \cdot 2} = 420$$

   (a) How many permutations are there of the letters in "meter"?
   (b) In how many ways (according only to the manufacturer) can six cars place in a stock-car race, if three of the cars are Fords, one is a Chevrolet, one is a Plymouth, and one is a Dodge?
   (c) In how many ways can the television director of Exercise 6 schedule the six commercials, if the sponsor has three commercials, each of which is to be shown twice?

(d) In its cookbook section, a bookstore has 4 copies of the *New York Times Cookbook*, 2 copies of *The Joy of Cooking*, 5 copies of the *Better Homes and Gardens Cookbook*, and one copy of *The Secret of Cooking for Dogs*. If these books are sold one at a time, in how many different ways (that is, in how many different sequences), can they be sold?

9. Multiplying and dividing the respective expressions by $(n - r)!$, verify that

   (a) the formula for the number of permutations can be written as

   $$_nP_r = \frac{n!}{(n - r)!}$$

   (b) the formula for the number of combinations can be written as

   $$\binom{n}{r} = \frac{n!}{(n - r)! \cdot r!}$$

10. Verify the *identity* $\binom{n}{r} = \binom{n}{n - r}$, which is important in connection with the use of Table VIII. [*Hint:* express each of the binomial coefficients in terms of factorials in accordance with the formula of part (b) of Exercise 9.]

11. Without referring to Table VIII, find the number of ways in which a restaurant chain can choose 2 of 13 locations for the construction of new restaurants.

12. Without referring to Table VIII, find the number of ways in which a 4-man committee can be chosen from among the 16 teachers of a private school.

13. Without referring to Table VIII, find the number of ways in which the Internal Revenue Service can choose 6 of 15 income tax returns for a special audit.

14. In Exercise 12 on page 94, the reader was asked for the total number of ways in which a student can check off his answers in a true-false test consisting of 20 questions. In how many different ways can a student get
   (a) 10 right and 10 wrong;
   (b) 12 right and 8 wrong;
   (c) 15 right and 5 wrong?

15. To fill a number of vacancies, the personnel manager of a company has to choose 4 secretaries from among 8 applicants and 2 file clerks from among 5 applicants. What is the total number of ways in which he can make his selection?

16. In hiring his staff, the director of a laboratory has to choose 2 chemists from among 9 applicants, 3 physicists from among 6 applicants, 4 mathematicians from among 12 applicants, and 6 technicians from among 17 applicants. What is the total number of ways in which he can make his choice?

17. A shipment of 12 transistor radios contains one that is defective. In how many ways can we choose three of these radios for inspection so that
     (a) the defective radio is not included;
     (b) the defective radio is included?

18. Suppose that among the 12 radios of Exercise 17 there are two defectives. In how many ways can we choose three of the radios for inspection so that
     (a) neither of the defective radios is included;
     (b) both of the defective radios are included;
     (c) only one of the defective radios is included?

19. The quantity $\binom{n}{r}$ defined on page 97 is referred to as a *binomial coefficient* because it is, in fact, the coefficient of a^r in the binomial expansion of $(a + b)^n$. Verify that this is true for $n = 2$, 3, and 4, by expanding $(a + b)^2$, $(a + b)^3$, and $(a + b)^4$, and comparing the coefficients with the corresponding values of $\binom{n}{r}$ given in Table VIII.

20. The number of combinations of n objects taken r at a time can easily be determined by means of the following arrangement called *Pascal's triangle:*

$$
\begin{array}{ccccccc}
 & & & 1 & 1 & & \\
 & & 1 & 2 & 1 & & \\
 & 1 & 3 & 3 & 1 & & \\
 1 & 4 & 6 & 4 & 1 & & \\
1 & 5 & 10 & 10 & 5 & 1 & \\
\end{array}
$$

. . . .

where each row begins with a 1, ends with a 1, and each other entry is given by the sum of the nearest two entries in the row immediately above.
     (a) Verify that the third row of the triangle contains the values of $\binom{3}{r}$ for $r = 0$, 1, 2, and 3.
     (b) Verify that the fourth row of the triangle contains the values of $\binom{4}{r}$ for $r = 0$, 1, 2, 3, and 4.

(c) Making use of the fact that in general the nth row of Pascal's triangle contains the values of $\binom{n}{r}$ for $r = 0, 1, 2, \ldots$, and n, calculate the values of $\binom{6}{r}$ for $r = 0, 1, 2, 3, 4, 5$, and 6.

21. Verify the *identity* $\binom{n+1}{r} = \binom{n}{r} + \binom{n}{r-1}$ by expressing each of the binomial coefficients in terms of factorials. Explain why this identity justifies the method used in the construction of Pascal's triangle in Exercise 20.

5.4 Probabilities and Odds

So far, we have studied only *what is possible* in a given situation. In some instances we listed all possibilities, and in others we merely determined how many different possibilities there are. Now we shall go one step further and judge also *what is probable* and *what is improbable* (or as we put it on page 87, what is likely and what is unlikely, or what is credible and what is incredible).

Historically, the oldest way of measuring probabilities applies when all the possible outcomes are *equally likely*, as is presumably the case in many games of chance. We can then say that

> *If there are n equally likely possibilities, of which one must occur and s are regarded as favorable, namely, as a "success,"*
>
> *then the probability of a "success" is given by the ratio $\frac{s}{n}$.*

In the application of this rule, it should be understood that the terms "favorable" and "success" are used rather loosely—what is favorable to one player is unfavorable to his opponent, and what is a success from one point of view is a failure from another. Thus, the terms "favorable" and "success" can be applied to any particular kind of outcome with which we happen to be concerned, even if "favorable" means that a television set does not work, or a "success" is a person's coming down with the flu. This usage dates back to the days when probabilities were quoted only in connection with games of chance.

Following the rule of the preceding paragraph, we find, for example, that the probability of drawing an ace from an ordinary deck of 52 playing cards is $\frac{4}{52}$ (there are four aces), the probability of getting *heads* with a

balanced coin is $\frac{1}{2}$, and the probability of rolling a 5 or a 6 with a die is $\frac{2}{6}$.
To consider a somewhat more complicated example, let us find the probability that two cards drawn from an ordinary deck of 52 playing cards will both be black. According to what we learned in the preceding section, the total number of possibilities is $\binom{52}{2}$, the number of favorable possibilities is $\binom{26}{2}$ since half of the 52 playing cards are black and the other half are red, and it follows that the probability of getting two black cards is

$$\frac{\binom{26}{2}}{\binom{52}{2}} = \frac{\dfrac{26 \cdot 25}{2}}{\dfrac{52 \cdot 51}{2}} = \frac{25}{102}$$

Although equally likely possibilities are found mostly in games of chance, the probability concept we have discussed applies also in a great variety of situations where gambling devices are used to make so-called *random selections*—say, when offices are assigned to teaching assistants by lot, when laboratory animals are chosen for an experiment (perhaps, by the method which we shall discuss in Chapter 9) so that each one has the same chance of being selected, when each family in a town has the same chance of being included in a sample survey, or when each ball bearing has the same chance of being chosen for inspection. Note, however, that the rule on page 102 does *not* provide us with a *definition* of "probability," since the idea of possibilities being equally likely (namely, being *equiprobable*, or *having the same probability*) is assumed to be "intuitively understood." And this is not the only shortcoming of the *classical* probability concept which we have discussed; as we shall see in the remainder of this chapter (and, for that matter, throughout the whole book), there are many situations in which the various possibilities cannot be regarded as equally likely. For example, we are not dealing with equally likely possibilities when we are concerned with the question of whether there will be rain, sunshine, snow, or hail, when we are wondering whether or not a person will recover from a disease, or when we want to predict the outcome of an election or the score of a football game.

Among the various other probability concepts, most widely held is the *frequency* interpretation, according to which *the probability of an event (happening, or outcome) is interpreted as the proportion of the time that similar events will occur in the long run.* Thus, if we say that there is a probability of 0.84 that a jet from New York to Chicago will arrive on time, this means that such flights actually arrive on time about 84 per cent of the time. Also, if the weather bureau predicts that there is a 40 per cent chance for rain, namely, a probability of 0.40 this means that under similar weather conditions it will rain about 40 per cent of the time. More generally, we say that an event has a probability of, say, 0.90, in the same sense in which we might say that in cold weather our car will start on the first try about 90 per cent of the time. *We cannot guarantee what will happen on any particular occasion—the car may start on the first try and then it may not—but if we kept records for a long time, we should find that the proportion of "successes" is very close to 0.90.*

In accordance with the frequency interpretation of probability, we *estimate* the probability of an event by observing how often (what part of the time) similar events have occurred in the past. For instance, if data kept by the F.A.A. show that (over a period of time) 630 of 750 jets from New York to Chicago arrived on time, we estimate the probability that any one jet from New York to Chicago (perhaps, the next one due to arrive) will get there on time as $\frac{630}{750} = 0.84$. Similarly, if 504 of 813 automatic dishwashers sold by a large dealer required repairs within the first year, we estimate the probability that any such dishwasher will require repairs within the first year as $\frac{504}{813} = 0.62$, and if (over a number of years) 3,435 of the 9,270 freshmen who entered a given college managed to graduate, we estimate as $\frac{3,435}{9,270} = 0.37$ the probability that a freshman entering this college will graduate.

When probabilities are thus estimated, it is only reasonable to ask whether the estimates are any good. The answer, which is "Yes," is supported by a remarkable law called the *Law of Large Numbers*, which will be discussed further in Chapter 8. Informally, this law can be stated as follows:

If the number of times a situation is repeated becomes larger and larger, the proportion of successes will tend to come closer and closer to the actual probability of success.

To illustrate how this works, let us refer to the familiar example of repeatedly flipping a coin. If we observe the accumulated proportion of successes, say, *heads*, after every fifth flip and plot it graphically as in Figure 5.3, we should find that although it fluctuates, the proportion comes closer and closer to $\frac{1}{2}$, the *probability* of getting heads for each flip

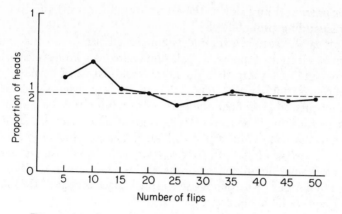

Figure 5.3. Graph illustrating the law of large numbers.

of the coin. A good way to develop an understanding of the Law of Large Numbers is through experimentation (that is, repeatedly flipping a coin, rolling a die, etc.) and this the reader will be asked to do in Exercises 1 and 2 on pages 241 and 242.

Having defined probabilities in terms of what happens to similar events in the long run, let us check for a moment whether it is at all meaningful to talk about the probability of an event which *cannot occur more than once.* Can we ask for the probability that Mrs. Barbara Smith's broken arm will heal within a month, or the probability that a certain major-party candidate will win an upcoming presidential election? If we put ourselves in the position of Mrs. Smith's doctor, we could check medical records, discover that such fractures have healed within a month in 39 per cent of thousands of cases, and apply this figure to Mrs. Smith's arm. This may not be of much comfort to Mrs. Smith, but it does provide a *meaning* for a probability statement concerning her arm—the probability that it will heal within a month is 0.39. Thus, *when we make a probability statement about a specific (non-repeatable) event, the frequency concept of probability leaves us no choice but to refer to a set of similar events.* This can lead to complications, for the choice of "similar" events is often neither obvious nor easy. With reference to Mrs. Smith's arm, for example, we might consider as "similar" only those cases where the fracture was in the same (left or right) arm, we might consider only those cases in which the patients were just as old, or we might consider only those cases in which the patients were also of the same height and the same weight as Mrs. Smith. Ultimately, this is a matter of choice, and *it is by no means contradictory that we can thus arrive at different probabilities concerning Mrs. Smith's arm.* It should be observed, however, that the more we narrow things down, the less information we have to estimate the corresponding probability.

So far as the second example is concerned, the one concerning the presidential election, suppose we ask some persons who have conducted a poll "how sure" they are that the given candidate will actually win. If their answer is "99 per cent sure," that is, if they assign the candidate's election a probability of 0.99, they are not implying that he would win 99 per cent of the time if he ran for office a great many times. *No, it means that the persons who conducted the poll based their conclusion (judgment, or decision) on methods which (in the long run) will "work" 99 per cent of the time.* In this sense, many of the probabilities which we use to express our faith in predictions or decisions are simply "success ratios" that apply to the methods we have employed.

An alternate point of view, which is currently gaining favor among quite a few statisticians, is to interpret probabilities as *personal* or *subjective*. To illustrate, suppose that a businessman feels that the *odds* for the success of a new venture, say, a new shoe store, are 3 to 2. This means that he would be willing to bet (or consider it fair to bet) $300 against $200, or perhaps $3,000 against $2,000, that the venture will succeed. In this way he expresses the *strength of his belief* regarding the uncertainties connected with the success of the new store. This method of dealing with uncertainties works well (and is certainly justifiable) in situations where there is very little direct evidence; in that case one may have no choice but to consider pertinent collateral information, "educated" guesses, and perhaps intuition and other subjective factors. Thus, the businessman's odds concerning the success of a new shoe store may well be based on his ideas about business conditions in general, the opinion of an expert, and his own subjective evaluation of the whole situation, including, perhaps, a small dose of optimism.

Regardless of how we interpret probabilities and odds, subjectively or in terms of frequencies or proportions, *the mathematical relationship between probabilities and odds is always the same.* It is given by the following rule:

> *If somebody considers it fair or equitable to bet a dollars against b dollars that a given event will occur, he is, in fact, assigning the event the probability* $\dfrac{a}{a+b}$.

Thus, the businessman who is willing to give odds of 3 to 2 that the new shoe store will succeed is actually assigning its success a probability of $\dfrac{3}{3+2} = 0.60$. Also, if the odds are 18 to 7 that a student *will not* get an

A in an algebra course, then the probability that he *will not* get an A is $\frac{18}{18 + 7} = 0.72$; correspondingly, the odds that he *will* get an A are 7 to 18, the probability that he *will* get an A is $\frac{7}{7 + 18} = 0.28$, and it should be observed that the sum of these two probabilities is $0.72 + 0.28 = 1.00$.

To illustrate how probabilities are converted into odds, let us refer back to the example on page 104, where we indicated that a probability of 0.90 implies that in cold weather our car will start about 90 per cent of the time. As we said, we cannot guarantee what will happen on any particular occasion, but we could have added that it would have been *fair* to bet \$9.00 against \$1.00 (or 90 cents against a dime) that the car will start on any given try. These odds of 9 to 1 would have been "fair," "reasonable," or "equitable," for we would win nine times as often as we would lose. Generalizing from this example, we can, thus, state the following rule:

If the probability of an event is p, then the odds for its occurrence are p to 1 − p and the odds against its occurrence are 1 − p to p.

For instance, if the probability that an item lost in a department store will never be claimed is 0.15, then the odds are 0.15 to 0.85, or 3 to 17, that a lost item will never be claimed (and they are 17 to 3 that it will be claimed). Also, if the probability that we shall have to wait for a table at our favorite restaurant is 0.25, then the odds are 0.25 to 0.75, or 1 to 3 that we shall have to wait for a table (and they are 3 to 1 that we shall not have to wait). Note that in both of these illustrations we followed the common practice of quoting odds as *ratios of positive integers* (having no common factors).

EXERCISES

1. When we roll a balanced die, what are the probabilities of getting (a) a 6; (b) an even number; (c) a 2, 3, 4, or 5?

2. When one card is drawn from a well-shuffled deck of 52 playing cards, what are the probabilities of getting (a) a black king; (b) a queen, king, or ace; (c) a red card; (d) a 5, 6, 7, or 8 of any suit?

3. If H stands for *heads* and T for *tails*, the four possible outcomes for two successive flips of a coin are HH, HT, TH, and TT. Assuming that

these four possibilities are equally likely, what are the respective probabilities of getting 0, 1, or 2 heads?

4. If H stands for *heads* and T *tails*, the eight possible outcome for three successive flips of a coin are HHH, HHT, HTH, THH, HTT, THT, TTH, and TTT. Assuming that these eight possibilities are equally likely, what are the respective probabilities of getting 0, 1, 2, or 3 heads?

5. A bowl contains 17 red beads, 10 white beads, 20 blue beads, and 3 black beads. If one of these beads is drawn at random, what are the probabilities that it will be (a) red; (b) blue or white; (c) black; (d) neither white nor black?

6. Assuming that in Exercise 17 on page 101 the three radios are chosen at random (that is, each possible set of three has the same chance of being selected), find the probabilities that (a) the defective radio will not be included; (b) the defective radio will be included.

7. Assuming that in Exercise 18 on page 101 the three radios are chosen at random (that is, each possible set of three has the same chance of being selected), find the probabilities that
    (a) neither of the defective radios will be included;
    (b) both of the defective radios will be included;
    (c) only one of the defective radios will be included.

8. With reference to Exercise 13 on page 100, suppose that only one of the 15 income tax returns contains an error. If the 6 returns are chosen in such a way that each possible choice has the same chance, what is the probability that the one with an error will be included?

9. A hoard of medieval silver coins discovered in Belgium included 60 struck for Charles the Bold and 40 struck for Philip the Good.
    (a) If a person is allowed to pick one of these coins at random, what is the probability that he will get a coin struck for Philip the Good?
    (b) If a person is allowed to pick two of these coins at random, what is the probability that he will get two coins struck for Philip the Good?

10. If data compiled by the manager of a department store show that 1,564 of 1,840 women who entered the store on a Saturday afternoon made at least one purchase, estimate the probability that a woman who enters the store on a Saturday afternoon will make at least one purchase.

11. In a sample of 400 cans of mixed nuts (taken from a very large shipment), 124 contained no pecans. Estimate the probability that there will be no pecans in a can of mixed nuts which is randomly selected from this shipment.

12. A study made by a traffic engineer showed that 1,375 of 2,870 cars which approached a certain intersection from the South made a right turn. Estimate the probability that a car approaching this intersection from the South will make a right turn.

13. Weather bureau statistics show that in a certain community it has rained 12 times in the last 60 years on the first Sunday in May, on which a service club holds its annual picnic.
    (a) Estimate the probability that it will rain this year on the day which the service club has chosen for its annual picnic.
    (b) What are the odds that it will *not* rain on that day?
    (c) If somebody offered to bet us $10 against our $3 that it will *not* rain on that day, whom would that bet favor?

14. Among the 351 times that Ken has gone fishing, he has come back empty-handed (that is, without a single catch) 117 times.
    (a) Estimate the probability that he will come back empty-handed from his next fishing trip.
    (b) What are the odds that he will catch at least one fish on his next fishing trip?
    (c) If someone offered Ken *even money* (that is, odds of 1 to 1) that he will *not* make a single catch on his next fishing trip, who would be favored by this bet?

15. In a sample of 741 cars stopped at a roadblock, the drivers of 247 did not have their seat belt fastened.
    (a) Estimate the probability that the driver of a car stopped at this roadblock will have his seat belt fastened.
    (b) What are the odds that the driver of a car stopped at this roadblock will have his seat belt fastened?
    (c) If we offered a friend a bet of $10 against his $5 that the driver of the next car stopped at the roadblock will have his seat belt fastened, who would be favored by this bet?

16. If a college senior feels that the odds are 5 to 3 that he will be admitted to law school, what is his subjective probability that this will be the case?

17. One insurance salesman offers another insurance salesman a bet of $11 against $9 that he will not be able to sell a policy to a difficult client. If the second insurance salesman considers this fair, what subjective probability is he, thus, assigning to his selling a policy to this client?

18. If somebody claims that the odds are 33 to 17 that a certain shipment will arrive on time, what probability does he assign to the shipment's arriving on time?

19. A sportswriter feels that 4 to 1 are fair odds that the home team will lose an upcoming football game. What probability is he, thus, assigning to the home team's losing this game?

20. If a stockbroker is *not* willing to bet $60 against $180 that the price of a certain stock will go up within a month, what does this tell us about his subjective probability that this will be the case (namely, that the price of the stock will go up within a month). (*Hint:* the answer should read "less than...".)

21. Suppose that a student is willing to bet $5 against $1, but not $7 against $1 that he will get a passing grade in a certain course. What does this tell us about the personal probability he assigns to his getting a passing grade in the course? (*Hint:* the answer should read "at least...but less than...".)

22. A television executive is willing to bet $500 against $4,500, but not $600 against $4,400 that a new dramatic show will be a success. What does this tell us about the executive's subjective probability that the new show will *not* be a success? (*Hint:* the answer should read "greater than...but at most...".)

23. Some philosophers have argued that if we have absolutely no information about the likelihood of the different possibilities, it is reasonable to regard them all as equally likely. This is sometimes referred to as the "Principle of *Equal Ignorance.*" Discuss the argument that human life either does or does not exist elsewhere in the universe, and since we really have no information one way or the other, the probability that human life exists elsewhere in the universe is $\frac{1}{2}$.

24. The following illustrates how one's intuition can be misleading in connection with probabilities or odds: A box contains 100 beads, some red and some white. One bead will be drawn, and you are asked to call beforehand whether it is going to be red or white. At what odds would you be willing to bet on this game if
    (a) you have no idea how many of the beads are red and how many are white;
    (b) you are told that 50 of the beads are red and 50 are white?
    It is a curious fact that most people seem to be much more willing to gamble under condition (b) than under condition (a).

BIBLIOGRAPHY

Informal introductions to probability, written essentially for the layman, may be found in

LEVINSON, H. C., *Chance, Luck, and Statistics.* New York: Dover Publications, Inc., 1963.

WEAVER, W., *Lady Luck—The Theory of Probability.* Garden City, N.Y.: Doubleday & Company, Inc., 1963.

and subjective probabilities, in particular, are discussed in

BOREL, E., *Elements of the Theory of Probability*. Englewood Cliffs, N.J.: Prentice-Hall, Inc., 1965.

KYBURG, H. E. JR., and SMOKLER, H. E., *Studies in Subjective Probability*. New York: John Wiley & Sons, Inc., 1964.

LINDLEY, D. V., *Introduction to Probability and Statistics from a Bayesian Viewpoint. Part 1, Probability*. Cambridge: Cambridge University Press, 1965.

An interesting discussion of various philosophical questions concerning probabilities is given in

NAGEL, E., *Principles of the Theory of Probability*. Chicago: University of Chicago Press, 1939.

6

Rules of Probability

6.1 Introduction

In the study of probability there are basically *three kinds of questions*. First, there is the question of what we *mean*, for example, when we say that the probability for rain is 0.80, when we say that the probability for the success of a new venture is 0.35, or when we say that the probability for a candidate's election is 0.63; then there is the question of how probabilities are *measured* (namely, how their values are determined in actual practice); and finally there is the question of *how probabilities "behave,"* namely, what mathematical rules they have to obey.

The first and second kinds of question have already been discussed to some extent in Section 5.4. In the *classical concept* we determine probabilities by means of the formula $\frac{s}{n}$ on page 102, but circumvent all questions of meaning by accepting "equally likely" as something which is *intuitively understood*. Be that as it may, it certainly limits the applicability of this concept. In the *frequency concept*, which is very widely held, a probability is interpreted as a proportion, or percentage, in the long run, and its value is obtained (estimated) by observing what proportion of the time similar events have occurred in the past. Such probabilities are also referred to as "objective," in contrast to *subjective*, or *personal*, probabilities, which are meant to express the strength of a person's belief. Subjective probabilities could be evaluated by simply asking a person what he considers "fair odds" that an event will occur, and then converting these odds into a probability as on page 106. More realistic, perhaps, would be to make a person "put up or shut up," namely, to see how he would react if there were really something at stake. How this can be done will be explained in Chapter 7.

In this chapter we shall study the third kind of question, namely, that about the rules according to which probabilities "behave." As we shall see in Section 6.3, there are essentially three basic rules on which all other rules of probability are based, and it is important to keep in mind that these basic rules are supposed to apply regardless of whether we interpret probabilities objectively, subjectively, or in the classical sense.

6.2 Sample Spaces and Events

In statistics it is customary to refer to any process of observation or measurement as an *experiment*. Thus, using the term in a very wide sense, an experiment may consist of the simple process of noting whether a light is on or off; it may consist of determining the number of imperfections in a piece of cloth; or it may consist of the very complicated process of finding the mass of an electron. The results one obtains from an experiment, whether they be simple "yes" or "no" answers, instrument readings, or whatever, are called the *outcomes* of the experiment.

In most problems in which uncertainties are connected with the various outcomes of an experiment, it is convenient to represent the outcomes by means of points. This has the advantage that we can treat such problems mathematically, where we would otherwise have to verbalize about the various outcomes. For example, if an experiment consists of determining whether a person interviewed in a poll favors a proposed civil rights legislation, is undecided, or opposes the legislation, we might represent these three outcomes by means of the three points of Figure 6.1, to which we arbitrarily assigned the code numbers 1, 2, and

Figure 6.1. Outcomes of one-person interview.

3. Note that we could just as well have used any other configuration of points, and that we could have assigned to them any other arbitrary set of numbers (or letters).

Had we been interested in the reactions of two persons interviewed in the poll, we could have presented the outcomes by means of the points

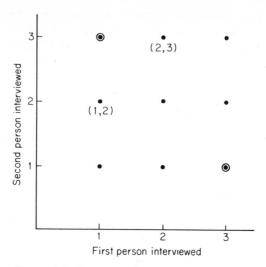

Figure 6.2. Outcomes of two-person interview.

of Figure 6.2, where 1, 2, and 3 stand again for a response favoring the legislation, indecision, and a response opposing the legislation. Thus, the point (1, 2) represents the outcome where the first person interviewed is for the legislation while the second is undecided, and the point (2, 3) represents the outcome where the first person is undecided and the second opposes the legislation. Had we been interested in the response of n persons in this experiment, there would have been 3^n possible outcomes and as many points. For instance, for 4 persons there would have been $3 \cdot 3 \cdot 3 \cdot 3 = 81$ points, among which (2, 1, 1, 3), for example, represents the outcome where the first person interviewed is undecided, the second and third persons are for the legislation, while the fourth person is against it. (Since each point is now determined by four numbers, we cannot picture these points as readily as those of Figures 6.1 and 6.2; in fact, this would require a space of four dimensions.)

It is customary to refer to the total set of points representing all the possible outcomes of an experiment as the *sample space* of the experiment and to denote it by the letter S. Thus, the three points of Figure 6.1 constitute the sample space for interviewing one person, and the nine points of Figure 6.2 constitute the sample space for interviewing two. In any discussion of an experiment, no matter how simple, *it is always important to specify the sample space with which we are concerned*, and as we shall see, this may depend partly on what we look upon as an individual outcome. For example, if in the two-person-interview case we had been interested only in the *total number of persons* favoring the legislation, being undecided, or opposing the legislation, we could have used the *three-dimensional* sample space of Figure 6.3 instead of the *two-dimensional* sample space of Figure 6.2. Here the first coordinate gives the number of persons favoring the legislation, the second coordinate gives the number of persons who are undecided, while the third coordinate gives the number of persons opposing the legislation. Thus, the point

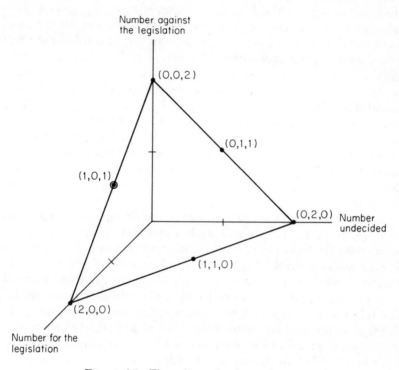

Figure 6.3. Three-dimensional sample space.

(2, 0, 0) of Figure 6.3 represents the outcome where both persons interviewed favor the legislation, while (1, 0, 1) represents the case where one person is for the legislation while the other is against it. Note that the two points (1, 3) and (3, 1) circled in Figure 6.2 *together* represent the same outcome as the single point (1, 0, 1) circled in Figure 6.3, and that the sample spaces of Figures 6.2 and 6.3 consist of 9 points and 6 points, respectively, even though they pertain to the same experiment. Generally speaking, it is desirable to use sample spaces whose points represent outcomes which cannot be further "subdivided"; that is, each individual point should not represent two or more possibilities which are distinguishable in some fashion. Unless there is a special reason for not following this rule, we would thus prefer the sample space of Figure 6.2 to that of Figure 6.3.

Earlier we described the sample spaces of Figures 6.2 and 6.3 as two-dimensional and three-dimensional, and we could similarly describe the sample space of Figure 6.1 as one-dimensional. Although it may be useful to know the number of dimensions in which we happen to picture the points of a sample space, it is more common to classify sample spaces according to the number of points which they contain. All the sample spaces mentioned so far have been *finite;* that is, they consisted of a finite, or fixed, number of points. Other examples of finite, though much larger, sample spaces are the one representing all possible 5-card poker hands one can deal with an ordinary deck of 52 playing cards (there are 2,598,960 possibilities), and the one representing all the possible ways in which the 10 schools in the Big Ten Conference can finish the football season (there are 3,628,800 possibilities, not counting ties). In the remainder of this chapter, we shall consider only finite sample spaces, although in later chapters we shall consider also sample spaces that are *infinite* (for instance, when dealing with measurements assuming values on a continuous scale).

Having explained what we mean by a sample space, let us now state what we mean by an *event*. This is important because probabilities always refer to the occurrence or nonoccurrence of events. For instance, we may assign a probability to the event that there will be anywhere from 24 to 30 drop-outs among 85 high school students, the event that a 60-year-old person will live to be 70, the event that there will be at least 6 heads in 10 flips of a coin, the event that a tire will last at least 18,000 miles before it has to be recapped, and so forth. Generally speaking, *all events with which we are concerned in probability theory are represented by subsets of appropriate sample spaces* (by *subset* we mean any part of a set, including the set as a whole, and, trivially, the empty set which has no elements at all). In other words, "event" is the nontechnical term and "subset of a sample space" is the corresponding mathematical counterpart. Thus, in Figure 6.2 the subset which consists of the point (2, 2) represents the event that both persons interviewed are undecided; the subset which consists of the two points circled in Figure 6.2 represents the event that one of the persons interviewed is for the legislation while the other is against it; and the subset which consists of the five points (3, 1), (3, 2), (3, 3), (1, 3), and (2, 3) represents the event that at least one of the two persons interviewed is against the legislation.

Still referring to the two-person-interview example, let us now suppose that X stands for the event that at least one of the two persons interviewed is undecided, Y stands for the event that both persons interviewed respond in the same way, and Z stands for the event that one of the persons interviewed is for the legislation and the other one is against it. Referring to Figure 6.4, which shows the same sample space as Figure

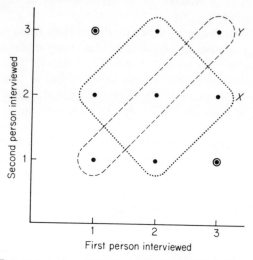

Figure 6.4. Sample space for two-person interview.

6.2, we find that event X is represented by the five points inside the dotted line, event Y is represented by the three points inside the dashed line, and event Z is represented by the two points which are circled. Note that the subsets which represent events X and Z or Y and Z (but not X and Y) have no points in common—they are referred to as *mutually exclusive events*, which means that they cannot both occur in the same experiment.

In many situations we are interested in events which are actually combinations of two or more simpler kinds of events. For instance in our example we might be interested in the event that "either both persons interviewed respond in the same way *or* one person is for the legislation while the other is against it," or we might be interested in the event that "both persons interviewed respond in the same way *and* at least one of them is undecided." In the first case we are interested in the event that *either Y or Z occurs*, which is represented by the subset consisting of the points (1, 1), (2, 2), (3, 3), (1, 3), and (3, 1); in the second case, we are interested in the event that *Y and X both occur*, which is represented by the subset consisting of the single point (2, 2).

In general, if A and B are any two events we define their *union* $A \cup B$

as the event which consists of all the individual outcomes contained either in A, in B, or in both. It is customary to read $A \cup B$ as "A *cup* B" or simply as "A *or* B." Referring again to the example of the preceding paragraph, we can now say that we found that $Y \cup Z$ consists of the points $(1, 1)$, $(2, 2)$, $(3, 3)$, $(1, 3)$, and $(3, 1)$, and it will be left to the reader to check that $X \cup Z$ consists of all of the points of the sample space except $(1, 1)$ and $(3, 3)$.

If A and B are any two events we define their *intersection* $A \cap B$ as the event which consists of all the individual outcomes contained in both A and B. Here $A \cap B$ reads "A *cap* B" or simply "A *and* B." Again referring to the same example, we can now say that we found that $Y \cap X$ consists of the single point $(2, 2)$; earlier we saw that $X \cap Z$ has no elements at all, and if we use the symbol $\emptyset$ to denote the *empty set*, we can now write $X \cap Z = \emptyset$.

To complete our notation, let us define A', the *complement* of event A with respect to a given sample space S, as the subset which consists of all the individual outcomes of S that are *not* contained in A. Thus, with reference to our example X' represents the event that neither person is undecided, and Y' represents the event that the two persons interviewed

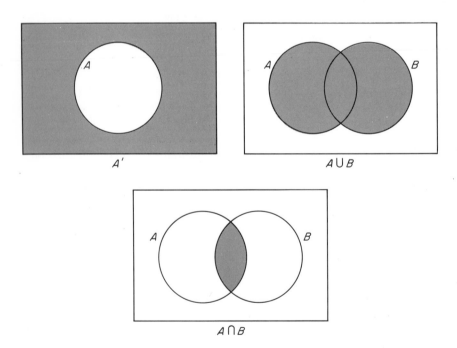

Figure 6.5. Venn diagrams.

did not respond the same way. It will be left to the reader to verify that Z' is, in fact, identical with $X \cup Y$.

Sample spaces and events, particularly relationships among events, are often depicted by means of *Venn diagrams* like those of Figures 6.5, 6.6, and 6.7. In each case the sample space is represented by a rectangle, while subsets (or events) are represented by regions within the rectangles, usually circles or parts of circles. Thus, the shaded regions of the three

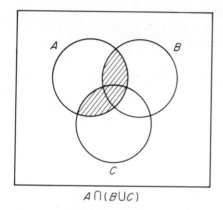

$$A \cap (B \cup C)$$

Figure 6.6. Venn diagram.

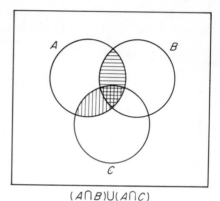

$$(A \cap B) \cup (A \cap C)$$

Figure 6.7. Venn diagram.

Venn diagrams of Figure 6.5 represent, respectively, the complement of event A, the union of events A and B, and the intersection of A and B. When dealing with three events, it is customary to draw the respective circles as in Figure 6.6. Note that the shaded region represents the event $A \cap (B \cup C)$, namely, the event which consists of all outcomes belonging to A and also to either B or C (or both).

Venn diagrams are often used to verify relationships among sets, thus making it unnecessary to give formal proofs based on the Algebra of Sets. To illustrate, let us verify that $A \cap (B \cup C) = (A \cap B) \cup (A \cap C)$, one of the so-called *distributive laws*. As we already pointed out, the event $A \cap (B \cup C)$ is represented by the shaded region of Figure 6.6. Now, $A \cap B$ is represented by the region of Figure 6.7 which is ruled horizontally, $A \cap C$ is represented by the region which is ruled vertically, and hence $(A \cap B) \cup (A \cap C)$ is represented by the region which is ruled in either (or both) directions. As can be seen by inspection, this region is identical with the shaded region of Figure 6.6, and this proves (or at least demonstrates) that $A \cap (B \cup C)$ is the same event as $(A \cap B) \cup (A \cap C)$.

EXERCISES

1. Referring to the sample space of Figure 6.2, described *in words* the event which is represented by each of the following sets of points:
   (a) (1, 2), (2, 2), and (3, 2);
   (b) (1, 1), (1, 2), and (1, 3);
   (c) (1, 1), (2, 1), (3, 1), (1, 2), (2, 2), and (3, 2);
   (d) (1, 3), (2, 2), and (3, 1).

2. Referring to the sample space of Figure 6.4, describe each of the following events *in words* and list the points which it contains:
   (a) X';   (c) $X \cup Y'$;
   (b) Y';   (d) $X \cap Y'$.

3. Referring to the sample space of Figure 6.2, let T be the event that the first person interviewed is for the legislation, let U be the event that the second person interviewed is for the legislation, and let V be the event that neither person interviewed is against the legislation. List the points which represent these three events.

4. With reference to Exercise 3, describe each of the following events *in words* and list the points which it contains:
   (a) T';   (c) $U \cap V$;   (e) $T' \cap U'$;
   (b) $U \cup V$;   (d) $T \cap V$;   (f) $T' \cup U'$.

5. Referring to Exercise 3, which of the following pairs of subsets represent *mutually exclusive* events:
   (a) T and U;   (c) T and V';
   (b) U and V;   (d) V' and $T \cap U$?

6. A company providing shuttle service between two nearby airports has two helicopters which leave the respective airports every hour on the hour; the larger of the two helicopters can carry 4 passengers while the smaller one can carry only 3 passengers.
   (a) Using two coordinates so that (1, 3), for example, represents the event that when the helicopters take off at a given hour the larger helicopter has one passenger while the smaller helicopter has three, and (2, 0) represents the event that the larger helicopter has two passengers while the smaller helicopter is empty, draw a diagram (similar to that of Figure 6.2) showing the 20 points of the corresponding sample space.
   (b) Describe *in words* the event which is represented by each of the following sets of points of the sample space: the event Q which consists of the points (2, 3), (3, 2), (3, 3), (4, 1), (4, 2), and (4, 3), the event R which consists of the points (0, 0), (1, 1), (2, 2), and (3, 3), the event T which consists of the points (0, 1), (0, 2), (0, 3), (1, 2), (1, 3), and (2, 3), and the event U which consists of the points (0, 3), (1, 2), (2, 1), and (3, 0).

   (c) Referring to part (b), list the points of the sample space which belong to each of the following subsets, and describe *in words* the events which they represent: (i) $Q \cup U$; (ii) $Q \cap T$; (iii) $R \cup T$; and (iv) $Q \cap R$.

   (d) With reference to part (b), which of the following pairs of subsets of the sample space represent mutually exclusive events:

      (i) R and T;   (iii) Q and T;

      (ii) R and U';   (iv) Q and U?

7. Having been transferred to a new city, an engineer is looking for a new 2-, 3-, or 4-bedroom house, and he finds that none of the houses that are available have fewer bedrooms than baths but have, of course, at least one bath.

   (a) Draw a diagram similar to that of Figure 6.2 (with the x-coordinate denoting the number of bedrooms and the y-coordinate the number of baths) which shows the nine different ways in which his choice can be made. For instance, (3, 1) represents the event that he chooses a 3-bedroom house with one bath, and (2, 2) represents the event that he choses a 2-bedroom house with two baths.

   (b) Describe *in words* the event which is represented by each of the following sets of points: event E which consists of the points (3, 2) and (4, 1), event F which consists of the points (3, 1), (4, 1), and (4, 2), and event G which consists of the points (2, 1), (2, 2), (3, 1), (3, 2), and (3, 3). Also indicate these three sets on the diagram of part (a) by enclosing the respective points by means of a solid line, a dotted line, and a dashed line.

   (c) Referring to part (b), describe each of the following events *in words* and list the points which it contains:

      (i) F';   (iii) $E \cap F$;   (v) $E \cap G$;

      (ii) G';   (iv) $F \cup G$;   (vi) $E \cup G'$.

   (d) Referring to part (b), which of the following pairs of subsets represent mutually exclusive events:

      (i) E and G;   (iii) F' and G;

      (ii) E and $F \cap G$;   (iv) G' and $E \cap F$.

8. Mr. Green has four friends in San Francisco, whom he may or may not have the time to call on a two-day visit to this city. He will not call any one of these friends more than once.

   (a) Using two coordinates so that (2, 1), for example, represents the event that he will call two of these friends on the first day and one on the second day, and (0, 2) represents the event that he will not call any of these friends on the first day and two on the second day, draw a diagram similar to Figure 6.2 which shows the 15 possibilities.

   (b) List the points of the sample space of part (a) which constitute the following events: event R that he will call all four of his friends, event M that he will call more of his friends on the first day than on the second day, event T that he will call at least 3 of these friends on the second day, and event U that he will call only one of these friends.

(c) Referring to part (b), describe each of the following events *in words* and list the points which it contains:

(i) R';   (iv) $R \cap M$;

(ii) M';   (v) $U \cap M$;

(iii) $R \cup T$;   (vi) $R' \cap T$.

(d) Referring to part (b), which of the following pairs of subsets represent mutually exclusive events:

(i) R and T;   (iii) U and M;

(ii) M and T;   (iv) U and T?

9. Suppose that in the example in the text three persons had been interviewed, and that 1, 2, and 3 stand, as before, for "favors the legislation," "is undecided," and "is against the legislation."

(a) Using three coordinates so that $(2, 1, 3)$, for example, represents the event that the first person interviewed is undecided, the second is for the legislation, and the third is against it, list the coordinates of the other 26 points of the corresponding sample space.

(b) Describe *in words* the event J which consists of the points $(1, 1, 1)$, $(1, 1, 2)$, and $(1, 1, 3)$, the event K which consists of the points $(1, 1, 1)$, $(2, 2, 2)$, and $(3, 3, 3)$, the event L which consists of the points $(1, 2, 1)$, $(1, 2, 3)$, $(3, 2, 1)$, and $(3, 2, 3)$, and the event M which consists of the points $(1, 2, 3)$, $(1, 3, 2)$, $(2, 1, 3)$, $(2, 3, 1)$, $(3, 1, 2)$, and $(3, 2, 1)$.

(c) Referring to part (b), list the points of the sample space which belong to each of the following subsets, and describe *in words* the events which they represent:

(i) $J \cap K$;   (iii) $L \cap M$;

(ii) $K \cup M$;   (iv) $J \cap M'$.

(d) With reference to part (b), which of the following pairs of subsets of the sample space represent mutually exclusive events:

(i) J and K;   (iii) K and M;

(ii) J and L;   (iv) L and M?

10. Suppose that a group of students are traveling through Europe and that R is the event that they will visit Rome, T is the event that they will have a good time, and C is the event that they will run out of cash. With reference to the Venn diagram of Figure 6.8 list (by numbers) the regions or combinations of regions which represent the following events:

(a) The event that they will visit Rome, have a good time, but will not run out of cash.

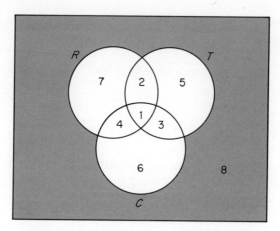

Figure 6.8. Venn diagram for Exercise 10.

   (b) The event that they will visit Rome, but will not have a good time and not run out of cash.

   (c) The event that they will not visit Rome and run out of cash.

   (d) The event that they will have a good time.

   (e) The event that they will not visit Rome, not have a good time, or not run out of cash.

11. With reference to Exercise 10 and the Venn diagram of Figure 6.8, explain *in words* what events are represented by the following regions;

   (a) Region 1;   (d) Regions 3 and 5 together;

   (b) Region 5;   (e) Regions 3, 5, and 6;

   (c) Regions 1 and 2 together;   (f) Regions 1, 2, 4, and 7.

12. Which of the following pairs of events are mutually exclusive? Explain your answers.

   (a) Having rain and sunshine on the 4th of July, 1975.

   (b) Being under 25 years of age and being President of the U.S.

   (c) One and the same person wearing black shoes and green socks.

   (d) A driver getting a ticket for speeding and his getting a ticket for going through a red light.

   (e) A person leaving Los Angeles by jet at 11:45 p.m. and arriving in Washington, D.C. on the same day.

   (f) A baseball player getting a walk and hitting a home run in the same game.

   (g) A baseball player getting a walk and hitting a home run in the same time at bat.

13. Use Venn diagrams to verify that the following relationships hold for any two sets A and B:

   (a) $A \cup (A \cap B) = A$;

(b) $A \cap (A \cup B) = A$;

(c) $(A \cap B) \cup (A \cap B') = A$;

(d) $A \cup B = (A \cap B) \cup (A' \cap B) \cup (A \cap B')$.

14. Use Venn diagrams to verify that $A \cup (B \cap C) = (A \cup B) \cap (A \cup C)$.

6.3 Rules of Probability

The rules according to which "mathematical objects" must behave are generally called *axioms* or *postulates*, where the first term applies when these rules are regarded as *self-evident truths*, while the second is used when they are more in the nature of *assumptions*; thus, we shall speak of the *postulates of probability*.

To formulate the postulates of probability and some of their immediate consequences, we shall continue the practice of denoting events by means of capital letters, and we shall write the probability of event A as $P(A)$, the probability of event B as $P(B)$, and so forth. Furthermore, we shall follow the common practice of denoting the set of all possible outcomes, namely, the *sample space*, by the letter S. As we shall formulate it here, the third postulate of probability applies only when the sample space S is *finite*; a modification that is required when S is *infinite* will be mentioned on page 180.

> POSTULATE 1: *The probability of any event is a positive real number or zero; symbolically, $P(A) \geq 0$ for any subset A of any given sample space S.*

> POSTULATE 2: *The probability of any sample space is equal to 1; symbolically, $P(S) = 1$ for any sample space S.*

It is important to note that both of these postulates are satisfied by the various probability concepts which we studied in the preceding chapter. So far as the first postulate is concerned, fractions or proportions of successes are always positive or zero, and so long as a and b (the amounts bet for and against the occurrence of an event) are positive, the probability $\dfrac{a}{a + b}$ cannot be negative.

The second postulate states indirectly that *certainty* is identified with a probability of 1; after all, it is always assumed that *one of the possibilities included in S must occur*, and it is to this certain event that we assign a probability of 1. In the classical concept $\dfrac{n}{n} = 1$ and so far as the frequency interpretation is concerned, a probability of 1 implies that the event will occur 100 percent of the time, or in other words, that it is certain to occur. So far as subjective probabilities are concerned, the surer we are that an event will occur, the "better" odds we should be willing to give— say, 100 to 1, 1,000 to 1, or perhaps even 1,000,000 to 1. The corresponding probabilities are $\dfrac{100}{100+1}$, $\dfrac{1,000}{1,000+1}$, and $\dfrac{1,000,000}{1,000,000+1}$ (or approximately 0.99, 0.999, and 0.999999), and it can be seen that *the surer we are that an event will occur, the closer its subjective probability will be to 1*.

In actual practice, we also assign a probability of 1 to events of which we are "practically certain" that they will occur. For instance, we would assign a probability of 1 to the event that at least one person will vote in the next presidential election, and we would assign a probability of 1 to the event that among all the new cars sold during any one model year at least one will be involved in an accident before it has been driven 10,000 miles.

The third postulate of probability is especially important, and it is not quite so "obvious" as the other two:

POSTULATE 3: *If two events are mutually exclusive, the probability that one or the other will occur equals the sum of their probabilities. Symbolically,*

$$P(A \cup B) = P(A) + P(B)$$

for any two mutually exclusive events A and B.

For instance, if the probabilities that a person will order steak or chicken in a certain restaurant are, respectively, 0.34 and 0.23, then the probability that he (or she) will order one or the other is $0.34 + 0.23 = 0.57$. Also, if the probabilities that a student will get an A or a B in a course are, respectively, 0.13 and 0.29, then the probability that he (or she) will get either an A or a B is $0.13 + 0.29 = 0.42$.

All this agrees with the first two of the probability concepts of Chapter 5. In the classical concept, if s_1 of n equally likely possibilities constitute event A and s_2 others constitute event B, then all these $s_1 + s_2$ equally likely possibilities constitute event $A \cup B$, and the respective probabilities are $P(A) = \dfrac{s_1}{n}$, $P(B) = \dfrac{s_2}{n}$, and $P(A \cup B) = \dfrac{s_1 + s_2}{n}$; this satisfies the third postulate. So far as the frequency interpretation is concerned, the postulate is satisfied, for if one event occurs, say, 38 per cent of the time, another event occurs 43 per cent of the time, and *they cannot both occur at the same time* (that is, they are mutually exclusive), then one or the other will occur $38 + 43 = 81$ per cent of the time. When it comes to subjective probabilities, the third postulate does *not* follow from our discussion in Section 5.4; however, proponents of the subjective point of view generally impose the third postulate as what they call the "consistency criterion" (see also Exercises 7 and 8 on page 132).

By using the three postulates of probability, we can derive many further rules according to which probabilities must "behave"—some of them are easy to prove and some are not, but they all have important applications. Among the immediate consequences of the three postulates we find that *probabilities can never be greater than 1*, that *an event which cannot occur has the probability 0*, and that *the respective probabilities that an event will occur and that it will not occur always add up to 1*. Symbolically,

$$P(A) \leq 1 \quad \textit{for any event } A$$

$$P(\emptyset) = 0$$

and

$$P(A) + P(A') = 1 \quad \textit{or} \quad P(A') = 1 - P(A).$$

The first of these results simply expresses the fact that there cannot be more favorable outcomes than there are outcomes, that an event cannot occur more than 100 per cent of the time, and that $\dfrac{a}{a + b}$ cannot exceed 1 when a and b are *positive amounts* bet for and against the occurrence of an event. The second result expresses the fact that when an event is impossible there are $s = 0$ favorable possibilities, that an impossible event happens 0 per cent of the time, and that a person would not bet *at any odds* on an event which cannot occur. In actual practice, we also assign 0 probabilities to events which are *so unlikely* that we are "practically certain" that they will not occur. Thus, we would assign a probability of 0 to the event that a monkey set loose on a typewriter will by chance type Plato's *Republic* word for word without a single mistake.

The third result can be derived formally from the three postulates of probability, but it also follows from the various probability concepts of Section 5.4. In the classical concept, if there are s "successes" then there are $n - s$ "failures," the respective probabilities are $\dfrac{s}{n}$ and $\dfrac{n - s}{n}$, and their sum is $\dfrac{s}{n} + \dfrac{n - s}{n} = 1$. So far as the frequency concept is concerned, if the probability that the home team will win a certain football game is 0.42, then the probability that it will lose or tie is 0.58, and if the probability that a student will pass a given test is 0.77, then the probability that he (or she) will not pass the test is 0.23, where all these figures are looked upon as proportions. Subjectively speaking, if a person considers it *fair*, or equitable, to bet a dollars against b dollars that a given event will occur, he is actually assigning the event the probability $\dfrac{a}{a+b}$ and its non-occurrence the probability $\dfrac{b}{b+a}$; evidently, these two probabilities also add up to 1.

The third postulate of probability applies only to *two* mutually exclusive events, but it can easily be generalized; repeatedly using this postulate, it can be shown that

> *If A_1, A_2, ..., and A_k are k mutually exclusive events (that is, only one of them can occur), then the probability that one of them will occur is*
>
> $$P(A_1 \cup A_2 \cup \ldots \cup A_k) = P(A_1) + P(A_2) + \ldots + P(A_k)$$

where $\cup$ may again be read as "or." For instance, if the probabilities that a given high school graduate will enroll at U.C.L.A., San Diego State, or the University of Arizona are, respectively, 0.07, 0.45, and 0.23, then the probability that he will enroll at one of these schools is $0.07 + 0.45 + 0.23 = 0.75$. Also, if Mr. Brown is planning to buy a new color television set and the probabilities are 0.14, 0.28, 0.11, and 0.09 that he will buy a Zenith, R.C.A., Motorola, or Magnavox, then the probability that he will buy one of these kinds of sets is $0.14 + 0.28 + 0.11 + 0.09 = 0.62$. Furthermore, the probability that the given high school graduate will not enroll at one of the three schools is 0.25, and if the probability that Mr. Brown will decide not to buy a new color television set is 0.16, then the probability that he will buy some other kind of set (that is, other than Zenith, R.C.A., Motorola, or Magnavox) is 0.22. Why?

The job of assigning probabilities to all the events that are possible in a given situation can be a very tedious task, indeed. For a sample space with as few as 5 points representing, say, the events that a person traveling to Europe visits Paris, London, Berlin, Rome, and Amsterdam, there are already $2^5 = 32$ possibilities; he may visit only London, or London and Rome, or Paris, Rome, and Amsterdam, ..., he may visit all five of these cities, and then he may not visit any of them at all. Things get worse very rapidly when a sample space has more than 5 points—for a sample space with as few as 20 individual outcomes, or points, there are already over a million different subsets or events, 1,048,576 to be exact.

Fortunately, it is seldom necessary to assign probabilities to all possible events, and the following rule (which is a direct application of the "generalized addition formula" given above) makes it easy to determine

the probability of any event on the basis of the probabilities which are
assigned to the individual outcomes (points) of the corresponding sample
space:

> *The probability of any event A is given by the sum of the prob-*
> *abilities of the individual outcomes comprising A.*

To illustrate this rule, let us refer again to the two-person-interview
example on page 114, and let us suppose that the nine points of the sample
space (shown originally in Figure 6.2) are assigned the probabilities given
in Figure 6.9. Altogether, there are $2^9 = 512$ different subsets or events,

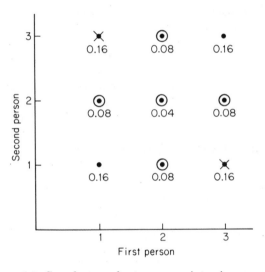

Figure 6.9. Sample space for two-person-interview example.

since there are two possibilities for each point depending on whether it is
included or excluded, and according to the above rule we can find the probability of any of them by simply adding the probabilities assigned to the
points which they contain. For instance, if we want to know the probability that *at least one of the two persons interviewed will be undecided*, we
have only to add the probabilities of the points circled in Figure 6.9, and
we get

$$0.08 + 0.08 + 0.04 + 0.08 + 0.08 = 0.36$$

Similarly, if we want to know the probability that one of the two persons
interviewed will be for the legislation while the other one will be against

it, we have only to add the probabilities of the points marked **X** in Figure
6.9, and we get $0.16 + 0.16 = 0.32$. The situation is even simpler when
the individual outcomes are all *equiprobable*, for the "generalized addition
rule" on page 127 will then lead to the formula $\dfrac{s}{n}$, which we introduced in
Chapter 5 in connection with the classical probability concept.

Since the third postulate applies only to mutually exclusive events, it
cannot be used, for example, to find the probability that at least one of
two roommates will pass a final exam in economics, the probability that a
person will break an arm or a rib in an automobile accident, or the probability that a customer will buy a shirt or a tie while shopping at Macy's
department store. In the first case, both roommates can pass the exam,
in the second case the person can break an arm as well as a rib, and in the
third case the customer can buy a shirt as well as a tie. To obtain a
formula for $P(A \cup B)$ which holds regardless of whether the events A
and B are mutually exclusive, let us consider the situation illustrated by
means of the Venn diagram of Figure 6.10; it concerns an insurance

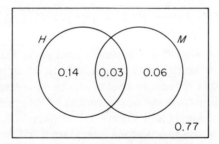

Figure 6.10. Venn diagram.

salesman's luck with a potential customer. The letter H stands for the
event that he will sell him a homeowner's policy, M stands for the event
that he will sell him a major medical policy, and it can be seen that

$$P(H) = 0.14 + 0.03 = 0.17$$
$$P(M) = 0.06 + 0.03 = 0.09$$

and

$$P(H \cup M) = 0.14 + 0.03 + 0.06 = 0.23$$

We were able to add the respective probabilities since they referred to
mutually exclusive events (namely, regions of the Venn diagram which

have no points in common), but had we *erroneously* used the third postulate of probability to calculate $P(H \cup M)$, we would have obtained

$$P(H) + P(M) = 0.17 + 0.09 = 0.26$$

which exceeds the *correct value* by 0.03. What happened is that $P(H \cap M) = 0.03$ was added in *twice*, once in $P(H) = 0.17$ and once in $P(M) = 0.09$, and we could correct for this by *subtracting* $P(H \cap M) = 0.03$ from the final result, namely, by writing

$$\begin{aligned} P(H \cup M) &= P(H) + P(M) - P(H \cap M) \\ &= 0.17 + 0.09 - 0.03 \\ &= 0.23 \end{aligned}$$

Since this kind of argument holds for any two events A and B, we can now state the following *general addition rule*, which applies regardless of whether A and B are mutually exclusive events:

▲ $$P(A \cup B) = P(A) + P(B) - P(A \cap B)$$ ▲

Note that when A and B *are* mutually exclusive, then $P(A \cap B) = 0$ (since *by definition* the two events cannot both occur at the same time), and the new formula reduces to that of the third postulate of probability on page 125. To illustrate the new formula, let us refer again to the two-person-interview example and Figure 6.9 on page 128. If A is the event that at least one of the two persons interviewed will be undecided and B is the event that both will respond in the same way, we already know from page 128 that $P(A) = 0.36$. Since B consists of the points $(1, 1)$, $(2, 2)$, and $(3, 3)$, and $A \cap B$ consists only of the point $(2, 2)$, it follows that $P(B) = 0.16 + 0.04 + 0.16 = 0.36$, $P(A \cap B) = 0.04$, and, hence, that

$$P(A \cup B) = 0.36 + 0.36 - 0.04 = 0.68$$

As a check, we could calculate $P(A \cup B)$ directly, namely, add the probabilities assigned to the seven points comprising $A \cup B$ in Figure 6.9.

To give another example, suppose that if a person visits his dentist the probability that he will have a cavity filled is 0.46, the probability that he will have a tooth extracted is 0.19, and the probability that he will have a cavity filled and a tooth extracted is 0.06. Substituting these values into the formula, we find that the probability that he will have a cavity filled or a tooth extracted is $0.46 + 0.19 - 0.06 = 0.59$.

EXERCISES

1. Suppose that in a study of juvenile delinquents in a certain metro-
    politan area, R stands for the event that the delinquent has dropped
    out of school and Q stands for the event that the delinquent's parents
    are on welfare. State *in words* what probability is expressed by
    (a) $P(R')$;   (c) $P(Q \cap R)$;   (e) $P(Q' \cap R')$;
    (b) $P(Q')$;   (d) $P(Q \cup R')$;   (f) $P(Q' \cup R)$.

2. If D is the event that a new novel will get good reviews, E is the event
    that it will be a best-seller, and F is the event that it will be made
    into a movie, write each of the following probabilities in symbolic
    form:
    (a) The probability that the new novel will be a best-seller and be
        made into a movie.
    (b) The probability that the new novel will not get good reviews
        but will be made into a movie.
    (c) The probability that the novel will neither get good reviews
        nor be a best-seller.
    (d) The probability that the new novel will get good reviews and
        be made into a movie.
    (e) The probability that the novel will be a best-seller, but will
        neither get good reviews nor be made into a movie.

3. Analyzing research done at their schools, three college presidents
    made the following claims: The first college president claims that the
    probabilities for more federal support, less federal support, or no
    change are, respectively, 0.07, 0.25, and 0.65; the second college presi-
    dent claims that these probabilities are, respectively, 0.14, 0.38, and
    0.48; and the third college president claims that these probabilities
    are, respectively, 0.09, 0.38, and 0.56. Comment on these claims.

4. Given the mutually exclusive events A and B for which $P(A) = 0.27$
    and $P(B) = 0.54$, find
    (a) $P(A')$;   (c) $P(A \cup B)$;   (e) $P(A' \cup B')$;
    (b) $P(B')$;   (d) $P(A \cap B)$;   (f) $P(A' \cap B')$.
    (*Hint:* draw a Venn diagram and fill in the probabilities associated
    with the various regions.)

5. Given the mutually exclusive events C and D for which $P(C) = 0.21$
    and $P(D) = 0.33$, find
    (a) $P(C')$;   (c) $P(C \cup D)$;   (e) $P(C' \cap D)$;
    (b) $P(D')$;   (d) $P(C \cap D')$;   (f) $P(C' \cup D')$.

6. Explain why there must be a mistake in each of the following:
    (a) $P(A) = 0.48$ and $P(A') = 0.42$;
    (b) $P(B) = 1.02$;
    (c) $P(C) = -0.03$;

(d) $P(A) = 0.45$ and $P(A \cap B) = 0.53$;

(e) $P(A) = 0.87$ and $P(A \cup B) = 0.79$.

7. A football coach claims that the odds are 3 to 2 that his team will win an upcoming game, while the odds against his team losing or tieing are, respectively, 4 to 1 and 9 to 1. Discuss the *consistency* of the corresponding probabilities.

8. A teacher feels that the odds are 2 to 1 against his getting a raise of $500 and 5 to 1 against his getting a raise of $1,000, but that it is an even bet (the odds are 1 to 1) that he will get either a $500 raise or a $1,000 raise. Discuss the *consistency* of the corresponding probabilities.

9. A stockbroker predicts that the odds are 2 to 1 that the value of GM stock will go up during the coming month and 3 to 1 that it will go down. Can these odds be right? Explain.

10. The probabilities that a consumer-testing service will rate a new anti-pollution device for cars poor, fair, good, or excellent are, respectively, 0.12, 0.23, 0.45, and 0.20. Find the probabilities that they will rate the device

   (a) poor or fair;

   (b) at least fair;

   (c) good or excellent;

   (d) fair or good.

11. The probabilities that a doctor's answering service will receive 0, 1, 2, 3, 4, 5, 6, or 7 *or more* calls for him during the lunch hour are, respectively, 0.001, 0.006, 0.022, 0.052, 0.091, 0.128, 0.149, and 0.551. What is the probability that he will receive

   (a) fewer than 5 calls;

   (b) at least 3 calls;

   (c) anywhere from 2 to 6 calls, inclusive?

12. A waiter knows from experience that the probabilities are, respectively, 0.13, 0.24, 0.09, 0.11, 0.05, and 0.07 that a customer will order chocolate cake, cherry pie, ice cream, rice pudding, sherbet, or watermelon for dessert. What are the probabilities that a customer (who can order only one of these desserts) will order

   (a) cherry pie or ice cream;

   (b) chocolate cake, rice pudding, or watermelon;

   (c) cherry pie, rice pudding, sherbet, or watermelon;

   (d) none of these desserts?

13. With reference to Exercise 6 on page 120, suppose that each point of the sample space has the same probability 1/20.

   (a) What are the respective probabilities of events Q, R, T, and U?

   (b) What are the probabilities of the four events of part (c) of that exercise?

14. With reference to Exercise 7 on page 121, suppose that the points (2, 1), (2, 2), (3, 1), (3, 2), (3, 3), (4, 1), (4, 2), (4, 3), and (4, 4), of the sample space have the respective probabilities 0.12, 0.08, 0.04, 0.26, 0.20, 0.01, 0.06, 0.17, and 0.06.
    (a) What are the corresponding probabilities of events E, F, and G?
    (b) What are the respective probabilities that the engineer will choose a 2-, 3-, or 4-bedroom house?
    (c) What are the probabilities of the six events of part (c) of that exercise?

15. Suppose that in Exercise 8 on page 121 the points (0, 0), (0, 1), (0, 2), (0, 3), (0, 4), (1, 0), (1, 1), (1, 2), (1, 3), (2, 0), (2, 1), (2, 2), (3, 0), (3, 1), and (4, 0) of the sample space are assigned the respective probabilities 0.14, 0.12, 0.05, 0.03, 0.04, 0.12, 0.10, 0.06, 0.05, 0.05, 0.06, 0.06, 0.03, 0.05, and 0.04.
    (a) What are the corresponding probabilities of events R, M, T, and U?
    (b) What are the respective probabilities that he will make 0, 1, 2, 3, or 4 calls on the first day?
    (c) What are the respective probabilities that he will make 0, 1, 2, 3, or 4 calls on the second day?
    (d) What are the respective probabilities that he will call 0, 1, 2, 3, or all 4 of his friends?

16. Given two events A and B for which $P(A) = 0.56$, $P(B) = 0.43$, and $P(A \cap B) = 0.18$, find
    (a) $P(A')$;   (c) $P(A \cup B)$;   (e) $P(A' \cup B)$;
    (b) $P(B')$;   (d) $P(A' \cap B)$;   (f) $P(A \cap B')$.
    (*Hint:* draw a Venn diagram and fill in the probabilities associated with the various regions.)

17. The probability that a person stopping at a gas station will ask to have his tires checked is 0.12, the probability that he will ask to have his oil checked is 0.29, and the probability that he will ask to have them both checked is 0.07.
    (a) What is the probability that a person stopping at this gas station will have either his tires or his oil checked?
    (b) What is the probability that a person stopping at this gas station will have neither his tires nor his oil checked?
    (*Hint:* draw a Venn diagram and fill in the probabilities associated with the various regions.)

18. For married couples living in a suburb, the probability that the husband will vote in a school bond election is 0.19, the probability that his wife will vote is 0.26, and the probability that they will both vote is 0.15.
    (a) What is the probability that either a husband or his wife will vote?

(b) What is the probability that neither a husband nor his wife will vote?

(*Hint:* draw a Venn diagram and fill in the probabilities associated with the various regions.)

19. The probability that a certain movie will get an award for good acting is 0.16, the probability that it will get an award for good directing is 0.30, and the probability that it will get awards for both is 0.09.
    (a) What is the probability that the movie will get either or both awards?
    (b) What is the probability that the movie will get only one of the two awards?
    (c) What is the probability that the movie will get neither award?

20. The following is a proof of the rule on page 126 which states that $P(A) \leq 1$ for any event A. Making use of the fact that *by definition* A and A' represent mutually exclusive events, and that $A \cup A' = S$ (since A and A' together contain all of the points of the sample space S), we can write $P(A \cup A') = P(S)$, and, hence,

$$\begin{array}{ll} P(A) + P(A') = P(S) & \textit{Step 1} \\ P(A) + P(A') = 1 & \textit{Step 2} \\ P(A) = 1 - P(A') & \textit{Step 3} \\ P(A) \leq 1 & \textit{Step 4} \end{array}$$

State which of the three postulates of probability justify the first, second, and fourth steps of this proof; the third step is simple arithmetic. Note also that in Step 2 we have actually proved the third of the rules given on page 126.

6.4 Conditional Probability

Very often, it is meaningless (or at least very confusing) to speak of the probability of an event without specifying the sample space with which we are concerned. For instance, if we ask for the probability that a lawyer makes more than $15,000 a year, we may well get many different answers *and they can all be correct.* One of these might apply to all lawyers in the United States, another might apply to lawyers handling only divorce cases, a third might apply only to lawyers employed by corporations, another might apply to lawyers handling only tax cases, and so on. Since the choice of the sample space (namely, the set of all possibilities under consideration) is by no means always self-evident, it is helpful to use the symbol $P(A \mid S)$ to denote the *conditional probability* of event A relative to the sample space S, or as we often call it "the probability of A given

S." The symbol $P(A \mid S)$ makes it explicit that we are referring to the sample space S (that is, a *particular* sample space S), and it is generally preferable to the abbreviated notation $P(A)$ unless the tacit choice of S is clearly understood. It is also preferable when we have to refer to *different* sample spaces in one and the same problem, as in the examples which follow.

To elaborate on the idea of a *conditional probability*, let us consider the following experiment: In a study of the effectiveness of an anti-allergy drug, 80 patients with allergy problems received an injection of the drug and two hours later only 16 of them exhibited allergic symptoms when exposed to ragweed pollen; as a control, 20 patients with allergy problems were injected with a placebo (that is, a preparation *not* containing the drug) and two hours later 8 of them exhibited allergic symptoms when exposed to ragweed pollen. Schematically, this information can be presented as follows:

	Received the Drug	Received Placebo
Allergic Symptoms	16	8
No Allergic Symptoms	64	12
	80	20

To examine these patients further, the scientist conducting the study draws their names by lot, so that each of them has a probability of $\frac{1}{100}$ of being chosen first. As can be seen from the above table, the chances that the first patient thus chosen will have received the drug and showed allergic symptoms are rather slim—the probability is $\frac{16}{100} = 0.16$ to be exact. Letting D denote the selection of a patient who received the drug and A the selction of a patient who exhibited allergic symptoms, we can write this probability as

$$P(D \cap A) = 0.16$$

Furthermore, it can be seen that the probability that the first patient thus chosen will have received the drug is

$$P(D) = \frac{16 + 64}{100} = 0.80$$

and the probability that he will have exhibited allergic symptoms is

$$P(A) = \frac{16 + 8}{100} = 0.24$$

where all these probabilities were obtained by means of the special formula for equiprobable events, which, as we saw on page 129, follows from the postulates of probability.

Since the value of the probability $P(D \cap A)$ is fairly low, suppose that the scientist conducting the study decides to limit the selection at first to patients who actually received the drug. The number of possibilities is thus reduced to 80, and if we assume that each of these 80 patients still has an equal chance, we find that

$$P(A \mid D) = \frac{16}{80} = 0.20$$

This is the *conditional probability* of the scientist selecting one of the patients who exhibited allergic symptoms *given that the patient actually received the drug*. Note that this conditional probability can also be written as

$$P(A \mid D) = \frac{16/100}{80/100} = \frac{P(D \cap A)}{P(D)}$$

namely, as the *ratio* of the probability of the selection of a patient who received the drug *and* exhibited allergic symptoms to the probability of the selection of a patient who received the drug.

Generalizing from this example, let us now make the following definition which applies to any two events A and B belonging to a given sample space S:

> *If $P(B)$ is not equal to zero, then the conditional probability of A relative to B, namely, the "probability of A given B," is given by*
>
> $$P(A \mid B) = \frac{P(A \cap B)}{P(B)}$$

Had the scientist limited the selection to patients who received the placebo, we can now argue that the probability of his first selecting one who exhibited allergic symptoms is

$$P(A \mid D') = \frac{P(A \cap D')}{P(D')} = \frac{0.08}{0.20} = 0.40$$

Of course, this result could have been obtained directly by observing that among the 20 patients who received a placebo, 8 $\left(\text{or } \dfrac{4}{10}\right)$ exhibited allergic symptoms. Note that $P(A \mid D')$ is much greater than $P(A \mid D)$, which suggests that the drug is probably quite effective.

Although we justified the formula for $P(A \mid B)$ by means of an example in which all outcomes were *equiprobable*, this is *not* a requirement for its use. To consider an example in which we are *not* dealing with equiprobable events, let us refer again to the two-person-interview example and Figure 6.9. If we let A denote the event that the first person interviewed is undecided (about the legislation) while B denotes the event that the second person is against it, we find that

$$P(A) = 0.08 + 0.04 + 0.08 = 0.20$$
$$P(B) = 0.16 + 0.08 + 0.16 = 0.40$$

and

$$P(A \cap B) = 0.08$$

Then, if we substitute the last two values into the formula for $P(A \mid B)$, we get

$$P(A \mid B) = \frac{P(A \cap B)}{P(B)} = \frac{0.08}{0.40} = 0.20$$

and this serves to illustrate the use of the formula for $P(A \mid B)$ when the outcomes are *not all equiprobable*.

Note also that $P(A \mid B) = 0.20 = P(A)$ in the preceding example, which means that the probability of event A is the same regardless of whether event B has occurred (occurs, or will occur), and we say that *event A is independent of event B*. Intuitively speaking, this means that *the occurrence of A is in no way affected by the occurrence or non-occurrence of B*, and this is something we should really have expected in this example—there should be no relationship (dependence) between the responses of two persons interviewed in a scientifically conducted "impartial" survey. Note that in the allergy example of this section event A was *not* independent of event D—whereas $P(A)$ was 0.24, $P(A \mid D)$ was 0.20.

As it can be shown in general that *event B is independent of event A whenever event A is independent of event B*, namely $P(B) = P(B \mid A)$ whenever $P(A) = P(A \mid B)$, it is customary to say simply that *A and B are independent* whenever one is independent of the other. As the reader will be asked to verify in Exercise 12 on page 146 for the two-person-interview example, the independence of A and B implies also that A is independent of B' and B is independent of A', namely, that $P(A \mid B') = P(A)$ and $P(B \mid A') = P(B)$. If two events A and B are *not independent*, we say that they are *dependent*.

So far we have used the formula $P(A \mid B) = \dfrac{P(A \cap B)}{P(B)}$ only to calculate conditional probabilities, which, of course, was the reason for which it was introduced. However, if we multiply the expressions on both sides of this equation by $P(B)$, we get

$$\blacktriangle \qquad P(B){\cdot}P(A \mid B) = P(A \cap B) \qquad \blacktriangle$$

and this provides us with a formula, sometimes referred to as a *multiplication rule*, which enables us to calculate the probability that two events will both occur. In words, the formula states that *the probability that two events will both occur is the product of the probability that one of the events will occur and the conditional probability that the other will occur given that the first event has occurred (occurs, or will occur)*. As it does not matter which event is referred to as A and which event is referred to as B, the above formula can also be written as

$$\blacktriangle \qquad P(A){\cdot}P(B \mid A) = P(B \cap A) \qquad \blacktriangle$$

and, of course, $P(A \cap B) = P(B \cap A)$, since $A \cap B$ and $B \cap A$ denote the same set.

To illustrate the use of these formulas, suppose we want to determine the probability of having the *bad luck* of randomly picking 2 defective television sets from a shipment of 15 sets among which 3 are defective. Assuming equal probabilities for each selection (which is what we mean by "randomly picking" the two sets), we find that the probability that the first one is defective is 3/15, and that the probability that the second one is defective *given that the first set was defective* is 2/14. Clearly, there are only 2 defectives among the 14 sets which remain after one defective set has been picked. Hence, the probability of choosing two sets which are *both defective* is

$$\frac{3}{15}{\cdot}\frac{2}{14} = \frac{1}{35}$$

A similar argument leads to the result that the probability of choosing two sets which are *not defective* is

$$\frac{12}{15}\cdot\frac{11}{14} = \frac{22}{35}$$

and it follows, by subtraction, that the probability of getting one good set and one defective set is $1 - \dfrac{1}{35} - \dfrac{22}{35} = \dfrac{12}{35}$. An alternate way of handling problems of this kind will be discussed in Chapter 8.

When A and B are *independent events,* we can substitute $P(A)$ for $P(A \mid B)$ into the first form of the multiplication rule on page 138, or $P(B)$ for $P(B \mid A)$ into the second, and we obtain the *special multiplication rule*

$$\blacktriangle \qquad\qquad P(A \cap B) = P(A)\cdot P(B) \qquad\qquad \blacktriangle$$

This formula can be used, for example, to find the probability of getting two *heads* in a row with a balanced coin or the probability of drawing two aces in a row from an ordinary deck of 52 playing cards *provided the first card is replaced before the second is drawn.* For the two flips of the coin we get $\dfrac{1}{2}\cdot\dfrac{1}{2} = \dfrac{1}{4}$ and for the two aces we get $\dfrac{4}{52}\cdot\dfrac{4}{52} = \dfrac{1}{169}$, since there are 4 aces among the 52 cards. (Had the first card not been replaced before the second card was drawn, the probability of getting two aces in a row would have been $\dfrac{4}{52}\cdot\dfrac{3}{51} = \dfrac{1}{221}$; this distinction will be discussed further in Chapter 8, as it is important in *statistics,* where we speak of "sampling with or without replacement.") The following are two further applications of the special multiplication rule: if the probability that a person will make a mistake in his income tax return is 0.12, then the probability that two totally unrelated persons (who do not use the same accountant) will both make a mistake is $(0.12)(0.12) = 0.0144$; if the probability that a person will choose blue as his favorite color is 0.24, then the probability that neither of two totally unrelated persons will choose blue is $(0.76)(0.76) = 0.5776$.

The special multiplication rule can easily be extended so that it applies to the occurrence of three or more independent events—*we simply multiply all of the respective probabilities.* For instance, the probability of getting 4 *heads* in a row with a balanced coin is $\dfrac{1}{2}\cdot\dfrac{1}{2}\cdot\dfrac{1}{2}\cdot\dfrac{1}{2} = \dfrac{1}{16}$, and the probability of first rolling two 1's and then some other number in three rolls of a balanced die is $\dfrac{1}{6}\cdot\dfrac{1}{6}\cdot\dfrac{5}{6} = \dfrac{5}{216}$. For dependent events the formulas become somewhat more complicated, as is illustrated in Exercise 16 on page 147.

6.5 Bayes' Rule

Although the two symbols $P(A \mid B)$ and $P(B \mid A)$ look very much alike, there is a great difference between the corresponding probabilities. In the allergy example of the preceding section, $P(A \mid D)$ is the probability of the scientist's choosing a patient who exhibited allergic symptoms *given that he received the drug*, and $P(D \mid A)$ is the probability of the scientist's choosing a patient who received the drug *given that he exhibited allergic symptoms*. As we already saw, $P(A \mid D) = 0.20$, and as can easily be verified, $P(D \mid A) = \dfrac{2}{3}$. Similarly, if C represents the event that a certain person committed a crime and G represents the event that he is judged guilty, then $P(G \mid C)$ is the probability that the person will be judged guilty *given that he actually committed the crime*, and $P(C \mid G)$ is the probability that the person actually commited the crime *given that he has been judged guilty*—clearly, there is a big difference between these two conditional probabilities.

Since there are many problems which involve such pairs of conditional probabilities, let us try to find a formula which expresses $P(B \mid A)$ in terms of $P(A \mid B)$ for any two events A and B. Fortunately, we do not have to look very far; all we have to do is equate the two expressions for $P(A \cap B)$ on page 138, and we get

$$P(A)\cdot P(B \mid A) = P(B)\cdot P(A \mid B)$$

and, hence,

▲
$$P(B \mid A) = \frac{P(B)\cdot P(A \mid B)}{P(A)}$$
▲

after dividing the expressions on both sides of the equation by $P(A)$. To illustrate the use of this formula, suppose that if a person with tuberculosis is given a chest X-ray, the probability is 0.98 that his condition will be detected; also, if a person without tuberculosis is given a chest X-ray, the probability that he will *erroneously* be diagnosed as having tuberculosis is 0.0001. What we would like to know is the probability that in a community where 0.3 per cent of all residents have tuberculosis, *a person thus diagnosed as having tuberculosis actually has this disease.* If we let A

denote the event that a person is thus diagnosed as having tuberculosis, while B denotes the event that a person (in this community) actually has the disease, the given information can be written as $P(A \mid B) = 0.98$, $P(A \mid B') = 0.0001$, and $P(B) = 0.003$. Before we can calculate $P(B \mid A)$ by means of the above formula, we first have to find $P(A)$, and to this end it is best to look at a tree diagram like that of Figure 6.11. Here A

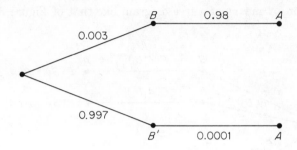

Figure 6.11. Tree diagram for tuberculosis-diagnosis example.

is reached either along the branch which passes through B or along the branch which passes through B', and the respective probabilities are $(0.003)(0.98) = 0.00294$ and $(1 - 0.003)(0.0001) = 0.0000997$, so that $P(A) = 0.00294 + 0.0000997 = 0.0030397$ (or approximately 0.00304). Actually, what we have done here is make use of the formula

$$\blacktriangle \qquad P(A) = P(B) \cdot P(A \mid B) + P(B') \cdot P(A \mid B') \qquad \blacktriangle$$

which is an immediate consequence of part (c) of Exercise 13 on page 124 and the multiplication rules on page 138. Having obtained the value of $P(A)$, we can now substitute into the above formula for $P(B \mid A)$, getting

$$P(B \mid A) = \frac{(0.003)(0.98)}{0.00304} = \frac{0.00294}{0.00304} = 0.97$$

for the probability that a person who is thus diagnosed as having tuberculosis actually has the disease.

The formula which we have used here to determine $P(B \mid A)$ is a very simple version of the somewhat controversial *Rule of Bayes*. There is no question about its *validity* (in fact, it is quite easy to prove), but arguments have been raised about its *applicability*. This is due to the fact that it involves a "backward" or "inverse" sort of reasoning—namely, *reasoning from effect to cause*. In our numerical example we asked for the probability that the tuberculosis diagnosis was "caused" by the person actually having the disease, in Exercise 18 on page 148 the reader will be asked to find the probability that the crash of an airplane was "caused" by structural failure, and in Exercise 17 on page 147 the reader will be asked to determine the probability that a company's success in getting a government contract was "caused" by its major competitor's failure to bid.

When there are more than two possible "causes," it is best to analyze the situation by means of a tree diagram like that of Figure 6.12, where

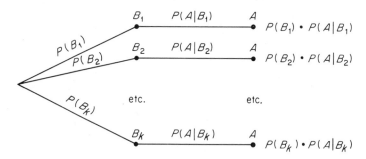

Figure 6.12. Tree diagram for Bayes' Rule.

the various possible "causes" of A are labeled B_1, B_2, ..., and B_k. With reference to this diagram we can say that $P(B_i \mid A)$ is the probability that event A was reached via the ith branch of the tree, for $i = 1, 2, \ldots$, or k, and that its value is given by the *ratio* of the probability associated with the ith branch, namely, $P(B_i) \cdot P(A \mid B_i)$, to the *sum* of the probabilities associated with *all* of the branches of the "tree." Symbolically,

$$\blacktriangle \quad P(B_i \mid A) = \frac{P(B_i) \cdot P(A \mid B_i)}{P(B_1) \cdot P(A \mid B_1) + P(B_2) \cdot P(A \mid B_2) + \ldots + P(B_k) \cdot P(A \mid B_k)} \quad \blacktriangle$$

for $i = 1, 2, \ldots$, or k. Note that the expression in the denominator actually equals $P(A)$—it is the sum of the probabilities of reaching A via the k branches of the tree diagram of Figure 6.12.

To illustrate this more general rule, suppose that in a cannery, assembly lines I, II, and III account respectively, for 50 per cent, 30 per cent, and 20 per cent of the total output. If 0.4 per cent of the cans from assembly line I are improperly sealed, while the corresponding percentages for assembly lines II and III are, respectively, 0.6 per cent and 1.2 per cent, what we would like to know is the probability that an improperly sealed can (discovered in the final inspection of outgoing products) came from assembly line I.

Picturing this situation as in Figure 6.13, where A represents the event that a can is improperly sealed, we find that the probabilities asso-

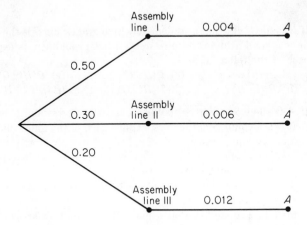

Figure 6.13. Tree diagram for illustration of Bayes' Rule.

ciated with the three branches of the tree diagram are, respectively, $(0.50)(0.004) = 0.002$, $(0.30)(0.006) = 0.0018$, and $(0.20)(0.012) = 0.0024$, and that they add up to 0.0062. Thus, the probability that an improperly sealed can came from assembly line I is

$$\frac{0.0020}{0.0062} = 0.32 \text{ (approximately)}$$

even though this assembly line porduced half of all the cans.

To solve this problem by means of *Bayes' formula* (that is, without reference to a tree diagram like that of Figure 6.13), we let B_1, B_2, and B_3 denote the respective events that a can comes from assembly lines I, II, and III, so that the given information can be written as $P(B_1) = 0.50$, $P(B_2) = 0.30$, $P(B_3) = 0.20$, $P(A \mid B_1) = 0.004$, $P(A \mid B_2) = 0.006$, and $P(A \mid B_3) = 0.012$, where A represents, as before, the event that a can is improperly sealed. Thus, substitution into the formula yields

$$P(B_1 \mid A) = \frac{(0.50)(0.004)}{(0.50)(0.004) + (0.30)(0.006) + (0.20)(0.012)} = \frac{0.0020}{0.0062}$$

and the result is, of course, the same as before. In Exercise 19 below the reader will be asked to find the corresponding probabilities that an improperly sealed can came from assembly lines II and III.

EXERCISES

1. If H is the event that a student gets high grades and G is the event that he is a good athlete, state *in words* what probability is expressed by each of the following:
   - (a) $P(H \mid G)$;
   - (c) $P(H \mid G')$;
   - (e) $P(H' \mid G')$;
   - (b) $P(G \mid H)$;
   - (d) $P(H' \mid G)$;
   - (f) $P(G' \mid H')$.

2. If A is the event that a play is an artistic success and C is the event that it is a commercial success, express each of the following probabilities in symbolic form:
   - (a) The probability that a play which is an artistic success is also a commercial success.
   - (b) The probability that a play which is a commercial success is not an artistic success.
   - (c) The probability that a play which is not an artistic success is also not a commercial success.
   - (d) The probability that a play which is not a commercial success is an artistic success.

3. If A is the event that a car which a dealer has in stock has air conditioning, Q is the event that it has power steering, and B is the event that it has bucket seats, state *in words* what probabilities are represented by
   - (a) $P(A \mid B)$;
   - (c) $P(B \mid A')$;
   - (e) $P(Q' \mid B')$;
   - (b) $P(A \mid Q')$;
   - (d) $P(B \cap Q \mid A)$;
   - (f) $P(B \mid A \cap Q)$.

4. With reference to Exercise 3, express each of the following probabilities in symbolic form:
   - (a) The probability that one of the dealer's cars with bucket seats also has power steering.
   - (b) The probability that one of the dealer's cars without air conditioning will have either power steering or bucket seats.
   - (c) The probability that one of the dealer's cars without power steering will have air conditioning but no bucket seats.
   - (d) The probability that one of the dealer's cars with neither bucket seats nor power steering will not have air conditioning either.

5. As part of a promotional scheme in California and Oregon, a company making cake mixes will award a grand prize of $50,000 (and several other prizes) to persons sending in their names on entry blanks, with the option of also including a box top of one of the company's products.

The breakdown of the 60,000 entries which they received is as shown in the following table:

	With Box Top	Without Box Top
California	32,000	11,000
Oregon	8,000	9,000

If the winner of the grand award is chosen by lot, C represents the event that it will be won by an entry from California, and B represents the event that it will be won by an entry which included a box top, find each of the following probabilities:

(a) $P(C)$;

(b) $P(C')$;

(c) $P(B)$;

(d) $P(B')$;

(e) $P(C \cap B)$;

(f) $P(C' \cap B')$;

(g) $P(C \mid B)$;

(h) $P(B \mid C)$;

(i) $P(C' \mid B')$;

(j) $P(B' \mid C')$.

6. Use the results of Exercise 5 to verify that

(a) $P(C \mid B) = \dfrac{P(C \cap B)}{P(B)}$;

(b) $P(C' \mid B') = \dfrac{P(C' \cap B')}{P(B')}$;

(c) $P(B \mid C) = \dfrac{P(C \cap B)}{P(C)}$;

(d) $P(B' \mid C') = \dfrac{P(C' \cap B')}{P(C')}$.

7. Suppose that in Exercise 5 the drawing for the grand prize is rigged so that by including a box top the chances of an entry are doubled. Recalculate the ten probabilities asked for in that exercise.

8. Referring to Exercise 17 on page 133, find
   (a) the probability that a person who has his tires checked will also have his oil checked;
   (b) the probability that a person who has his oil checked will also have his tires checked.

9. Referring to Exercise 18 on page 133, find
   (a) the probability that a husband will vote provided his wife is going to vote;
   (b) the probability that a wife will vote provided her husband is going to vote.

10. Referring to Exercise 19 on page 134, find
   (a) the probability that the movie will not get the award for good acting given that it will get the award for good directing;

    (b) the probability that the movie will get the award for good directing given that it will not get the award for good acting.

(*Hint:* draw a Venn diagram and fill in the probabilities associated with the various regions.)

11.   The director of a research laboratory has the following information: The probability that the equipment needed for a project will be delivered on time is 0.80, and the probability that the equipment will be delivered on time *and* the project will be completed on time is 0.45.

    (a) Find the probability that the project will be completed on time, given that the equipment was delivered on time.

    (b) If the probability that the project will be completed on time is 0.50, find the probability that it will be completed on time, given that the equipment was *not* delivered on time.

(*Hint:* draw a Venn diagram and fill in the probabilities associated with the various regions.)

12.   Referring to the two-person-interview example on page 137, verify that

    (a) event B is also independent of event A, namely, that $P(B \mid A) = P(B)$;

    (b) event A is also independent of event B', namely, that $P(A \mid B') = P(A)$;

    (c) event B is also independent of event A', namely, that $P(B \mid A') = P(B)$.

13.   With reference to Exercise 13 on page 132, are the two events Q and R independent?

14.   Which of the following pairs of events are independent and which are dependent?

    (a) Being intoxicated while driving and having an accident.

    (b) Getting threes in two successive rolls of a die.

    (c) Being a banker and having green eyes.

    (d) Having a flat tire and being late for work.

    (e) Being born in July and having flat feet.

    (f) Having a driver's license and owning a car.

    (g) Living in Colorado and being a stamp collector.

    (h) Any two mutually exclusive events.

15.   As we indicated on page 139, the probability that any number of independent events will occur is given by the product of their respective probabilities. Use this rule to find

    (a) the probability of getting 8 tails in a row with a balanced coin;

    (b) the probability of getting first 4 heads and then 4 tails with a balanced coin;

    (c) the probability of rolling either a 5 or a 6 in four successive rolls of a die;

    (d) the probability of drawing (with replacement) 3 hearts in a row from an ordinary deck of 52 playing cards;

    (e) the probability that a fairly good marksman will hit the target five times in a row, given that the probability of his hitting the target on any one try is 0.90 and that we can assume independence.

16. The problem of determining the probability that any number of events will occur becomes more complicated when the events are *not independent*. For three events A, B, and C, for example, the probability that they will all occur is obtained by multiplying the probability of A by the probability of B *given* A, and then multiplying the result by the probability of C *given* $A \cap B$. For instance, the probability of drawing (without replacement) 3 aces in a row from an ordinary deck of 52 playing cards is

$$\frac{4}{52} \cdot \frac{3}{51} \cdot \frac{2}{50} = \frac{1}{5,525}$$

Clearly, there are only 3 aces among the 51 cards which remain after the first ace has been drawn, and only 2 aces among the 50 cards which remain after the first two aces have been drawn.

    (a) Referring to the illustration on page 135, what is the probability that the first three patients drawn by lot for further study had all received the drug?

    (b) Symbolically or in words, give a rule for the probability that four events A, B, C, and D will all occur.

    (c) In a certain city, the probability of passing the test for a driver's license on the first try is 0.75; after that the probability of passing becomes 0.60, regardless of how often a person has failed. What is the probability of finally getting one's license on the fourth try?

    (d) In the Fall, the probability that a rainy day will be followed by a rainy day is 0.70 and the probability that a sunny day will be followed by a rainy day is 0.50. Assuming that each day is classified as being either rainy or sunny and that the weather on any given day depends only on the weather the day before, find the probability that a rainy day is followed by three more rainy days, then two sunny days, and finally another rainy day.

17. There is a fifty-fifty chance that Firm A will bid for the construction of a new city hall. Firm B submits a bid and the probability that it will get the job is 2/3 provided Firm A does not bid; if Firm A submits a bid, the probability that Firm B will get the job is only 1/5. If Firm B gets the job, what is the probability that Firm A did not bid?

18. The probability that an airplane accident due to structural failure is diagnosed correctly is 0.72 and the probability that an airplane accident which is *not* due to structural failure is diagnosed incorrectly as being due to structural failure is 0.12. If 40 per cent of all airplane accidents are due to structural failure, find the probability that an accident which is diagnosed as being due to structural failure is actually due to that cause.

19. Referring to the example on page 143, find the probabilities that an improperly sealed can came (a) from assembly line II, and (b) from assembly line III.

20. There is a fifty-fifty chance that Tom's ex-wife will show up at the New Year's Eve party to which he is planning to go. If she does not show up at the party the odds are 3 to 1 that Tom will have a good time, but if she does show up the odds are 4 to 1 that he will *not* have a good time. If we hear later on that Tom did have a good time at the party, what is the probability that his ex-wife did not show up?

21. A study made by the dean of a university shows that among all freshmen having difficulties, 70 per cent have academic difficulties while the other 30 per cent have difficulties adjusting socially to campus life. If the study also shows that the odds are 4 to 1 that a freshman with academic difficulties will not return for the sophomore year while the odds are only 2 to 1 that a freshman who has difficulties adjusting socially to campus life will not return for the sophomore year, *what is the probability that a student who had difficulties as a freshman and did not return for the sophomore year could not adjust socially to campus life?* (Assume, for the sake of argument, that these two kinds of difficulties are mutually exclusive.)

22. The probabilities that a brewery will decide to sponsor the television of football games, a soap opera, or a news program are, respectively, 0.50, 0.30, and 0.20. If they decide on the football games, the probability that they will get a high rating is 0.75; if they decide on a soap opera, the probability that they will get a high rating is 0.30; and if they decide on a news program, the probability that they will get a high rating is 0.15. If it turns out that they do get a high rating, what is the probability that they chose a soap opera?

23. (From Hans Reichenbach's *The Theory of Probability*, University of California Press, 1949.) Mr. Smith's gardener is not dependable; the probability that he will forget to water the rosebush during Smith's absence is 2/3. The rosebush is in questionable condition anyhow; if watered, the probability for its withering is 1/2; if it is not watered, the probability for its withering is 3/4. Upon returning Smith finds that the rosebush has withered. What is the probability that the gardener did not water the rosebush?

BIBLIOGRAPHY

More detailed, though still elementary, treatments of probability may be found in

FREUND, J. E., *Introduction to Probability*. Encino, Calif.: Dickenson Publishing Company, Inc., 1973.

GOLDBERG, S.. *Probability—An Introduction*. Englewood Cliffs, N.J.: Prentice-Hall, Inc., 1960.

MOSTELLER, F., ROURKE, R. E. K., and THOMAS, G. B., *Probability with Statistical Applications*, 2nd ed. Reading, Mass.: Addison-Wesley Publishing Company, Inc., 1970.

7

Expectations and Decisions

7.1 Introduction

When we are faced with uncertainties, our decisions are seldom based on probabilities alone, for in most cases we must also know something about the consequences (namely, the potential profits, losses, penalties, or rewards) to which we are exposed. For instance, if we must decide whether or not to buy a new car, knowing the chances that our old car will soon require repairs is not enough—to make an intelligent decision we would also have to know, among other things, the cost of the repairs and the trade-in value of our old car. To give another example, suppose that a road builder has to decide whether to bid on a job which promises a profit of $80,000 with a probability of 0.20, or a loss of $18,000 (due, perhaps, to bad estimates, strikes, or the late delivery of materials) with a probability of 0.80. Clearly, the probability of his making a profit is not very high, but, on the other hand, the amount he stands to gain is much greater than the amount he stands to lose. Both of these examples illustrate the need for a method of combining probabilities and consequences, and it is for this purpose that we shall introduce the concept of a *mathematical expectation*.

7.2 Mathematical Expectation

When we say that in Florida a couple can expect to have 1.46 children, that a person living in the United States can expect to eat 161.5 pounds of meat and 20.1 apples a year, or that a resident of Geneva, Switzerland, can expect to go to the movies 25.2 times a year, it must be obvious that we are not using the word "expect" in its colloquial sense. Some of these events cannot possibly occur, and it would certainly be very surprising if a person actually did eat 161.5 pounds of meat during a calendar year. So far as the Florida couples are concerned, some of them will have no children, some will have one child, some will have two children, some will have three, ..., and the 1.46 figure must be interpreted as an *average*, or as we shall call it here, a *mathematical expectation*.

Originally, the concept of a mathematical expectation arose in connection with games of chance, and in its simplest form it is given by the *product* of the amount a player stands to win and the probability that he will win. Thus, if we stand to win $5.00 if a balanced coin comes up *tails*, and nothing if it come up *heads*, our mathematical expectation is $5 \cdot \frac{1}{2} =$ $2.50. Similarly, if we consider buying one of 1,000 raffle tickets issued for a prize (say, a television set) worth $480.00, our mathematical expectation is $480(0.001) = 0.48$ or 48 cents; thus, it would be foolish to pay more than 48 cents for the ticket unless the proceeds of the raffle went to a worthy cause (or the difference could be credited to whatever pleasure a person might derive from placing a bet). Note that in this example 999 of the tickets will not pay anything at all, one ticket will pay $480.00 (or the equivalent in merchandize), so that altogether the 1,000 tickets pay $480, or *on the average* 48 cents per ticket.

So far, we have considered only examples in which there was a single "payoff," namely, one prize or a single payment. To demonstrate how the concept of a mathematical expectation can be generalized, let us change the raffle for the television set so that there is also a second prize (say, a record player) worth $120.00 and a third prize (say, a radio) worth $40.00. Now we can argue that 997 of the tickets will not pay anything at all, one ticket will pay the equivalent of $480, another will pay the equivalent of $120, while a third will pay the equivalent of $40; altogether, the 1,000 raffle tickets will thus pay $480 + $120 + $40 = $640, or *on the average* 64 cents per ticket—this is the *mathematical expectation* for each ticket. Looking at the problem in a different way, we could argue that if the raffle were repeated many times, we would lose 99.7 percent of the time and win each of the three prizes 0.1 per cent of the time. On the average we would thus win

$$0(0.997) + 480(0.001) + 120(0.001) + 40(0.001) = \$0.64$$

which is the sum of the products obtained by multiplying each amount by the corresponding probability. Generalizing from this example, let us now make the following definition:

> *If the probabilities of obtaining the amounts a_1, a_2, a_3, ..., or a_k, are, respectively, p_1, p_2, p_3, ..., and p_k, then the mathematical expectation is*

▲
$$E = a_1p_1 + a_2p_2 + a_3p_3 + \ldots + a_kp_k$$
▲

Each amount is multiplied by the corresponding probability, and the mathematical expectation, E, is given by the sum of all these products. So far as the a's are concerned, it is important to remember that they are *positive* when they represent profits, winnings, or gains (namely, amounts which we receive), and that they are *negative* when they represent losses, penalties, or deficits (namely, amounts which we have to pay). For instance, if we bet $5.00 on the flip of a coin (that is, we either win $5.00 or lose $5.00 depending on the outcome), the amounts a_1 and a_2 are $+5$ and -5, the probabilities are $p_1 = 0.50$ and $p_2 = 0.50$, and the mathematical expectation is

$$E = 5 \cdot \frac{1}{2} + (-5) \cdot \frac{1}{2} = 0$$

This is what the expectation should be in an *equitable game*, namely, in a game which does not favor either player.

Although we referred to the quantities a_1, a_2, ..., and a_k as "amounts," they need not be *cash* winnings, losses, penalties, or rewards. When we said that a Florida couple can *expect* to have 1.46 children, we referred to a result which was obtained by multiplying 0, 1, 2, ..., by the respective probabilities of a Florida couple's having that many children and then adding all these products. Similarly, if we say that a person over 65 can *expect* to visit a doctor 6.8 times a year, the a's in the formula for the mathematical expectation are 0, 1, 2, 3, ..., and they represent the corresponding numbers of visits.

To consider another example, suppose that a businessman is interested in investing in a piece of property for which the probabilities are 0.22, 0.36, 0.28, and 0.14 that he will sell it at a profit of $2,500, that he will sell it at a profit of $1,000, that he will break even, or that he will sell it at a loss of $1,500. If we substitute all these figures into the formula for E, we get

$$E = 2{,}500(0.22) + 1{,}000(0.36) + 0(0.28) + (-1{,}500)(0.14)$$
$$= \$700$$

and this is his *expected gross profit*. (Whether a profit of $700 makes the transaction worthwhile is another matter; this would have to depend on such things as sales commissions, money spent on advertising, the length of time the cash investment will be tied up, and so on.)

In the last example the a's were again sums of money, but they are not in the following situation pertaining to the annual number of hurricanes reported in a certain county: If it is known on the basis of weather bureau records that the probabilities for 0, 1, 2, 3, 4, 5, 6, or 7 hurricanes are, respectively, 0.09, 0.22, 0.26, 0.21, 0.13, 0.06, 0.02, and 0.01, we find that they can expect

$$0(0.09) + 1(0.22) + 2(0.26) + 3(0.21) + 4(0.13) + 5(0.06)$$
$$+ \, 6(0.02) + 7(0.01) = 2.38$$

hurricanes in the given county per year.

It has been suggested that a person's behavior is *rational* if in situations involving uncertainties and risks he always chooses the alternative having the highest mathematical expectation. Although this may look like a reasonable criterion for rational behavior, it involves a number of difficulties which we shall discuss later in this chapter. For the moment, let us merely point out that we tacitly used this criterion in our study of subjective probabilities (see Exercises 20 through 22 on page 110), where we assumed that it is rational to accept a bet when the odds are in our favor, or reject it when the odds are against us. (Of course, when the mathematical expectation is zero, there is no definite advantage or disadvantage to accepting the bet, leaving aside all considerations of the pleasure one might possibly get from making a bet.) Thus, if we feel that Candidate X has a probability of 0.30 of winning an election and someone offers us odds of 4 to 1 (say, $4.00 to $1.00) that he will not win, it would be "rational" for us to accept the bet since our expectation would be

$$4(0.30) + (-1)(0.70) = \$0.50$$

On the other hand, had we been offered odds of only 2 to 1 (say, $2.00 to $1.00), our expectation would have been

$$2(0.30) + (-1)(0.70) = -\$0.10$$

and it would not have been rational to accept the bet. Had we been offered odds of 7 to 3, our expectation would have been 0, and the bet would have been fair.

Although we presented in Section 5.3 a straightforward method of determining subjective probabilities in terms of "fair," "reasonable," or "acceptable" odds, it should be noted that this method also has some serious shortcomings. For one thing, if the amount of money involved is very small, most of us are rather careless in judging odds; also, some persons object to any form of gambling, or betting, on moral grounds, while others, for one reason or another, refuse to risk any part of their capital regardless of the odds. Of course, we cannot force anyone to gamble with his own money, but there is a way around this difficulty, which is illustrated by the following example: Suppose that a writer stands to get $2,500 if a story which he wrote is accepted by a magazine, but nothing if the manuscript is turned down. What we would like to know is *the probability which the writer assigns to the acceptance of his story by the magazine*. Instead of asking him for "fair" odds, as in Chapter 5, suppose that we give him the option of settling now, say, for $1,500 *in lieu of all future claims*. Although this may not provide us with a definite answer, we can learn a lot about the probability p (which he assigns to the acceptance of his story by the magazine) from his decision whether or not to accept the cash settlement of $1,500. If he *rejects* it, we can argue that $1,500 is less than his expectation of $2,500 \cdot p$, namely, that $1,500 < 2,500p$ and, hence $p > \dfrac{1,500}{2,500}$ (or 0.60). Correspondingly, if he *accepts* the cash settlement, $1,500 \geq 2,500p$ and, hence, $p \leq \dfrac{1,500}{2,500}$ (or 0.60). Similarly, as the reader will be asked to verify in Exercise 16 on page 156, we would find that $0.60 < p \leq 0.80$, if the writer *rejected* a cash settlement of $1,500 but accepted one of $2,000. To narrow it down even further, we might ask the writer directly what he would consider a "fair" settlement, and if he said $1,800 we could equate this amount to $2,500 \cdot p$ getting $1,800 = 2,500p$ and $p = \dfrac{1,800}{2,500}$ (or 0.72).

It should be observed that all these methods of determining subjective probabilities are based on the premise that a person will be relatively consistent in making decisions concerning risk-taking situations, that he will take such gambles seriously, and that he will act (intuitively, at least, or subconsciously) according to the principle of always maximizing one's expectations (see Section 7.3). Whether all this is actually the case is debatable, but it is of critical importance to the subjective probability concept.

EXERCISES

1. If a charitable organization raises funds by selling 2,000 raffle tickets for a painting worth $400, what is the mathematical expectation of a person who buys one of these raffle tickets?

2. As part of a promotional scheme, a soap manufacturer offers a first prize of $80,000 and a second prize of $30,000 to persons willing to try a new product (distributed without charge) and send in their names on the label. The winners will be drawn at random in front of a large television audience.
    (a) What would be each entrant's mathematical expectation, if 1,500,000 persons were to send in their names?
    (b) Would this make it worthwhile to spend the 8 cents postage it costs to send in an entry?

3. A jeweler wants to "unload" 5 watches that cost him $60.00 each and 45 watches that cost him $15.00 each. If he wraps these watches in identically-shaped unmarked boxes and lets each customer take his pick, find
    (a) each customer's mathematical expectation;
    (b) the jeweler's expected profit per customer, if he charges $22.00 for the privilege of taking a pick.

4. The two finalists in a golf tournament play 18 holes, with the winner getting $20,000 and the runner-up getting $12,000. What are the two player's mathematical expectations if
    (a) they are evenly matched;
    (b) the better player should be favored by odds of 3 to 1?

5. If someone were to give us $3.00 each time we roll a 6 with a balanced die, how much should we have to pay him when we roll a 1, 2, 3, 4, or 5 to make the game equitable?

6. If it is extremely cold in the East a guest ranch in Arizona will have 120 guests during the Christmas season; if it is cold (but not extremely cold) in the East they will have 104 guests, and if the weather is moderate in the East they will have only 75 guests. How many guests can they expect if the probabilities for extremely cold, cold, or moderate weather in the East are, respectively, 0.34, 0.54, and 0.12?

7. If the two league champions are evenly matched, the probabilities that a "best of seven" basketball play-off will take 4, 5, 6, or 7 games are, respectively, 1/8, 1/4, 5/16, and 5/16. Under these conditions, how many games can we expect such a play-off to last?

8. Referring to the illustration on page 150, what is the road builder's mathematical expectation?

9. An importer is offered a shipment of bananas for $6,000, and the probabilities that he will be able to sell them for $7,000, $6,500, $6,000, or $5,500 are, respectively, 0.25, 0.46, 0.19, and 0.10. If he buys the shipment, what is his expected gross profit?

10. A wage negotiator of a labor union feels that the odds are 3 to 1 that the members of the union will get a raise of 60 cents in their hourly wage, the odds are 17 to 3 against their getting a raise of 40 cents

in their hourly wage, and the odds are 9 to 1 against their getting a raise of 80 cents in their hourly wage. What is the corresponding expected raise in their hourly wage?

11.   A teacher knows that the probabilities for 0, 1, 2, 3, 4, or 5 of her pupils to be absent on any given day are, respectively, 0.21, 0.35, 0.27, 0.11, 0.05, and 0.01. How many of her pupils can she expect to be absent on any one day?

12.   The following table gives the probabilities that a woman who enters a given dress shop will buy 0, 1, 2, 3, or 4 dresses:

Number of Dresses	0	1	2	3	4
Probability	0.11	0.37	0.35	0.12	0.05

How many dresses can a woman entering this shop be expected to buy?

13.   Mr. Jones has the option of accepting a gift of $10.00 or of gambling on the outcome of a football game, where he is to receive $25.00 if the home team wins, nothing if it loses or ties. What can we say about the subjective probability he assigns to the home team's winning if he prefers to accept the $10.00 gift?

14.   With reference to Exercise 13, what can we say about the probability which Mr. Jones assigns to the home team's winning if
      (a) he would prefer the gamble on the game to an outright gift of $5.00;
      (b) he cannot make up his mind whether to gamble on the game or accept an outright gift of $8.00?

15.   A recent college graduate is faced by a decision which cannot wait, namely, that of accepting or rejecting a job paying $8,100 a year. What can we say about the probability which he assigns to his only other prospect, a job paying $10,800 a year, if he decides to reject the $8,100-a-year job?

16.   With reference to the example on page 154, show that if the writer rejects a cash settlement of $1,500 but accepts one of $2,000, then $0.60 < p \le 0.80$, where p is the probability that his story will be accepted by the magazine.

17.   To handle a liability suit, a lawyer has to decide whether to charge a straight fee of $600 or a contingent fee of $1,800 which he will get only if his client wins. What does the lawyer think about his client's chances if
      (a) he prefers the straight fee of $600;
      (b) he prefers the contingent fee of $1,800;
      (c) he cannot make up his mind?

18. The manufacturer of a new battery additive has to decide whether to sell his product for $1.00 a can, or for $1.25 with a "double-your-money-back-if-not-satisfied guarantee." How does he feel about the chances that a person will actually ask for double his money back if
    (a) he decides to sell the product for $1.00;
    (b) he decides to sell the product for $1.25 with the guarantee;
    (c) he cannot make up his mind?

19. Mr. Green has the choice of staying home and reading a good book or going to a party. If he goes to the party he might have a terrible time (to which he assigns a utility of 0), or he might have a wonderful time (to which he assigns a utility of 20 units). If he feels that the odds against his having a good time are 8 to 2 and he decides not to go, what can we say about the utility which he assigns to staying home and reading a good book?

20. Mr. Jones would love to beat Mr. Brown in an upcoming tennis tournament, but his chances are nil unless he takes $500.00 worth of extra lessons, which (according to the tennis pro at his club) will give him a fifty-fifty chance. If Mr. Jones assigns the utility U to his beating Mr. Brown and the utility $-\frac{1}{5}U$ to his losing to Mr. Brown, find U if Mr. Jones decides that it is just about worthwhile to spend the $500.00 on extra lessons.

7.3 Decision Making

When we are faced by uncertainties, mathematical expectations can often be used to a great advantage in making decisions. Generally speaking, if we have to choose between several alternatives, it is considered "rational" to select the one with the "most promising" mathematical expectation: the one which *maximizes expected profits, minimizes expected costs, maximizes expected tax advantages, minimizes expected losses,* and so on.

Although this approach to decision making has great intuitive appeal and sounds very logical, it is not without complications—there are many problems in which it is hard, if not impossible, to assign values to all of the a's (amounts) and all of the p's (probabilities) in the formula for E on page 152. To illustrate some of these difficulties, let us consider the following situation: The manager of the research division of a drug manufacturer has to decide whether to continue experimenting with a new anti-allergy drug. He figures that if the project is continued and the drug proves to be effective, this will be worth $1,500,000 to his company; if it is continued but proves to be unsuccessful, this will entail a loss of $900,000; if the project is discontinued but another company successfully develops the new drug, this will entail a loss of $330,000 (partly, for being put at a competitive disadvantage); and if the project is discontinued and nobody else has success with the new drug, there is a gain of $15,000 (accounted for by funds allocated to the project which remain unspent). Knowing that his competitors are also working on the drug, the research manager can summarize all this information as in the following table:

	The New Drug is Effective	The New Drug is Not Effective
He Decides to Continue the Project	$1,500,000	−$900,000
He Decides to Discontinue the Project	−$330,000	$15,000

Evidently, it will be advantageous to continue the project only if the new anti-allergy drug is actually effective, and the research manager's decision will, therefore, have to depend on the chances that this will be the case. Suppose, for instance, he feels that there is a fifty-fifty chance that the new drug will prove to be effective. He can then argue that *if the project is continued,* his company's *expected gain* is

$$1,500,000 \cdot \frac{1}{2} + (-900,000) \cdot \frac{1}{2} = \$300,000$$

and *if the project is not continued,* his company's *expected gain* is

$$(-330,000) \cdot \frac{1}{2} + 15,000 \cdot \frac{1}{2} = -\$157,500$$

Since an expected gain of $300,000 is obviously preferable to an *expected loss* (negative gain) of $157,500, it stands to reason that the project should be continued.

Of course, the conclusion at which we have arrived is based on the assumption that the research manager's appraisal of the chances for success is correct. *What if he was a bit hasty in assessing the odds?* What if the odds against the drug's proving to be effective should have been 2 to 1, or perhaps even 4 to 1? As the reader will be asked to verify in Exercise 1 on page 160, the research manager should discontinue the project if the odds against success are 4 to 1, and the whole situation is a toss-up when the odds against success are 2 to 1. The point we are trying to make is that *when decisions are based on mathematical expectations, one must be fairly certain that one's estimates of the probabilities p in the formula on page 152 are "correct"* (or at least reasonably close).

The way in which we have studied this problem is referred to as a *Bayesian analysis*. In this kind of analysis, probabilities are assigned to the possibilities about which uncertainties exist (the so-called "States of Nature"), which in our example were the anti-allergy drug's being effective or ineffective; then, *we choose whichever alternative promises the greatest expected gain or the smallest expected loss.* As the reader will see in Exercise 10 on page 163, a Bayesian analysis can also include the possibility of delaying any final action until further information is obtained. This is of special importance in statistics, where we generally deal with sample data obtained from surveys or experiments, and may have to decide how large a sample to take, whether a given sample is adequate for reaching a decision, or whether further observations will have to be made.

Let us now examine briefly what the research manager might do if he had no idea about the chances that the new drug will turn out to be effective. To suggest one possibility, suppose that he is a *confirmed optimist*. Looking at the situation through rose-colored glasses, he notes that if the project is continued his company might gain as much as $1,500,000, whereas the decision to discontinue the project can at best lead to a gain of $15,000. Always hoping for the best (perhaps, in the sense of wishful thinking), he would decide to continue the project, and we might say that by doing so he is *maximizing his company's maximum gain.* (In other words, he is choosing the alternative for which the company's greatest possible gain is a maximum.)

Now suppose that the research manager is a *confirmed pessimist*. Always looking for the worst that can happen, he finds that his company may lose $900,000 if the project is continued, but only $330,000 if it is discontinued. Always expecting the worst (perhaps, in the sense of fear or resignation), he would decide to discontinue the project, and we might say that by doing so he is *minimizing his company's maximum losses.* (In other words, he is choosing the alternative for which the company's greatest possible loss is the least, and we refer to this as the *minimax criterion.*)

There are several other criteria on which decisions can be based in the absence of any knowledge about the probabilities associated with the various alternatives (namely, the probabilities of the various "States of Nature"). One of these, based on the *fear of losing out on a good deal* is given in Exercise 7 on page 162. Of course, the research manager could always leave his decision to chance, say, by flipping a coin—*heads* the project is continued and *tails* it is discontinued—or perhaps by rolling a die, but this leads to further difficulties, as the reader will be asked to show in Exercises 11 through 13 on page 164.

Earlier in this section, we pointed out that it is essential to know the correct values of the p's in the formula for a mathematical expectation, and it may have occurred to the reader that the same applies also to the a's, the amounts. We did not worry about this in our example, assuming that the figures in the table on page 158 were given correctly by the company's accountants, but as the reader will discover from Exercise 2 below, this can be a source of further complications. Generally speaking, the problem of assigning "cash values" to the consequences of one's decisions can pose serious difficulties. This is true, especially, when the consequences involve such *intangibles* as the over-all effects of the choice of one's career, the possible side effects of a new drug, the emotional effects of a broken home, the satisfaction a person may get from making a scientific discovery or seeing his writings published, and so on.

Finally, let us also point out that there are situations in which the criteria we have discussed are outweighed by other considerations. For instance, if the research manager of our example knew that he will be fired if his company loses more than $400,000 as the result of his decision, he would be foolish (though perhaps unselfish) to recommend that the project be continued. The situation would be reversed if he knew that he will be fired unless his decision leads to a *substantial* gain. Further examples of situations where extraneous factors affect decisions are given in Exercises 4 and 17 on pages 161 and 165, and they should stress the point that *there is no universal rule which will always lead to the best possible decisions.*

EXERCISES

1. Referring to the example in the text, find the company's expected gains corresponding to the project being continued and being discontinued, when
   (a) the odds are 4 to 1 that the anti-allergy drug will not be effective;
   (b) the odds are 3 to 1 that the anti-allergy drug will not be effective;
   (c) the odds are 2 to 1 that the anti-allergy drug will not be effective.

2. Referring to the example in the text, suppose that the accounting department of the company discovers the following mistake: If the project is continued and the new drug turns out to be effective, the company stands to gain $3,000,000 instead of $1,500,000. Use the results obtained in the text and in Exercise 1 to check whether this would affect the research manager's decision when he feels that
   (a) there is a fifty-fifty chance that the new drug will be effective;
   (b) the odds are 4 to 1 that the new drug will not be effective;

    (c) the odds are 3 to 1 that the new drug will not be effective;

    (d) the odds are 2 to 1 that the new drug will not be effective.

3. A group of businessmen is faced with the problem of whether to put up the funds needed to build a new sports arena or continue to hold its sports promotions in a college gymnasium. They figure that if the new arena is built and they can get a professional basketball franchise, there will be a profit of \$1,025,000 during the next five years; if the new arena is built and they cannot get a professional basketball franchise, there will be a deficit of \$250,000 during the next five years; if the new arena is not built and they get a professional basketball franchise, they will make a profit of \$500,000 during the next five years; and if the new arena is not built and they cannot get a professional basketball franchise, there will be a profit of only \$50,000 during the next five years (from their other promotions).

    (a) Present all this information in a table like that on page 158.

    (b) If they believe an official of the professional basketball organization who tells them that the odds are 2 to 1 against their getting the franchise, what should the businessmen decide to do so as to maximize the expected profit for the next five years?

    (c) If they believe the sports editor of a local newspaper who tells them that the odds are only 3 to 2 against their getting the franchise, what should the businessmen decide to do so as to maximize the expected profit for the next five years?

    (d) If one of these businessmen is a confirmed pessimist, would he be inclined to vote for or against putting up the funds for the new arena? Explain your answer.

    (e) If one of these businessmen is a confirmed optimist, would he be inclined to vote for or against putting up the funds for the new arena? Explain your answer.

4. With reference to Exercise 3, what would one expect the businessmen to do if

    (a) they will be bankrupt unless they can make a profit of at least \$600,000 during the next five years;

    (b) they will be bankrupt unless they can make a profit of at least \$30,000 during the next five years.

5. Mr. Cooper is planning to attend a convention in Washington, D.C., and he must send in his room reservation immediately. The convention is so large that the activities are held partly in Hotel I and partly in Hotel II, and Mr. Cooper does not know whether the particular session he wants to attend will be held at Hotel I or Hotel II. He is planning to stay only one day, which would cost him \$25.20 at Hotel I and \$21.60 at Hotel II, but it will cost him an extra \$6.00 for cab fare if he stays at the wrong hotel.

    (a) Present all this information in a table like the one on page 158.

    (b) Where should he make his reservation if he wants to minimize his expected expenses and feels that the odds are 2 to 1 that the session he wants to attend will be held at Hotel I?

(c) Where should he make his reservation if the odds quoted in (b) should have been 5 to 1 instead of 2 to 1?

(d) Where should he make his reservation if the odds quoted in (b) should have been 4 to 1?

(e) Where should he make his reservation if he were a confirmed pessimist?

(f) Where should he make his reservation if he were a confirmed optimist?

6. Mrs. Green, who lives in a suburb, plans to spend an afternoon shopping in downtown Boston, and she has some difficulty deciding whether or not to take along her raincoat. If it rains, she will be inconvenienced if she does not bring it along, and if it does not rain, she will be inconvenienced if she does. On the other hand, it will be convenient to have the raincoat if it rains, and it would seem reasonable to say that she is neither convenienced nor inconvenienced if she does not bring her raincoat and it does not rain. To express all this numerically, suppose that the numbers in the following table are in *units of inconvenience,* so that the negative value reflects *convenience:*

	It Rains	It Does Not Rain
She Takes the Raincoat Along	−25	15
She Does Not Take the Raincoat Along	50	0

(a) What should Mrs. Green do to *minimize her expected inconvenience,* if she feels that the odds against rain are 6 to 1?

(b) What should Mrs. Green do to *minimize her expected inconvenience,* if she feels that the odds against rain are 4 to 1?

(c) Does it matter whether or not Mrs. Green takes her raincoat, if she feels that the odds are 5 to 1 that it will not rain?

(d) What will Mrs. Green do if she is a confirmed optimist?

(e) What will Mrs. Green do if she is a confirmed pessimist?

7. Suppose that the research manager of the illustration in the text is the kind of person who always worries about *missing out on a good deal.* Looking at the table on page 158, he would argue that if he decides to continue the project and it turns out that the drug is not effective, he would have been better off by $915,000 if he had decided not to continue the project; this is the *difference* between $15,000 (the company's gain if he had made the right decision) and −$900,000 (the

amount which his company lost due to his making the wrong decision).
Referring to this difference as the *opportunity loss* (or *regret*) associated
with this situation, verify that
   (a) the opportunity loss is $1,830,000 when he decides to discon-
       tinue the project and it turns out that the new anti-allergy drug
       is effective;
   (b) the opportunity loss is zero in the other two cases.
Also, what action should the research manager take to *minimize the
maximum loss of opportunity?*

8. With reference to Exercise 3 and the terminology of the preceding
   exercise, find the opportunity loss of the businessmen when
   (a) they decide to put up the money for the new arena and it turns
       out that they cannot get the franchise;
   (b) they decide not to put up the money for the new arena and it
       turns out that they get the franchise.
   Making use of the fact that the opportunity loss is zero in the other
   two cases, decide what the businessmen should do so as to *minimize
   the maximum loss of opportunity.*

9. With reference to the terminology introduced in Exercise 7, what
   should Mr. Cooper of Exercise 5 do so as to *minimize the maximum
   loss of opportunity?*

10. With reference to the illustration in the text, suppose that the research
    manager has the option of continuing the project for three months at a
    cost of $75,000, after which he will be *certain* (for all practical pur-
    poses) whether the new drug is effective. Of course, this raises the
    question whether it is worthwhile to spend the $75,000. To answer it,
    let us take the research manager's original fifty-fifty chance that the
    drug will turn out to be effective. If he knew *for sure* whether or not
    the drug will turn out to be effective, the right decision would yield
    his company either $1,500,000 or $15,000 (the corresponding entries
    in the table on page 158), and since the corresponding probabilities
    are supposedly $\frac{1}{2}$ and $\frac{1}{2}$, we find that the company could thus expect
    to gain

$$1,500,000 \cdot \frac{1}{2} + 15,000 \cdot \frac{1}{2} = \$757,500$$

    and this is what is called the *expected value of perfect information*. In
    general, *this is the amount one can expect to gain, profit, or win in a
    given situation if one always makes the right decision.* Since the value
    which we obtained here exceeds by $457,500 the $300,000 which the
    company can expect to gain if the project is continued, and by
    $915,000 the $157,500 which the company can expect to lose if the

project is not continued (see page 158), it stands to reason that the $75,000 would be well spent.

(a) With reference to Exercise 3 and the odds of part (b), what is the expected value of perfect information? Would it be worthwhile to the businessmen to spend $25,000 to find out for sure whether they will be able to get the franchise before they make their decision about the arena?

(b) With reference to Exercise 5 and the odds of part (c), what is the expected value of perfect information? Would it be worthwhile to spend $1.80 on a long-distance call to find out for certain in which hotel the session (Mr. Cooper wants to attend) will be held?

11. With reference to the illustration in the text, suppose that the research manager bases his decision on the flip of a balanced coin, and the project is continued if and only if the coin comes up *heads*. What is the company's expected gain in the case that

(a) the anti-allergy drug is effective;

(b) the anti-allergy drug is not effective?

12. With reference to the illustration in the text, suppose that the research manager bases his decision on the roll of a balanced die and decides to discontinue the project if and only if it comes up 5 or 6. What is the company's expected gain in the case that

(a) the anti-allergy drug is effective;

(b) the anti-allergy drug is not effective.

Compare the results obtained here with those of Exercise 11, and explain why a confirmed *pessimist* would prefer the gambling scheme of Exercise 11 to that of this exercise.

13. With reference to the illustration in the text, suppose that the research manager bases his decision on a gambling scheme which will make him decide to continue the project with the probability p, and discontinue it with the probability $1 - p$. Express his company's expected gains corresponding to whether or not the drug is actually effective in terms of p, and show that they are equal when $p = \dfrac{23}{183}$. Explain why a *confirmed pessimist* would prefer this gambling scheme (label 23 slips of paper "continue," 160 slips of paper "discontinue," and draw one) to those of Exercises 11 and 12.

14. With reference to Exercise 3, suppose that the businessmen reach their decision by drawing a card from an ordinary deck of 52 playing cards, and decide to put up the money for the arena if and only if it is a heart (the odds against this are 3 to 1). What is their expected profit for the next five years if

(a) they get the franchise;

(b) they do not get the franchise?

15. Would a confirmed pessimist prefer the gambling scheme of Exercise 14 to the businessmen basing their decision on the flip of a coin?

16. A retailer has shelf space for 4 highly perishable items which are destroyed at the end of the day if they are not sold. The unit cost of the item is $3.00, the selling price is $6.00, and the profit is thus $3.00 per item sold. How many items should the retailer stock so as to maximize his expected profit, if he knows that the probabilities of the demand for 0, 1, 2, 3, or 4 items are, respectively, 0.10, 0.30, 0.30, 0.20, and 0.10?

17. A contractor has to choose between two jobs. The first job promises a profit of $120,000 with a probability of 3/4 or a loss of $30,000 (due to strikes and other delays) with a probability of 1/4; the second job promises a profit of $180,000 with a probability of 1/2 or a loss of $45,000 with a probability of 1/2.
     (a) Which job should the contractor choose if he wants to maximize his expected profit?
     (b) Which job would the contractor probably choose if his business is in fairly bad shape and he will go broke unless he can make a profit of at least $150,000 on his next job?

7.4 Statistical Decision Problems

The purpose of the minimax criterion of Section 7.3 was to protect one against the worst thing that can happen in a given situation. This may be desirable in some cases, but the application of this pessimistic criterion becomes highly questionable in most problems of statistical inference, where generalizations and decisions are based on samples. Suppose, for example, that a manufacturer is ready to ship a lot of 40,000 electric light bulbs, and that the department in charge of controlling outgoing quality tested a sample of 25 of these bulbs, all of which met specifications. The manufacturer is faced with a typical problem of statistical inference— he must make a decision about the quality (acceptability) of the whole lot on the basis of a sample. Now then, the worst that can possibly happen is that the 39,975 light bulbs which were not tested are all defective, and if this is what the manufacturer wants to protect himself against, he has no choice but to junk the whole lot.

Since this example is typical of many problems of statistics, it is clear that we must look for alternate criteria on which to base decisions, estimates, and predictions. To examine some possibilities, let us consider the following example, which also serves to bring out some interesting properties of the mean, the median, and some of the other measures of location discussed in Chapter 3: Suppose that the numbers 11, 15, 19, 20, and 20 are written on five slips of paper which are thoroughly mixed in a bowl; one slip is drawn at random, and *we are to predict the number that is on this slip*. Thus, we are faced with making a decision, and before we go any further, we shall have to know something about the consequences to which we are exposed. After all, if there are no penalties for being wrong, no rewards for being right or close, *there is noting at stake* and we might just as well predict that the number drawn will be 137 or 55.1 even though we know that there are no such numbers in the bowl. Thus, let us investigate what decision we might reach in each of the following situations:

1. *We are given a reward of $10.00 if the number drawn is exactly the one we predict, and we are fined $3.00 if a different number is drawn.*
2. *We are paid a fee of $5.00 for making the prediction, but fined an amount of money equal in dollars to the size of our error.*
3. *We are paid a fee of $15.00 for making the prediction, but fined an amount of money equal in dollars to the square of our error.*

To select in each case a "best possible" decision, we shall use the criterion of maximizing one's expected profit.

Investigating the first case, we find that if we decide on the number 20, the *mode* of the five numbers, we stand to make $10.00 with a probability of $\frac{2}{5}$, we stand to lose $3.00 with a probability of $\frac{3}{5}$, and our expected profit is

$$10 \cdot \frac{2}{5} + (-3) \cdot \frac{3}{5} = \$2.20$$

As can easily be verified, this is the best possible prediction; if we predicted 11, 15, or 19, the expected profit would be −$0.40, or a *loss* of $0.40, while any other prediction, say, 8 or 13, would entail a *sure* loss of $3.00. This demonstrates that in a situation where we have to pick the *exact value on the nose* (that is, where there is no reward for being close), the best decision is to use the mode.

In the second case it is the *median* which yields the most profitable predictions. If we predict that the number drawn will be 19, the *median* of the five numbers, the fine will be $8.00, $4.00, $0.00, or $1.00, depending on whether the number drawn is 11, 15, 19, or 20, and hence the *expected fine* is

$$8 \cdot \frac{1}{5} + 4 \cdot \frac{1}{5} + 0 \cdot \frac{1}{5} + 1 \cdot \frac{2}{5} = \$2.80$$

and the *expected profit* is $5.00 − $2.80 = $3.20. As can easily be verified, the expected fine would have been greater for any number other than the median. For instance, if we predicted that the number drawn will be 17, the *mean* of the five numbers, the fine would have been $6.00, $2.00, $2.00, or $3.00 depending on the number drawn, and the *expected fine* would have been

$$6 \cdot \frac{1}{5} + 2 \cdot \frac{1}{5} + 2 \cdot \frac{1}{5} + 3 \cdot \frac{2}{5} = \$3.20$$

The mean comes into its own right in the third case, where the fine goes up very rapidly with the size of the error. Predicting that the number drawn will be 17, the *mean* of the five numbers, we find that the fine will be $36.00, $4.00, $4.00, or $9.00, depending on whether the number drawn is 11, 15, 19, or 20, and hence the *expected fine* is

$$36 \cdot \frac{1}{5} + 4 \cdot \frac{1}{5} + 4 \cdot \frac{1}{5} + 9 \cdot \frac{2}{5} = \$12.40$$

It can easily be shown that the expected fine would have been greater for any other prediction, and it will be left to the reader to verify in Exercise 2 on page 168 that it would have been $16.40 if we had used the median, namely, if we had predicted the number 19.

The third case, where we are concerned with the *squares* of the errors, plays a very important role in the theory of statistics; it ties in closely with the *method of least squares* which we shall study in Chapter 15. The idea of trying to minimize the squared errors is justifiable on the grounds that in actual practice the seriousness of an error often increases very rapidly with the size of the error, *more* rapidly than the magnitude of the error itself.

One of the greatest difficulties in applying the methods of this chapter to realistic problems in statistics is that we seldom know the exact values of all the risks that are involved; that is, we seldom know the exact values of the "payoff" corresponding to the various eventualities. If a medical research group has to decide whether or not to recommend the general release of a new experimental drug, how can they put a cash value on the damage that might be done by not waiting for a more thorough analysis of possible side effects, or on the lives that might be lost by not making the drug available to the public right away? Similarly, if a committee of professors has to decide which of several applicants should be admitted to a medical school or, perhaps, receive a scholarship, how can they possibly forsee all the consequences that might be involved?

The fact that we seldom have adequate information about relevant probabilities also provides obstacles to finding suitable decision criteria; without them, is it "reasonable" to base decisions, say, on optimism or pessimism? Questions like these are difficult to answer, but their analysis at least serves the purpose of revealing some of the logic that underlies statistical thinking.

EXERCISES

1. Referring to the retailer of Exercise 16 on page 165, suppose he has no idea about the potential demand for the item. How many of the items should he stock so as to minimize the maximum losses to which he may be exposed. Discuss the reasonableness of this criterion in a problem of this kind.

2. Verify that in the third case listed on page 166, the expected fine would be $16.40 if we used the median of the five numbers to make our prediction.

3. To consider a variation of the example of this section, suppose that there are several slips of paper marked 12, several marked 15, several marked 18, several marked 26, and we do not know how many there are of each kind. If the payoff is as in the second case on page 166, what prediction would *minimize the maximum possible fine?* What name did we give to this statistic in Chapter 3?

4. Suppose we are asked to predict what proportion of the seniors of a very large high school will fail a college admission test, and all the information we have is that the proportion is not less than 0.06 and not greater than 0.14. What prediction would be "best" if we wanted to *minimize the maximum error?*

5. The ages of seven finalists in a beauty contest are 18, 18, 18, 19, 21, 22, and 24, and we are to predict the age of the winner.
   (a) If we assign each finalist the same probability and there is a reward for being exactly right, none for being close, what prediction maximizes the expected reward?
   (b) If we assign each finalist the same probability and there is a penalty proportional to the magnitude of the error, what prediction minimizes the expected penalty?
   (c) Repeat part (b) for the case where the penalty is proportional to the square of the size of the error.
   (d) What prediction would we make if we had no idea about the finalists' chances and we wanted to minimize the greatest possible error?

BIBLIOGRAPHY

More detailed treatments of the subject matter of this chapter may be found in

BROSS, I. D. J., *Design for Decision*. New York: The Macmillan Company, 1953.

JEFFREY, R. C., *The Logic of Decision*. New York: McGraw-Hill, Inc., 1965.

MORGAN, B. W., *An Introduction to Bayesian Statistical Decision Processes*. Englewood Cliffs, N.J.: Prentice-Hall, Inc., 1968.

8

Probability Distributions

8.1 Introduction

In most statistical problems we are interested only in *one aspect*, or in *a few aspects*, of the outcomes of experiments. If a student takes a true-false test, for example, his grade will depend only on the number of questions which he misses, and it does not matter whether he misses the 8th, 17th, 23rd, and 44th questions, or the 15th, 20th, 31st, and 42nd. Similarly, a geologist may be interested only in the age of a rock and not in its hardness, an agricultural research worker may be interested only in the effect which a fertilizer has on the total yield of a crop and not in the effect which it has on its flavor, and a sociologist may be interested only in a person's ancestry and not in his income or age. Two or more aspects of the outcomes would be of interest, for example, if an engineer investigated the durability and brightness of automobile headlights, and if a market research worker asked housewives about the size of their family, their husband's type of work, whether or not they like a certain product (and, perhaps, also about their age, income, and education).

In each of these examples the respective persons are interested in numbers that are associated with the outcomes of situations involving an element of chance, namely, in the values taken on by so-called *random variables*. (With reference to the examples of the preceding paragraph, this is true also for the sociologist who is interested in ancestry and the market research worker who asks about types of work, for such things can be assigned numerical codes.) To be more explicit, let us consider Figure 8.1, which (like Figure 6.9 on page 128) pictures the sample space

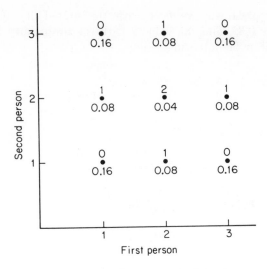

Figure 8.1. Sample space for two-person-interview example.

for the two-person-interview example. Note, however, that we have added another number to each point—the number 0 to the points (1, 1), (1, 3), (3, 1), and (3, 3), the number 1 to the points (1, 2), (2, 1), (2, 3), and (3, 2), and the number 2 to the point (2, 2). As should be apparent, we have thus associated with each point of the sample space the corresponding number of persons who are undecided about the proposed civil rights legislation. Since "associating a number with each point of a sample space" is just another way of saying that we are defining a *function* over the points of the sample space, random variables are, strictly speaking, functions. Conceptually, though, most beginners find it easiest to think of random variables simply as quantities which can take on different values depending on chance. Thus, the number of speeding tickets issued each day on the freeway from Phoenix to Tucson is a random variable, and so is the annual production of soybeans in the United States, the number of persons visiting Disneyland each week, the number of defectives produced each day by a machine, the number of mistakes a student makes in a test, and so forth.

In the study of random variables, we are usually interested mainly in the probabilities with which they take on the various values within their range. How such probabilities can be obtained, and displayed, is the subject matter of this chapter, and to give an example, let us refer again to Figure 8.1. Adding the probabilities associated with the respective points, we find that the random variable "the number of persons who are undecided" takes on the value 0 with the probability 0.16 + 0.16 +

$0.16 + 0.16 = 0.64$, the value 1 with the probability $0.08 + 0.08 + 0.08 + 0.08 = 0.32$, and the value 2 with the probability 0.04. All this is summarized in the following table:

Number of Persons Who are Undecided	Probability
0	0.64
1	0.32
2	0.04

8.2 Probability Functions

The above table, as well as the two which follow, serve to illustrate what we mean by a *probability function*, namely, *a correspondence which assigns probabilities to the values of a random variable.* The first of the two tables which follow was easily obtained on the basis of the assumption that each face of the die in question has a probability of $\frac{1}{6}$, and the second was obtained by considering as equally likely the eight possible outcomes HHH, HHT, HTH, THH, HTT, THT, TTH, and TTT of three flips of a coin, where H stands for *heads* and T for *tails*:

Number of Points Rolled With a Die	Probability		Number of Heads Obtained in Three Flips of a Coin	Probability
1	$\frac{1}{6}$		0	$\frac{1}{8}$
2	$\frac{1}{6}$		1	$\frac{3}{8}$
3	$\frac{1}{6}$		2	$\frac{3}{8}$
4	$\frac{1}{6}$		3	$\frac{1}{8}$
5	$\frac{1}{6}$			
6	$\frac{1}{6}$			

Note that in each of these examples *the sum of all the probabilities is 1*, and that by virtue of the second postulate of probability on page 124 this must always be the case. Note also that since the values of probability functions are probabilities, they must always be positive or zero, and cannot exceed 1.

Whenever possible, we try to express probability functions by means of formulas which enable us to calculate the probabilities associated with the various values of a random variable. With the usual *functional notation* we can thus write

$$f(x) = \frac{1}{6} \qquad \text{for } x = 1, 2, 3, 4, 5, 6$$

for the first of the above examples, where $f(1)$ represents the probability of rolling a 1, $f(2)$ represents the probability of rolling a 2, and so on. As the reader will be asked to verify in Exercise 1 on page 182, a corresponding formula for the number of heads obtained in three flips of a coin can be written as

$$f(x) = \frac{\binom{3}{x}}{8} \qquad \text{for } x = 0, 1, 2, 3$$

where $\binom{3}{x}$ is a binomial coefficient as defined on page 97.

8.3 The Binomial Distribution

There are many applied problems in which we are interested in the probability that an event will take place x times in n "trials," or in other words, x times out of n, while the probability that it will take place in any one trial is some fixed number p and the trials are independent. We may thus be interested in the probability of getting 24 responses to 80 mail questionnaires, the probability that in a sample of 50 voters 32 will favor Candidate A, the probability that 3 of 10 laboratory mice react positively to a new drug, and so on. Referring to the occurrence of any one of the individual events as a "success", we are thus interested in the probability of getting x *successes in n trials*. To handle problems of this kind, which incidentially include the second example on page 172, we use a special probability function, that of the *bionomial distribution*.

If p denotes the probability of a success on any given trial, the probability of getting x successes in n trials (and hence, x successes and $n - x$ failures) in some *specific order* is $p^x(1 - p)^{n-x}$. There is one factor p for each success, one factor $1 - p$ for each failure, and the x factors p

and $n - x$ factors $1 - p$ are all multiplied together by virtue of the assumption that the n trials are independent. Since this probability is the same for each point of the sample space where there are x successes and $n - x$ failures (it does not depend on the order in which the successes and failures are obtained), the desired probability for x successes in n trials *in any order* is obtained by multiplying $p^x(1 - p)^{n-x}$ by the number of points of the sample space (that is, individual outcomes) where there are x successes and $n - x$ failures. In other words, $p^x(1 - p)^{n-x}$ is multiplied by the number of ways in which the x successes can be distributed among the n trials, namely, by $\binom{n}{x}$. We have thus arrived at the following result:

> *The probability of getting x successes in n independent trials is given by*

$$\blacktriangle \qquad f(x) = \binom{n}{x} p^x(1 - p)^{n-x} \qquad for \ x = 0, 1, 2, \ldots, n \qquad \blacktriangle$$

> *where p is the constant probability of a success for each individual trial.*

It is customary to say that the number of successes in n trials is a random variable having the *binomial probability distribution*, or simply the *binomial distribution*. The terms "probability distribution" and "probability function" are often used interchangeably, although some persons make the distinction that the term "probability distribution" refers to *all* the probabilities associated with a random variable, and not only those given directly by its probability function. {Incidentally, we refer to this distribution as the binomial distribution because for $x = 0, 1, 2, \ldots$, and n, the values of its probability function are given by the successive terms of the *binomial expansion* of $[(1 - p) + p]^n$.}

To illustrate the use of the above formula, let us first calculate the probability of getting 5 heads and 7 tails in 12 flips of a balanced coin. Substituting $x = 5$, $n = 12$, $p = \dfrac{1}{2}$, and $\binom{12}{5} = 792$ (see Table VIII), we get

$$f(5) = 792 \left(\frac{1}{2}\right)^5 \left(1 - \frac{1}{2}\right)^{12-5} = \frac{99}{512}$$

or approximately 0.19. Similarly, to find the probability that 7 of 10 mice used in an experiment will react positively to a drug, when the probability that any one of them will react positively is $\dfrac{4}{5}$, we substitute

$x = 7$, $n = 10$, $p = \dfrac{4}{5}$, and $\dbinom{10}{7} = 120$ (see Table VIII), and we get

$$f(7) = 120 \left(\frac{4}{5}\right)^7 \left(1 - \frac{4}{5}\right)^{10-7} = \frac{1{,}966{,}080}{9{,}765{,}625}$$

or approximately 0.20

To give an example in which we calculate *all* of the values of a binomial distribution, suppose that a safety engineer claims that only 60 per cent of all drivers whose cars are equipped with seat belts use them on short trips. Assuming that this figure is correct, what are the probabilities that under such conditions 0, 1, 2, 3, 4, or 5 of 5 drivers will be using their seat belts? Substituting $n = 5$, $p = 0.60$, and, respectively, $x = 0, 1, 2, 3, 4$, and 5, we get

$$f(0) = \binom{5}{0} (0.60)^0 (1 - 0.60)^{5-0} = 0.010$$

$$f(1) = \binom{5}{1} (0.60)^1 (1 - 0.60)^{5-1} = 0.077$$

$$f(2) = \binom{5}{2} (0.60)^2 (1 - 0.60)^{5-2} = 0.230$$

$$f(3) = \binom{5}{3} (0.60)^3 (1 - 0.60)^{5-3} = 0.346$$

$$f(4) = \binom{5}{4} (0.60)^4 (1 - 0.60)^{5-4} = 0.259$$

$$f(5) = \binom{5}{5} (0.60)^5 (1 - 0.60)^{5-5} = 0.078$$

where all the answers are rounded to three decimals. A histogram of this distribution is shown in Figure 8.2.

In actual practice, problems pertaining to the binomial distribution are seldom solved by direct substitution into the formula. Sometimes we use approximations such as those of Sections 8.5 and 8.11, but more often we refer to special tables such as Table VI or the more detailed tables listed in the Bibliography on page 218. Table VI at the end of this book is limited to the binomial probabilities for $n = 2$ to $n = 15$, and $p = 0.05$, 0.1, 0.2, 0.3, 0.4, 0.5, 0.6, 0.7, 0.8, 0.9, and 0.95.

The probabilities which led to Figure 8.2 were actually read off Table VI, and so far as the two examples on page 174 are concerned, we can now check the results by looking up the values corresponding, respec-

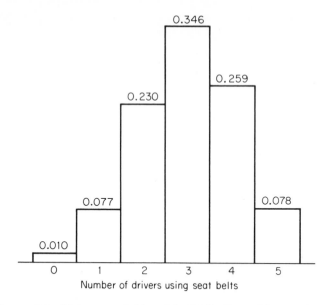

Figure 8.2. Histogram of binomial distribution with $n = 5$ and $p = 0.60$.

tively, to $x = 5$, $n = 12$, and $p = 0.50$, and $x = 7$, $n = 10$, and $p = 0.80$. The corresponding entries in Table VI are 0.193 and 0.201, and when we round these figures to two decimals we get the same results as before.

To consider some further problems in which Table VI can be used to advantage, suppose that in the example on page 174 we had wanted to know the probability of getting *at least* 7 heads in 12 flips of a coin. Evidently, the use of the formula would entail a good deal of work, but from the table it follows immediately that the answer rounded to three decimals is

$$0.193 + 0.121 + 0.054 + 0.016 + 0.003 = 0.387$$

which was obtained by adding the respective probabilities for $x = 7$, 8, ..., and 12. Suppose, also, that we want to know the probability that two of 15 persons crossing the English Channel by boat will get seasick, given that any one of them will get seasick with a probability of 0.05. Substituting $x = 2$, $n = 15$, and $p = 0.05$ into the formula for the binomial distribution, we get

$$f(2) = \binom{15}{2} (0.05)^2 (1 - 0.05)^{15-2}$$

and it must be evident that the evaluation of this quantity would require fairly extensive calculations (perhaps, based on logarithms). On the other hand, Table VI tells us directly that the answer is 0.135.

Before going on to the study of another probability distribution, let us remind the reader that the binomial distribution applies only when the probability of a "success" remains *constant* from trial to trial and the trials are, furthermore, *independent*. The formula on page 174 or Table VI cannot be used, therefore, to find the probability that it will rain, say, on 25 of 120 consecutive days (at a certain weather station). Not only does the probability for rain vary over such a lengthy period of time, but the "trials" are not even independent. Clearly, the probability that it will rain on any given day will depend to some extent on whether it did or did not rain on the day before.

8.4 The Hypergeometric Distribution

In Section 6.4 we mentioned *sampling with and without replacement* to illustrate the multiplication rules for independent and dependent events. Now, to introduce a distribution analogous to the binomial distribution which applies to *sampling without replacement*, let us consider the following example: A company ships automatic dishwashers in lots of 24. Before they are shipped, though, an inspector randomly selects 4 dishwashers from each lot, and the lot "passes" this inspection only if all four are in perfect condition; otherwise, each dishwasher is checked out individually (and repaired, if necessary) at a considerable cost. Clearly, this kind of *sampling inspection* involves certain risks—it is possible for a lot to "pass" this inspection even though 12, or even 20, of the 24 dishwashers have serious defects, and it is possible for a lot to "fail" this inspection even though only one of the dishwashers has a slight defect. This raises many questions. For instance, it may be of special interest to know the probability that a lot will "pass" the inspection when, say, 4 of the dishwashers are not in perfect condition. This means that we shall have to find the probability of 4 successes (perfect dishwashers) in 4 trials, and we might be tempted to argue that since 20 of the dishwashers are in perfect condition, the probability of getting such a dishwasher is $\dfrac{20}{24} = \dfrac{5}{6}$, and hence the probability of "4 successes in 4 trials" is

$$f(4) = \binom{4}{4}\left(\frac{5}{6}\right)^4\left(1 - \frac{5}{6}\right)^{4-4} = 1\left(\frac{5}{6}\right)^4 = \frac{625}{1{,}296}$$

or approximately 0.48 in accordance with the formula for the binomial distribution.

This result would be correct if each dishwasher were replaced before the next one is chosen for inspection, but if sampling is *without replacement*, the probability that the second dishwasher will be perfect *given that the first one was perfect* is $\frac{19}{23}$, the probability that the third dishwasher will be perfect *given that the first two were perfect* is $\frac{18}{22}$, and so forth. Thus, the trials are not independent and the binomial distribution does not apply. To obtain the correct result we might argue that 4 of the 24 dishwashers can be chosen in $\binom{24}{4} = 10{,}626$ ways, 4 of the 20 perfect dishwashers can be chosen in $\binom{20}{4} = 4{,}845$ ways, and hence that the probability of getting 4 perfect dishwashers is

$$\frac{\binom{20}{4}}{\binom{24}{4}} = \frac{4{,}845}{10{,}626}$$

or approximately 0.46 *provided the 10,626 possibilities can be regarded as equally* likely.

In general, if n objects are to be chosen from a set of a objects of one kind (call them "successes") and b objects of another kind (call them "failures"), and we are interested in the probability of getting "x successes and $n - x$ failures," we can similarly argue that the x successes can be chosen in $\binom{a}{x}$ ways, the $n - x$ failures can be chosen in $\binom{b}{n-x}$ ways, and hence x successes and $n - x$ failures can be chosen in $\binom{a}{x} \cdot \binom{b}{n-x}$ ways. Also, n objects can be chosen from the whole set of $a + b$ objects in $\binom{a+b}{n}$ ways, and if we regard these as equally likely, the probability of getting "x successes and $n - x$ failures" is

$$\blacktriangle \qquad f(x) = \frac{\binom{a}{x} \cdot \binom{b}{n-x}}{\binom{a+b}{n}} \qquad \text{for } x = 0, 1, 2, \ldots, \text{ or } n \qquad \blacktriangle$$

This is the formula for the *hypergeometric distribution*, and we should add that it applies only when x does not exceed a and $n - x$ does not exceed b; clearly, we cannot get more successes (or failures) than there are to begin with in the whole set. For instance, in the dishwasher example we had $a = 20$, $b = 4$, $n = 4$, and $x = 4$, and the probability was

$$\frac{\binom{20}{4} \cdot \binom{4}{4 - 4}}{\binom{20 + 4}{4}} = \frac{\binom{20}{4}}{\binom{24}{4}} = 0.46$$

To consider another application of the hypergeometric distribution suppose that as part of an air pollution survey an inspector decides to examine the exhaust of 8 of a company's 16 trucks. He suspects that 5 of the trucks emit excessive amounts of pollutants, and he wants to know the probability that *if his suspicion is correct*, his sample will catch *at least* 3 of these 5 trucks. The probability he wants to know is given by $f(3) + f(4) + f(5)$, where each term in this sum is to be calculated by means of the formula for the hypergeometric distribution with $a = 5$, $b = 11$, and $n = 8$. Substituting these values together with $x = 3$, $x = 4$, and $x = 5$, respectively, we get

$$f(3) = \frac{\binom{5}{3} \cdot \binom{11}{5}}{\binom{16}{8}} = \frac{10 \cdot 462}{12{,}870} = 0.359$$

$$f(4) = \frac{\binom{5}{4} \cdot \binom{11}{4}}{\binom{16}{8}} = \frac{5 \cdot 330}{12{,}870} = 0.128$$

$$f(5) = \frac{\binom{5}{5} \cdot \binom{11}{3}}{\binom{16}{8}} = \frac{1 \cdot 165}{12{,}870} = 0.013$$

and the probability that the inspector will catch at least 3 of the "bad" trucks is $0.359 + 0.128 + 0.013 = 0.500$. This suggests that the inspector should, perhaps, examine more than 8 of the trucks, and it will be left to the reader to show in Exercise 13 on page 184 that the probability of catching at least 3 of the "bad" trucks is increased to 0.76 if the inspector examines 10 of the trucks.

In the beginning of this section we introduced the hypergeometric distribution in connection with a problem in which we *erroneously* used the binomial distribution. Actually, when n is small compared to $a + b$, the binomial distribution often provides a very good *approximation* to the hypergeometric distribution. It is generally agreed that this approximation can be used so long as n constitutes less than 5 percent of $a + b$; this is good because the binomial distribution has been tabulated much more extensively than the hypergeometric distribution, and it is generally easier to use.

8.5 The Poisson Distribution

If n is large and p is small, binomial probabilities are often approximated by means of the formula

$$\blacktriangle \qquad f(x) = \frac{(np)^x \cdot e^{-np}}{x!} \qquad \text{for } x = 0, 1, 2, 3, \ldots \qquad \blacktriangle$$

which is that for the *Poisson distribution*. Here e is the number $2.71828\ldots$ used in connection with natural logarithms, and the values of e^{-np} may be obtained from Table XI at the end of the book. Note also that for this distribution the random variable x can take on the infinite set of values $x = 0, 1, 2, 3, \ldots$; practically speaking, though, this will not pose any problems, since the probabilities become negligible (very close to zero) after the first few values of $\dot{x}$.*

To illustrate the use of the Poisson formula, suppose that an insurance company writing major medical insurance has 1,860 policy holders. Assuming that the probability of any one of these policy holders filing a claim during any one year is $p = \dfrac{1}{600}$, let us find the probabilities that 0, 1, 2, 3, ..., of the 1,860 policy holders will file a claim during a given year. Ruling out the use of the formula for the binomial distribution for practical reasons, we substitute $n = 1,860$, $p = \dfrac{1}{600}$, and, hence, $np =$

---

* As formulated in Chapter 6, the postulates of probability apply only when the sample space S is *finite*. When the sample space is *countably infinite* (namely, when there are as many outcomes as there are whole numbers, as in our example), the third postulate will have to modified so that for any sequence of mutually exclusive events $A_1, A_2, A_3, \ldots$,

$$P(A_1 \cup A_2 \cup A_3 \cup \ldots) = P(A_1) + P(A_2) + P(A_3) + \ldots$$

$1860 \cdot \dfrac{1}{600} = 3.1$ into the formula for the Poisson distribution, getting

$$f(0) = \frac{(3.1)^{0} \cdot e^{-3.1}}{0!} = 0.045, \qquad f(1) = \frac{(3.1)^{1} \cdot e^{-3.1}}{1!} = 0.140,$$

$$f(2) = \frac{(3.1)^{2} \cdot e^{-3.1}}{2!} = 0.216, \qquad f(\dot{3}) = \frac{(3.1)^{3} \cdot e^{-3.1}}{3!} = 0.223$$

for the first four values of x. Note that $e^{-3.1} = 0.045$ is the only value we had to look up in Table XI. Continuing this way, we obtain the values shown in the following table, where we stopped with $x = 10$ since the probabilities for $x = 11$, $x = 12$, ..., are all 0.000 rounded to 3 decimals:

Number of Policy Holders Filing Claims	Probability
0	0.045
1	0.140
2	0.216
3	0.223
4	0.173
5	0.107
6	0.056
7	0.025
8	0.010
9	0.003
10	0.001

The Poisson distribution also has many important applications which have no direct connection with the binomial distribution. In that case np is replaced by the parameter λ (*lambda*) and we calculate the probability of getting x "successes" by means of the formula

▲ $$f(x) = \frac{\lambda^{x} \cdot e^{-\lambda}}{x!} \qquad \text{for } x = 0, 1, 2, 3, \ldots$$ ▲

As the reader will be asked to verify in Exercise 12 on page 194 (see also page 189), the parameter λ can be interpreted as the *expected*, or *average*, number of successes.

The above formula applies in many situations where we can expect a fixed number of "successes" per unit time (or for some other kind of unit),

say, when a bank can expect to receive 6 bad checks per day, when 1.6
accidents can be expected per day at a busy intersection, when 12 small
pieces of meat can be expected in a frozen meat pie, when 5.2 imperfections can be expected per roll of cloth, when 0.3 complaints per visitor
can be expected by the manager of a resort, and so on. For instance, the
probability that the above-mentioned bank will receive 4 bad checks on
any given day is

$$f(4) = \frac{6^4 \cdot e^{-6}}{4!} = \frac{1{,}296(0.0025)}{24} = 0.135$$

and the probability that there will be 3 accidents at the above-mentioned intersection on any given day is

$$f(3) = \frac{1.6^3 \cdot e^{-1.6}}{3!} = \frac{(4.096)(0.202)}{6} = 0.138$$

EXERCISES

1. Verify that if $x = 0$, 1, 2, and 3 is substituted into the formula

$$f(x) = \frac{\binom{3}{x}}{8}$$

the results are the probabilities shown in the second of the two tables
on page 172.

2. Check whether the following can be looked upon as probability functions (defined in each case only for the given values of x) and explain
your answers:

(a) $f(x) = \dfrac{1}{4}$ for $x = 0$, 1, 2, 3, or 4;

(b) $f(x) = \dfrac{x + 1}{10}$ for $x = 0$, 1, 2, or 3;

(c) $f(x) = \dfrac{x - 2}{5}$ for $x = 1$, 2, 3, 4, or 5;

(d) $f(x) = \dfrac{x^2}{30}$ for $x = 0$, 1, 2, 3, or 4.

3. In each case check whether the given values can be looked upon as the
values of the probability function of a random variable which can
take on only the values 1, 2, 3, and 4, and explain your answers:

    (a) $f(1) = 0.24, f(2) = 0.24, f(3) = 0.24, f(4) = 0.24$;
    (b) $f(1) = 1/6, f(2) = 2/6, f(3) = 2/6, f(4) = 1/6$;
    (c) $f(1) = 0.13, f(2) = 0.38, f(3) = 0.04, f(4) = 0.45$;
    (d) $f(1) = 1/2, f(2) = 1/4, f(3) = 1/8, f(4) = 1/16$.

4. Use the formula for the binomial distribution to find the probability of getting
    (a) exactly 3 heads in 8 flips of a balanced coin;
    (b) *at most* 3 heads in 8 flips of a balanced coin;
    (c) exactly 1 one in 3 rolls of a balanced die;
    (d) *at most* 1 one in 3 rolls of a balanced die.

5. A multiple-choice test consists of 8 questions and 3 answers to each question (of which only one is correct). If a student answers each question by rolling a balanced die and checking the first answer if he gets a 1 or a 2, the second answer if he gets a 3 or a 4, and the third answer if he gets a 5 or a 6, find (by means of the formula for the binomial distribution) the probability of getting
    (a) exactly 3 correct answers;
    (b) no correct answers;
    (c) at least 6 correct answers.

6. Use the formula for the binomial distribution to find the probability that exactly 2 of 6 hibiscus plants will survive a frost, given that any such plant will survive a frost with a probability of 0.30. Check the answer in Table VI.

7. Assuming that it is true that 2 in 10 industrial accidents are due to fatigue, find the probability that 2 of 8 industrial accidents will be due to fatigue
    (a) by using the formula for the binomial distribution;
    (b) by referring to Table VI.

8. The probability that Mr. Jones' secretary will not put the correct postage on a letter is 0.20. Find the probability that this secretary will not put the correct postage on 3 of 9 letters
    (a) by using the formula for the binomial distribution;
    (b) by referring to Table VI.

9. A doctor knows from experience that 10 per cent of the patients to whom he gives a certain drug will have undesirable side effects. Use Table VI to find the probabilities that among 12 patients to whom he gives the drug
    (a) none will have undesirable side effects;
    (b) at most 2 will have undesirable side effects;
    (c) at least 3 will have undesirable side effects.

10. A study has shown that in a large city 70 per cent of all families in a certain income group own at least one color television set. Use Table

VI to find the probabilities that among 15 such families (randomly selected for a survey)
  (a) all will have at least one color television set;
  (b) anywhere from 8 to 12, inclusive, will have at least one color television set;
  (c) at most 6 have at least one color television set;
  (d) at least 10 have at least one color television set.

11. A social scientist claims that only 60 per cent of all high school seniors who are capable of college work go to college. If this is so, use Table VI to find the probabilities that among 14 high school seniors capable of doing college work
  (a) exactly 9 will go to college;
  (b) at least 10 will go to college;
  (c) at most 8 will go to college;
  (d) anywhere from 7 through 11, inclusive, will go to college.

12. In connection with situations to which the binomial distribution applies, we are sometimes interested in the occurrence of the *first* success. For this to happen on the xth trial, it must be preceded by $x - 1$ failures, and it follows that the probability of getting the first success on the xth trial is given by

$$f(x) = p(1 - p)^{x-1} \quad \text{for } x = 0, 1, 2, 3, \ldots$$

This is the formula for the *geometric distribution* and it should be noted that, as in the case of the Poisson distribution, there is a countable infinity of possibilities. Use the above formula to solve the following problems:
  (a) Which is more likely, getting the first *heads* on the 5th flip of a balanced coin or getting the first *six* on the 5th roll of a balanced die?
  (b) If the probability that a burglar will get caught on any given "job" is 0.25, what is the probability that he will get caught for the first time on his fifth "job"?
  (c) Suppose the probability is 0.20 that any given person will believe a nasty rumor about the private life of a certain politician. What is the probability that the 6th person to hear the rumor will be the first one to believe it?
  (d) If the probabilities of having a male or female offspring are both 0.50, find the probability that a family's 4th child is their first son.

13. Verify the value given on page 179 for the probability that the inspector will catch at least 3 of the 5 "bad" trucks, if he examines 10 of the company's 16 trucks.

14. Referring to the illustration on page 177, where 4 of the 24 dishwashers were not in perfect condition, find the probabilities that among the 4 dishwashers randomly selected
    (a) exactly one is not in perfect condition;
    (b) exactly two are not in perfect condition;
    (c) exactly three are not in perfect condition;
    (d) none are in perfect condition.
    Combining these results with that obtained on page 178, verify that the sum of the probabilities is 1.

15. Find the probability that an I.R.S. auditor will catch 3 income tax returns with illegitimate deductions, if he randomly selects 5 returns from among 12 returns of which 6 contain illegitimate deductions.

16. A collection of 18 Spanish gold doubloons contains 4 counterfeits. If 2 of these coins are randomly selected to be sold at auction, what are the probabilities that
    (a) neither coin is a counterfeit;
    (b) one of the coins is a counterfeit;
    (c) both coins are counterfeits?

17. A secretary is supposed to send 3 of 8 letters by special delivery. If she gets them all mixed up and randomly puts special delivery stamps on 3 of them, what is the probability that
    (a) she puts all the special delivery stamps on the wrong letters;
    (b) she puts all the special delivery stamps on the right letters?

18. When she buys a dozen eggs, Mrs. Murphy always inspects 3 of the eggs carefully for cracks, and if at least one of them has a crack she looks for another carton. If she randomly selects the eggs which she inspects, what are the probabilities that Mrs. Murphy will buy a carton with
    (a) 2 cracked eggs;
    (b) 3 cracked eggs;
    (c) 4 cracked eggs?

19. A shipment of 120 burglar alarms contains 5 that are defective. If 3 of these burglar alarms are randomly selected and shipped to a customer, find the probability that he will get exactly one bad unit using
    (a) the formula for the hypergeometric distribution;
    (b) the binomial distribution with $p = \dfrac{5}{120}$ and $n = 3$ as an approximation.

20. Among the 200 employees of a company, 120 are union members while the others are not. If 6 of the employees are to be chosen by lot to serve on a committee which administrates the pension fund,

find the probability that three of them will be union members while the others are not, using

   (a) the formula for the hypergeometric distribution;

   (b) the binomial distribution with $p = \dfrac{120}{200}$ and $n = 6$ as an approximation.

21. If 4 per cent of the school children in a very large city have an I.Q. over 125, what is the probability that among 60 of them, randomly selected, exactly 2 will have an I.Q. over 125? Use the Poisson approximation.

22. If 1.8 per cent of the fuses delivered to an arsenal are defective, what is the probability that among 400 of them, randomly selected, exactly 6 are defective? Use the Poisson approximation.

23. If 6.4 per cent of all licensed drivers in a certain city get at least one speeding ticket a year, what is the probability that among 75 of them, randomly selected, exactly 71 will not get a speeding ticket in a given year? Use the Poisson approximation.

24. Given that the switchboard of a police station has on the average 4.2 incoming calls per minute, use the formula for the Poisson distribution with $\lambda = 4.2$ to find the probabilities that in a given minute there will be
   (a) no incoming calls;
   (b) exactly 2 incoming calls;
   (c) exactly 5 incoming calls.

25. If the number of golf balls a person loses while playing a certain course is a random variable having the Poisson distribution with $\lambda = 0.20$, find the probabilities that a person playing this course will
   (a) not lose any of his golf balls;
   (b) lose exactly one of his golf balls;
   (c) lose at most two of his golf balls.

26. If the number of complaints which a laundry receives per day is a random variable having the Poisson distribution with $\lambda = 3.5$, find the probabilities that on any given day the laundry will receive
   (a) no complaints;
   (b) exactly 2 complaints;
   (c) exactly 4 complaints.

27. If medical records show that there are on the average 0.3 cases of rabies reported per month, use the formula for the Poisson distribution with $\lambda = 0.3$ to find the probabilities that in a given month
   (a) no cases of rabies will be reported;
   (b) exactly one case of rabies will be reported;
   (c) exactly two cases of rabies will be reported;
   (d) exactly three cases of rabies will be reported.

8.6 The Mean of a Probability Distribution

On page 150 we claimed that a married couple in Florida can expect to have 1.46 children, and we pointed out that this figure is the sum of the products obtained by multiplying 0, 1, 2, 3, 4, ..., by the corresponding probabilities that a married couple in Florida will have that many children. A bit later we actually worked out such a problem (that is, we added the products obtained by multiplying each value of a random variable by the corresponding probability) and showed that in a certain county one can expect 2.38 hurricanes per year. If we now apply this sort of reasoning to the two examples which we gave in the beginning of Section 8.2, we find that the number of points we can expect in the roll of a die is

$$1 \cdot \frac{1}{6} + 2 \cdot \frac{1}{6} + 3 \cdot \frac{1}{6} + 4 \cdot \frac{1}{6} + 5 \cdot \frac{1}{6} + 6 \cdot \frac{1}{6} = 3.5$$

and that the number of heads we can expect in three flips of a coin is

$$0 \cdot \frac{1}{8} + 1 \cdot \frac{3}{8} + 2 \cdot \frac{3}{8} + 3 \cdot \frac{1}{8} = 1.5$$

Of course, we cannot actually roll 3.5 with a die or get 1.5 heads in three flips of a coin; like all mathematical expectations these figures have to be looked upon as *averages*, or as we referred to them in Chapter 3 as *means*.

Indeed, the expected values which we have calculated in these last two examples, and also in the hurricane example, are referred to as the *means* of the corresponding probability distributions. In general, if we are given a probability function, that is, the probabilities $f(x)$ which are associated with the various values of a random variable, we define the *mean of the distribution of this random variable*, or simply, the *mean of its distribution*, as

▲　　　　　　　　　　$\mu = \Sigma \, x \cdot f(x)$　　　　　　　　　　▲

where the summation extends over all values which the random variable can take on.

To give another example, let us refer back to the illustration on page 175, where we were concerned with the number of drivers who used their seat belts on short trips. If we now multiply the values of the random

variable by the corresponding probabilities, we find that among 5 drivers
on the average

$$\mu = 0(0.010) + 1(0.077) + 2(0.230) + 3(0.346) + 4(0.259) + 5(0.078)$$
$$= 3.001$$

use their seat belts on short trips. Of course, this is based on the assumption that the probability for any one of them to wear his seat belt on a
short trip is $p = 0.60$.

When a random variable assumes a very large set of values, the calculation of μ can be quite laborious. For instance, if we wanted to know how
many among 800 customers entering a store can be expected to make a
purchase, and the probability that any one of them will make a purchase
is 0.40, we would first have to calculate the 801 probabilities corresponding
to 0, 1, 2, ..., and 800 of them making a purchase. However, if we think
for a moment, we might argue that in the long run 40 per cent of the
customers make a purchase, 40 per cent of 800 is 320, and, hence, we can
expect 320 of the 800 customers to make a purchase. Similarly, if a balanced coin is flipped 1,000 times, we can argue that we should get heads
about 50 per cent of the time and, hence, that we can *expect* 1,000(0.50) =
500 heads. These two values are, indeed, correct, and we were able to
find them so easily because there exists the special formula

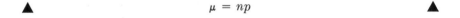

$$\mu = np$$

for the *mean of a binomial distribution*. In words, the mean of a binomial
distribution is simply the product of the number of trials and the probability of success on an individual trial.

Using this special formula, we can now verify the result which we obtained in the seat belt example and that which we obtained for three
flips of a coin. For the seat belt example we had $n = 5$ and $p = 0.60$, so
that we can now argue directly that $\mu = 5(0.60) = 3$. (The 0.001 difference between this result and the one obtained before is accounted for by
the fact that the probabilities on page 175 were rounded to three decimals.) For the number of heads which we can expect to get in three flips
of a balanced coin, we now get $3 \cdot \frac{1}{2} = 1.5$, since $n = 3$ and $p = \frac{1}{2}$, and
this agrees with the result we obtained before.

It is important to remember, of course, that the formula $\mu = np$
applies only to binomial distributions. Fortunately, though, there are

other special formulas for other special distributions. For the *hyper-geometric distribution* the formula is

▲
$$\mu = \frac{n \cdot a}{a + b}$$
▲

so that for the example on page 179 we find that the inspector can *expect* to catch

$$\mu = \frac{8 \cdot 5}{5 + 11} = 2.5$$

of the "bad" trucks. This should also not come as a surprise—he examines half of the trucks and catches half of the "bad" ones. Also, for the *Poisson distribution* $\mu = \lambda$, as we already indicated on page 181. Proofs of all these formulas may be found in any textbook on mathematical statistics.

8.7 The Variance of a Probability Distribution

In Chapter 4 we saw that there are many problems in which we must describe the *variability* of a distribution (that is, its spread or dispersion) as well as its mean or some other measure of location. *This is also the case when it comes to probability distributions.* As we pointed out in Chapter 4, the most widely used measure of variability is the *variance,* or its square root the *standard deviation,* which measures the spread of a set of data by averaging the squared deviations from the mean. When dealing with probability distributions, we measure variability in almost the same way, but instead of averaging the squared deviations from the mean, we find their *expected value.* If x is a value of some random variable and the mean of its probability distribution is equal to μ, then the deviation from the mean is $x - \mu$ and we define the *variance of the probability distribution* as the value we can expect for the squared deviation from the mean, namely, as

▲
$$\sigma^2 = \Sigma \ (x - \mu)^2 \cdot f(x)$$
▲

The summation extends over all values assumed by the random variable and, as in Chapter 4, the square root of the variance defines the *standard deviation* σ of the probability distribution.

To illustrate the calculation of the variance of a probability distribution, let us refer again to the seat belt example, for which the probabilities are given on page 175. Since the mean of this distribution was shown to be $\mu = 3$, we can arrange the calculations as in the following table:

Number of Drivers Using Seat Belts	Probability	Deviation From Mean	Square of Deviation From Mean	$(x - \mu)^2 f(x)$
0	0.010	−3	9	0.090
1	0.077	−2	4	0.308
2	0.230	−1	1	0.230
3	0.346	0	0	0.000
4	0.259	1	1	0.259
5	0.078	2	4	0.312

$$\sigma^2 = 1.199$$

The values in the last column are obtained by multiplying each squared deviation from the mean by the corresponding probability, and their sum gives the variance of the distribution. Also, $\sigma = \sqrt{1.199}$, which is approximately 1.1. Following these same steps, the reader will be asked to show in Exercise 6 on page 193 that the variance of the first distribution of Section 8.2 (the one for the number of points rolled with a die) is $\frac{35}{12}$, and in Exercise 7 on page 193 that the variance of the second distribution (the one for the number of heads in three flips of a coin) is 0.75. The calculations shown in the above table were quite easy, and this was due mainly to the fact that the deviations from the mean were whole numbers. When this is not the case, it may well be worthwhile to use the short-cut formula for σ^2 which will be given in Exercise 3 on page 193.

As in the case of the mean, the calculation of the standard deviation or the variance can generally be simplified when we deal with special kinds of probability distributions. For instance, for the *binomial distribution* we have the formula

▲ $\sigma^2 = np(1 - p)$ ▲

which we shall not prove, but which can easily be verified for our two examples. For the seat belt distribution we have $n = 5$ and $p = 0.60$, so that

$$\sigma^2 = 5(0.60)(1 - 0.60) = 1.2$$

and this differs from the result obtained before by the very small rounding error of $1.2 - 1.199 = 0.001$. For the number of heads in three flips of a coin, where $n = 3$ and $p = \dfrac{1}{2}$, the formula yields

$$\sigma^2 = 3 \cdot \frac{1}{2} \cdot \left(1 - \frac{1}{2}\right) = \frac{3}{4}$$

and this is the result which the reader should get by the long method in Exercise 7 on page 193. (There also exist special formulas for the variance of other special distributions, as is illustrated in Exercise 12 on page 194.)

Intuitively speaking, the variance or the standard deviation of a probability distribution measure its spread or its dispersion: when σ is small the probability is high that we will get a value close to the mean, and when σ is large we are more likely to get a value far away from the mean. This important idea is expressed rigorously in *Chebyshev's Theorem*, which we introduced on page 72 in connection with distributions of observed data. For probability distributions, the theorem reads as follows:

> *Given a probability distribution with the mean μ and the standard deviation σ, the probability of obtaining a value within k standard deviations of the mean is at least $1 - 1/k^2$.*

Thus, the probability of getting a value within two standard deviations of the mean (a value between $\mu - 2\sigma$ and $\mu + 2\sigma$) is at least 3/4, the probability of getting a value within five standard deviations of the mean is at least 24/25, and the probability of getting a value within ten standard deviations of the mean is at least 99/100. The quantity k in Chebyshev's Theorem can be any positive number, although the theorem becomes trivial when k is 1 or less. Changing around the argument, Chebyshev's Theorem can also be stated as follows:

> *Given a probability distribution with the mean μ and the standard deviation σ, the probability of obtaining a value which differs from the mean by at least k standard deviations is at most $1/k^2$.*

Thus, the probability of getting a value which differs from the mean by at least two standard deviations is at most $1/2^2 = 1/4$, the probability of getting a value which differs from the mean by at least four standard deviations is at most $1/4^2 = 1/16$, and the probability of getting a value which differs from the mean by at least ten standard deviations is at most $1/10^2 = 1/100$. Of course, when we say "differs" here, we mean "differs either way."

To consider some applications, suppose that the number of telephone calls which a doctor receives between 9 A.M. and 10 A.M. is a random variable whose distribution has the mean $\mu = 14$ and the standard deviation $\sigma = 3.74$. Thus, using Chebyshev's Theorem with $k = 2$, we can say that the probability is at least $3/4$ that he will receive between $14 - 2(3.74) = 6.52$ and $14 + 2(3.74) = 21.48$ calls, namely, that he will receive anywhere from 7 to 21 calls.

To give an example which pertains to the binomial distribution, suppose we actually obtained 148 heads and 252 tails in 400 flips of a coin and we are wondering whether this is sufficient evidence to raise the question whether the coin is really balanced. If the coin is balanced we are dealing with a binomial distribution having $n = 400$ and $p = 1/2$, and substitution into the special formulas on pages 188 and 190 yields

$$\mu = np = 400(1/2) = 200$$

and

$$\sigma = \sqrt{np(1 - p)} = \sqrt{400(1/2)(1/2)} = 10$$

For $k = 5$, which we chose more or less arbitrarily, Chebyshev's Theorem (in the second form) asserts that the probability of getting at most $200 - 5(10) = 150$ heads or at least $200 + 5(10) = 250$ heads is *at most* 0.04. Since the number of heads which we obtained is less than 150 and a probability of 0.04 is quite small, it may well be advisable to question the "honesty" (that is, balance) of the coin.

EXERCISES

1. The following table gives the probabilities that a man who enters a men's clothing store will buy 0, 1, 2, 3, or 4 shirts:

Number of Shirts x	0	1	2	3	4
Probability $f(x)$	0.29	0.49	0.16	0.05	0.01

   Calculate the mean and the variance of this distribution.

2. The following table gives the probabilities that a computer will malfunction 0, 1, 2, 3, 4, 5, or 6 times on any given day:

Number of Malfunctions x	0	1	2	3	4	5	6
Probability $f(x)$	0.17	0.29	0.27	0.16	0.07	0.03	0.01

   Calculate the mean and the standard deviation of this distribution.

3. By using the rules for summations given in Section 3.6, we can derive the following short-cut formula for the variance:

$$\sigma^2 = \Sigma\, x^2{\cdot}f(x) - \mu^2$$

The advantage of this formula is that we do not have to work with the deviations from the mean. Instead, we subtract μ^2 from the sum of the products obtained by multiplying the square of each value of the random variable by the corresponding probability.
   (a) Use this short-cut formula to find the variance of the probability distribution of Exercise 1.
   (b) Use this short-cut formula to find the standard deviation of the probability distribution of Exercise 2.

4. As can easily be verified by means of the formula for the binomial distribution (or by listing all 32 possibilities), the probabilities of getting 0, 1, 2, 3, 4, or 5 heads in five flips of a balanced coin are, respectively, 1/32, 5/32, 10/32, 10/32, 5/32, and 1/32. Find the mean and the variance of this binomial distribution using
   (a) the given probabilities and the basic formulas for μ and σ^2;
   (b) the special formulas for the mean and the variance of a binomial distribution.

5. As can easily be verified by means of the formula for the binomial distribution, the probabilities of getting 0, 1, 2, 3, or 4 *fives* in 4 rolls of a die are, respectively, 625/1296, 500/1296, 150/1296, 20/1296, and 1/1296. Find the mean of this binomial distribution using
   (a) the given probabilities and the basic formula for μ;
   (b) the special formula for the mean of a binomial distribution.
   Also find the standard deviation of this distribution using
   (c) the short-cut formula of Exercise 3;
   (d) the special formula for the standard deviation of a binomial distribution.

6. Using the probabilities in the table on page 172, verify that $\sigma^2 = \dfrac{35}{12}$ for the distribution of the number of points rolled with a balanced die.

7. Using the probabilities in the table on page 172, verify that $\sigma^2 = 0.75$ for the distribution of the number of heads obtained in three flips of a balanced coin.

8. Find the mean and the standard deviation of the distribution of each of the following (binomial) random variables:
   (a) the number of heads obtained in 600 flips of a balanced coin;
   (b) the number of *threes* obtained in 1,200 rolls of a balanced die;
   (c) the number of persons (among 400 invited) who will attend the

opening of a new branch of a bank, when the probability that
any one of them will attend is 0.85;

(d) the number of defectives in a sample of 2,400 parts made by a
machine, when the probability that any one of the parts is
defective is 0.04;

(e) the number of students (among 800 interviewed) who do not
like the food served at the university cafeteria, when the probability that any one of them will not like the food served there
is 0.30.

9. On page 179 we calculated the probabilities that an inspector will
catch 3, 4, or 5 of the company's 5 trucks which emit excessive
amounts of pollutants, if he examines 8 of the company's 16 trucks.

(a) Show that the corresponding probabilities for his catching 0, 1,
or 2 of the "bad" trucks are, respectively, 0.013, 0.128, and
0.359.

(b) Use the probabilities obtained on page 179 and those of part
(a) to find the mean and the variance of the distribution of
the number of "bad" trucks which the inspector will catch.

(c) Use the special formula on page 189 to check the value obtained
in (b) for the mean of the distribution.

10. Use the results of Exercise 14 on page 185 and the probability of
getting four perfect dishwashers which we calculated on page 178,
to find the mean and the standard deviation of the distribution of
the number of dishwashers that are not in perfect condition. Check
the result obtained for μ by means of the special formula given on
page 189.

11. Use the results of Exercise 16 on page 185 to determine the mean
and the variance of the distribution of the number of counterfeit
doubloons that will be selected to be sold at auction. Check the value
obtained for μ by means of the special formula on page 189.

12. Referring to the table on page 181, calculate the mean and the
variance of the distribution of the number of policy holders who file
a claim during any given year. Use these results to verify that the
mean and the variance of a Poisson distribution are given by the
formulas

$$\mu = \lambda \quad \text{and} \quad \sigma^2 = \lambda$$

13. If a random variable has the Poisson distribution with $\lambda = 4$, it can
be shown that it will take on the values 0, 1, 2, 3, 4, 5, 6, 7, 8, 9, 10,
11, or 12 with the respective probabilities 0.018, 0.073, 0.147, 0.195,
0.195, 0.156, 0.104, 0.060, 0.030, 0.013, 0.005, 0.002, and 0.001.

(a) Calculate the mean and the variance of this distribution.

(b) Use the results obtained in (a) to verify the two formulas given
in Exercise 12.

14. A student answers the 100 questions of a true-false test by flipping a balanced coin (*heads* is "true" and *tails* is "false").
    (a) Use the special formulas on pages 188 and 190 to calculate μ and σ for the number of correct answers he will get.
    (b) What does the first form of Chebyshev's Theorem with $k = 4$ tell us about the number of correct answers he should get?
    (c) According to the second form of Chebyshev's Theorem, with what probability can we assert that the student will get at most 20 or at least 80 correct answers?

15. The number of marriage licenses issued in a certain city during the month of June averages $\mu = 154$ with a standard deviation of $\sigma = 8.5$.
    (a) What does the first form of Chebyshev's Theorem with $k = 2$ tell us about the number of marriage licenses that should be issued in this city during the month of June?
    (b) What does the second form of Chebyshev's Theorem with $k = 3$ tell us about the number of marriage licenses that should be issued in this city during the month of June?

16. The daily number of customers to whom a restaurant serves breakfast on a weekday is a random variable with $\mu = 112$ and $\sigma = 12$. According to Chebyshev's Theorem, with what probability can we assert that on any given weekday the restaurant will serve lunch to
    (a) anywhere between 52 and 172 customers;
    (b) at least 160 customers?

17. Use Chebyshev's Theorem to verify that the probability is at least $\frac{35}{36} = 0.972$ that
    (a) in 900 flips of a balanced coin there will be between 360 and 540 heads, and hence the *proportion* of heads will be between 0.40 and 0.60;
    (b) in 10,000 flips of a balanced coin there will be between 4,700 and 5,300 heads, and hence the *proportion* of heads will be between 0.47 and 0.53;
    (c) in 1,000,000 flips of a balanced coin there will be between 497,000 and 503,000 heads, and hence the *proportion* of heads will be between 0.497 and 0.503.
    This exercise serves to illustrate the so-called *Law of Large Numbers*, according to which the proportion of successes approaches the probability of a success when the number of trials becomes larger and larger.

8.8 Continuous Distributions

Random variables are usually classified according to the sets of values they can assume. We speak of *discrete* random variables when they can assume only a finite set of values or as many values as there are whole numbers. The number of heads obtained in 12 flips of a coin is a discrete random variable as it cannot assume values other than 0, 1, 2, ..., and 12; if a coin is flipped until heads appears for the first time, the number of the flip on which this occurs is also a discrete random variable, but this time the random variable can assume as many values as there are whole numbers. This is true also for a random variable having the Poisson distribution which we introduced in Section 8.5.

In contrast to discrete random variables, we shall say that a random variable is *continuous* if it can assume values on a continuous scale. Such quantities as time, length, and temperature are measured on continuous scales and their measurements are referred to as values of continuous random variables. In order to associate probabilities with continuous random variables, we shall now introduce the concept of a *continuous distribution*, or *probability density*.

When we first discussed histograms in Chapter 2, we pointed out that the frequencies, percentages, and proportions (and we might now add probabilities) which are associated with the various classes are represented by the *areas* of the rectangles. For example, the areas of the rectangles of Figure 8.3 represent the probabilities of getting 0, 1, 2, ..., and 10 heads in 10 flips of a balanced coin or, better, they are equal to or proportional to these probabilities. If we now look carefully at Figure 8.4, which is an

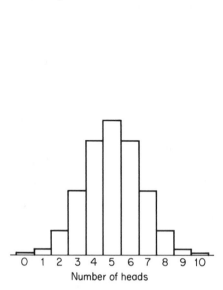

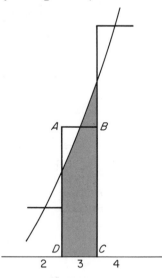

Figure 8.3. Histogram of the distribution of the number of heads in 10 flips of a coin.

Figure 8.4. Enlargement of part of Figure 8.3.

enlargement of a portion of Figure 8.3, it is apparent that the area of rectangle *ABCD* is nearly equal to the shaded area under the continuous curve which we have drawn to approximate the histogram. Since the area of rectangle *ABCD* is equal to (or proportional to) the probability of getting 3 heads in 10 tosses of a balanced coin, we can say that this probability is also given by the shaded area under the continuous curve. More generally, *if a histogram is approximated by means of a smooth curve, the frequency, percentage, or probability associated with any given class (or interval) is represented by the corresponding area under the curve.*

For example, if we approximated the distribution of 1967 family incomes in the United States with a smooth curve as in Figure 8.5, we can determine what proportion of the incomes fall into any given interval by looking at the corresponding area under the curve. By comparing the two shaded areas of Figure 8.5 with the total area under the curve (representing 100 per cent), we can judge that roughly 12 per cent of the families had incomes of \$15,000 or more and that roughly 41 per cent of the families had incomes under \$7,000.

Had we drawn Figure 8.5 so that the total area under the curve

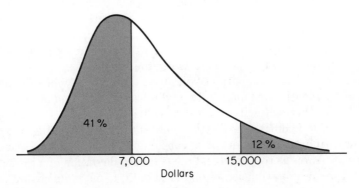

Figure 8.5. Distribution of 1967 family incomes in the United States.

actually equalled 1, the proportion of the families belonging to any income group would have been given directly by the corresponding area under the curve. Speaking in terms of probabilities rather than proportions, we refer to a function whose graph is like that of Figure 8.5 (with the total area under the curve equal to 1) as a *probability density*, and sometimes informally as a *continuous distribution*.* What characterizes a

---

* This terminology is borrowed from physics, where usage of the terms "weight" and "density" parallels usage of the terms "probability" and "probability density" in statistics.

probability density is the fact that *the area under the curve between two values a and b (see Figure 8.6) is equal to the probability that a random variable having this "continuous distribution" will assume a value between a and b.* Thus, when dealing with continuous random variables, probabilities are given by areas under appropriate curves.

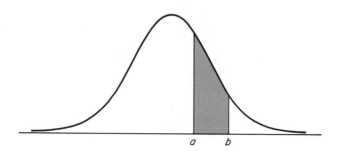

Figure 8.6. Probability density.

Since continuous distributions can always be looked upon as close approximations to histograms, we can define the mean and the standard deviation of continuous distributions in the following informal way: if a continuous distribution is approximated with a sequence of histograms having narrower and narrower classes, the means of the distributions represented by these histograms will approach a value which defines the mean of the continuous distribution. Similarly, the standard deviations of these distributions will approach a value which defines the standard deviation of the continuous distribution. Intuitively speaking, the mean and the standard deviation of a continuous distribution measure the identical features as the mean and the standard deviation of a probability distribution or a distribution of observed data, namely, its center and its spread. More rigorous definitions of these concepts cannot be given without the use of calculus.

8.9 The Normal Distribution

Among the many continuous distributions used in statistics, the *normal distribution* is by far the most important. Its study dates back to the eighteenth century and investigations into the nature of experimental errors. It was observed that discrepancies between repeated measurements of the same physical quantity displayed a surprising degree of regularity; their patterns (distribution), it was found, could be closely approximated by a certain kind of continuous distribution, referred to as the "normal curve of errors" and attributed to the laws of chance. The mathematical properties of this continuous distribution and its theoretical basis were first investigated by Pierre Laplace (1749–1827), Abraham de Moivre (1667–1745), and Carl Gauss (1777–1855).

The graph of a normal distribution is a bell-shaped curve that extends indefinitely in both directions. Although this may not be apparent from a small drawing like the one of Figure 8.7, the curve comes closer and closer

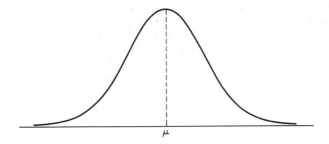

Figure 8.7. Normal distribution.

to the horizontal axis without ever reaching it, no matter how far we might go in either direction away from the mean. Fortunately, it is seldom necessary to extend the tails of a normal distribution very far because the area under that part of the curve lying more than 4 or 5 standard deviations away from the mean is for most practical purposes negligible.

An important feature of a normal distribution is that it is completely determined by its mean and its standard deviation. In other words, the mathematical equation for the normal distribution is such that we can determine the area under the curve between any two points on the horizontal scale if we are given its mean and its standard deviation. In practice, we obtain areas under the graph of a normal distribution, or simply a normal curve, by means of special tables, such as Table I at the end of the book. To be able to use this table, we shall first have to explain what is meant by the normal distribution in its *standard form* or, as it is also called, the *standard normal distribution*. Since the equation of the normal distribution depends on μ and σ, we get different curves and, hence, different areas for different values of μ and σ. For instance, Figure 8.8 shows the superimposed graphs of two normal distributions, one having $\mu = 10$ and $\sigma = 5$, and the other having $\mu = 20$ and $\sigma = 10$. As can be seen from this diagram, the area under the curve, say, between 10 and 12, is *not* the same for the two distributions.

As it would be physically impossible to construct separate tables of

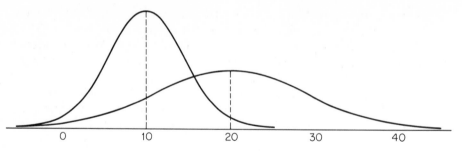

Figure 8.8. Two normal distributions with different means and standard deviations.

normal curve areas for each conceivable pair of values of μ and σ, we tabulate these areas only for the so-called *standard normal distribution* which has $\mu = 0$ and $\sigma = 1$. Then, we obtain areas under *any* normal distribution by performing the change of scale shown in Figure 8.9. All

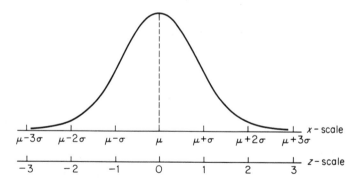

Figure 8.9. Change of scale.

we really do here is to convert the units of measurement into *standard units* (see page 74) by means of the formula

$$z = \frac{x - \mu}{\sigma}$$

To find areas under normal curves whose mean and standard deviation are not 0 and 1, we have only to convert the x's (the values to the left of which, to the right of which, or between which we want to determine areas under the curve) into z's and then use Table I. *The entries in this table are the areas under the standard normal distribution between the mean ($z = 0$) and $z = 0.00, 0.01, 0.02, \ldots, 3.08,$ and 3.09.* In other words, the entries in Table I are areas under the standard normal distribution like the one shaded in Figure 8.10. Note that Table I has no entries corresponding to negative values of z, but these are not needed by virtue of the *symmetry* of a normal curve about its mean. We can find the area under the stan-

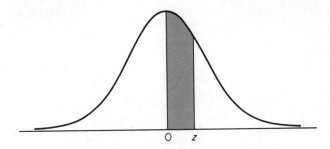

Figure 8.10. Tabulated normal curve areas.

dard normal distribution, say, between $z = -1.33$ and $z = 0$, by looking up instead the area between $z = 0$ and $z = 1.33$. As can be checked in Table I, the answer is 0.4082.

Questions concerning areas under normal distributions arise in various ways, and the ability to find any desired area quickly can be a big help. For instance, although the table gives only areas between the mean $z = 0$ and selected positive values of z, we often have to find areas to the right of a given value of z, to the right or left of $-z$, between two given values of z, and so forth. Finding any one of these areas is easy, provided we remember exactly what area corresponds to an entry in the table and we make use of the fact that the normal curve is symmetrical, so that the area to the left of $z = 0$ as well as the area to the right of $z = 0$ is equal to 0.5000. With this knowledge we find, for example, that the probability of getting a z less than 0.94 (the area under the curve to the left of $z = 0.94$) is $0.5000 + 0.3264 = 0.8264$, and that the probability of getting a z greater than -0.65 (the area under the curve to the right of $z = -0.65$) is $0.5000 + 0.2422 = 0.7422$ (see also Figure 8.11). Similarly, we find that the probability of getting a z greater than 1.76 is $0.5000 - 0.4608 = 0.0392$, and that the probability of getting a z less than -0.85 is $0.5000 - 0.3023 = 0.1977$ (see Figure 8.11). The probability of getting a z between 0.87 and 1.28 is $0.3997 - 0.3078 = 0.0919$, and the probability of getting a z between -0.34 and 0.62 is $0.1331 + 0.2324 = 0.3655$ (see Figure 8.11).

There are also problems in which we are given areas under the normal curve and we are asked to find the corresponding values of z. For instance,

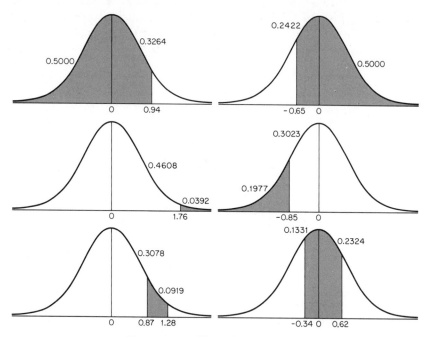

Figure 8.11. Normal distributions.

if we want to find a z which is such that the area to its right equals 0.1000, it is apparent from Figure 8.12 that this z will have to correspond to an entry of 0.4000 in Table I. Referring to this table, we find that the closest value is $z = 1.28$ (for which the entry in the table is 0.3997).

   Table I also enables us to verify the frequently-heard observation that for reasonably symmetrical bell-shaped distributions about 68 per cent of the data will fall within one standard deviation of the mean,

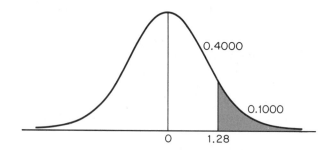

Figure 8.12. Normal distribution.

about 95 per cent of the data fall within two standard deviations of the mean, and over 99 per cent of the data fall within three standard deviations of the mean. These figures apply to normal distributions, and in Exercise 6 on page 208 the reader will be asked to verify that 0.6826 of the area under the standard normal curve falls between $z = -1$ and $z = 1$, that 0.9544 of the area falls between $z = -2$ and $z = 2$, and that 0.9974 of the area falls between $z = -3$ and $z = 3$. This last result also provides a basis for our earlier remark that (although the "tails" extend indefinitely in both directions) the area under a normal curve beyond 4 or 5 standard deviations from the mean is negligible.

To give an example in which we must first convert to standard units, let us suppose that a random variable has a normal distribution with $\mu = 24$ and $\sigma = 12$, and that we want to find the probability that it will assume a value between 17.4 and 58.8 (see Figure 8.13). Converting to standard units we obtain

$$z_1 = \frac{17.4 - 24}{12} = -0.55 \quad \text{and} \quad z_2 = \frac{58.8 - 24}{12} = 2.90$$

and since the areas corresponding to these z's are 0.2088 and 0.4981, respectively, the answer is $0.2088 + 0.4981 = 0.7069$.

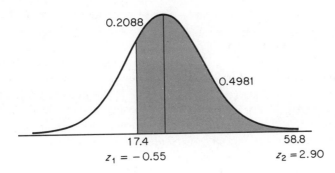

Figure 8.13. Normal distribution.

There are various ways in which we can test whether an observed distribution fits the over-all pattern of a normal curve. The one we shall discuss here is not the best; it is largely subjective, but it has the decided advantage that it is very easy to perform. To illustrate this technique, let us refer again to the distribution of the applicants for secretarial positions, which we used as an illustration in Chapters 2 through 4. If we divide each of the cumulative frequencies of the table on page 15 by 150 (the total number of scores), and then multiply by 100 to express the figures as percentages, we obtain the following *cumulative percentage distribution:*

Scores	Cumulative Percentage
Less than 9.5	0.0
Less than 19.5	0.7
Less than 29.5	4.7
Less than 39.5	10.7
Less than 49.5	31.3
Less than 59.5	59.3
Less than 69.5	80.7
Less than 79.5	92.0
Less than 89.5	98.7
Less than 99.5	100.0

Note that we substituted the class boundaries for the class limits in the column on the left.

Before we actually plot this cumulative percentage distribution on the special graph paper illustrated in Figure 8.14, let us briefly investigate its scales. As can be seen from Figure 8.14, the cumulative percentage scale is already marked off in the rather unusual pattern which makes the paper suitable for our purpose. The other scale consists of equal subdivisions that are not labeled; in our example they are used to indicate the class boundaries 19.5, 29.5, ..., and 89.5. The graph paper shown in Figure 8.14 is called *arithmetic probability paper,* and it can be obtained in the bookstores of most colleges and universities.

If we plot on this kind of graph paper the cumulative "less than" percentages which correspond to the class boundaries of a distribution and the points thus obtained follow the general pattern of a straight line, we can consider this as evidence that the distribution has roughly the same shape of a normal distribution. So far as our example is concerned, it would thus seem that we are justified to say that the distribution of the 150 scores has roughly the shape of a normal distribution. (Actually, in a graph like that of Figure 8.14 only large and obvious departures from a straight line are real evidence that the data do not follow the pattern of a normal distribution.)   Note that in Figure 8.14 we did not plot the percentages which correspond to 9.5 and 99.5; as we pointed out earlier, we never quite reach 0 or 100 per cent of the area under the curve no matter how far we go away from the mean in either direction.

It should be understood that the method we have described here is only a crude (and highly *subjective*) way of checking whether a distribution

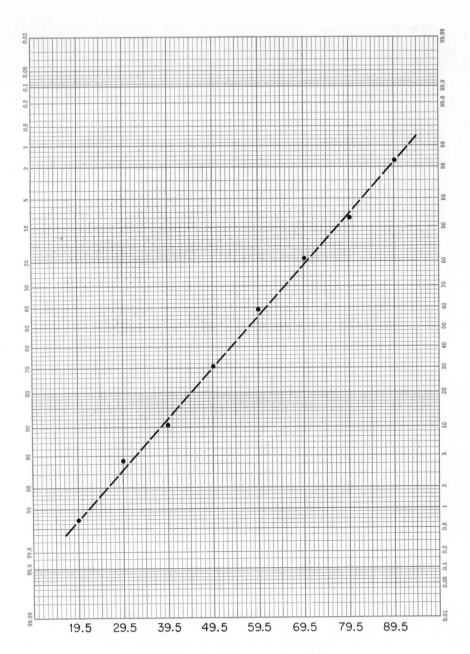

Figure 8.14. Arithmetic probability paper.

follows the pattern of a normal curve. A more *objective* method will be introduced in Chapter 12. Let us also point out that the special graph paper shown in Figure 8.14 can be used to get quick estimates of the mean and the standard deviation of a distribution having roughly the shape of a normal distribution. How this is done is explained in Exercise 11 below.

EXERCISES

1. Suppose that a continuous random variable can only take on values on the continuous interval from 2 to 5 and that the graph of its probability density (called the *uniform density*) is as shown in Figure 8.15.

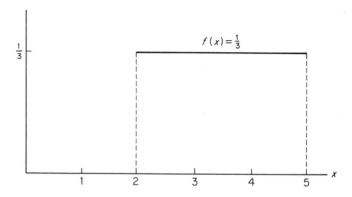

Figure 8.15. Uniform density.

What is the probability that the random variable will take on a value
   (a) between 3 and 4;
   (b) less than 2.8;
   (c) greater than 4.3;
   (d) between 2.6 and 4.4?
Would the answer have been different in part (b) if we had asked for the probability that the random variable will take on a value less than or equal to 2.8?

2. Suppose that a random variable can only take on values on the continuous interval from 0 to 4, and that the graph of its probability density (called a *triangular density*) is as shown in Figure 8.16. Verify that the total area under the curve (line) equals 1. Also, what is the probability that the random variable will take on a value
   (a) less than 2;
   (b) greater than 3;
   (c) between 1 and 2.5?

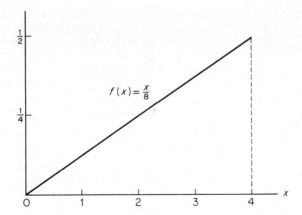

Figure 8.16. Triangular density.

[*Hint:* in parts (b) and (c) subtract the areas of appropriate triangles or use the formula for the area of a trapezoid; the area of a triangle is one-half times the product of its base and its height.]

3. The *exponential distribution* is another continuous distribution which has many applications in problems of statistical inference. If a random variable has an exponential distribution with the mean μ, the probability that it assumes a value between 0 and a given *positive* value x (see shaded area of Figure 8.17) is $1 - e^{-x/\mu}$. Here e is the constant

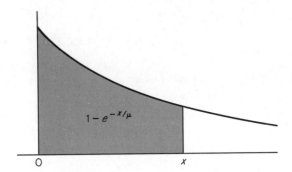

Figure 8.17. Exponential density.

$2.71828\ldots$ which also appears in the formula for the Poisson distribution; values of $e^{-x/\mu}$ can be obtained directly from Table XI.

(a) Find the probabilities that a random variable having an exponential distribution with $\mu = 10$ assumes a value between 0 and 4, a value greater than 6, and a value between 8 and 12.

(b) The lifetime of a certain kind of battery is a random variable which has an exponential distribution with a mean of 200 hours. What is the probability that such a battery will last at most 100 hours? Also find the probability that such a battery will last anywhere from 400 to 600 hours.

(c) Medical research has shown that the interval between successive reports of a rare contagious disease is a random variable having an exponential distribution with a mean of 110 days. What is the probability of not getting a report on an incidence of the disease for at least 275 days from the day the last case was reported?

4. Find the area under the standard normal distribution which lies
   (a) between $z = 0$ and $z = 0.85$;
   (b) between $z = -1.15$ and $z = 0$;
   (c) to the right of $z = 0.37$;
   (d) to the left of $z = -1.21$;
   (e) to the left of $z = 1.55$;
   (f) to the right of $z = -0.64$;
   (g) between $z = 0.85$ and $z = 1.35$;
   (h) between $z = -0.20$ and $z = -0.65$;
   (i) between $z = -1.24$ and $z = 1.24$;
   (j) between $z = -0.66$ and $z = 1.39$.

5. Find z if
   (a) the normal curve area between 0 and z is 0.4911;
   (b) the normal curve area to the right of z is 0.1093;
   (c) the normal curve area to the right of z is 0.6217;
   (d) the normal curve area to the left of z is 0.0217;
   (e) the normal curve area to the left of z is 0.8106;
   (f) the normal curve area between $-z$ and z is 0.9298.

6. Find the probability that a random variable having the standard normal distribution takes on a value between $-z$ and z if (a) $z = 1$; (b) $z = 2$; (c) $z = 3$.

7. In later chapters we shall use the symbols $z_{\alpha/2}$ and z_{α} to denote the values of z for which the area under the standard normal distribution *to their right* equals $\alpha/2$ and α, respectively. Find
   (a) $z_{0.10}$;   (c) $z_{0.025}$;   (e) $z_{0.01}$;
   (b) $z_{0.05}$;   (d) $z_{0.02}$;   (f) $z_{0.005}$.

8. A random variable has a normal distribution with the mean $\mu = 112.4$ and the standard deviation $\sigma = 3.6$. What is the probability that this random variable will take on a value
   (a) less than 117.8;
   (b) greater than 109.7;
   (c) between 116.9 and 120.5;
   (d) between 106.1 and 114.2?

9. A normal distribution has the mean $\mu = 77.0$. Find its standard deviation if 20 per cent of the area under the curve lies to the right of 90.0.

10. A random variable has a normal distribution with the standard deviation $\sigma = 10$. Find its mean if the probability that the random variable takes on a value less than 80.5 is 0.3264.

11. Arithmetic probability paper can be used to obtain crude estimates of the mean and the standard deviation of a distribution, provided that this distribution follows fairly closely the pattern of a normal curve. To find the mean, we have only to observe that since the normal distribution is symmetrical, 50 per cent of the area under the curve lies to the left of the mean. Hence, if we check the 50 per cent mark on the vertical scale and go horizontally to the line we fit to the points, then the corresponding point on the horizontal scale provides an estimate of the mean of the distribution. To obtain an estimate of the standard deviation, let us observe that the area under the standard normal distribution to the left of $z = -1$ is roughly 0.16 and that the area to the left of $z = +1$ is roughly 0.84. Hence, if we check 16 and 84 per cent on the vertical scale, we can judge by the straight line we have fitted to the points (representing the cumulative distribution of the data) what values on the horizontal scale correspond to $z = -1$ and $z = +1$; their difference divided by 2 provides an estimate of the standard deviation. Use this method to estimate the mean and the standard deviation of the distribution of the scores of the 150 applicants from Figure 8.14 and compare the results with the exact values previously obtained.

12. Convert the distribution of Exercise 3 on page 24, the one dealing with the weekly earnings of secretaries in the Phoenix area, into a cumulative "less than" percentage distribution, and use arithmetic probability paper to judge whether the shape of this distribution is close to that of a normal curve.

13. Convert the distribution of Exercise 5 on page 24, the one dealing with the weights of 50 bags of coffee, into a cumulative "less than" percentage distribution, and use arithmetic probability paper to judge whether this distribution has roughly the shape of a normal distribution.

14. Plot the cumulative "less than" percentage distribution of whichever data you grouped among those of Exercises 9, 10, or 11 on pages 25 and 26 on arithmetic probability paper, and judge whether the distribution of the data has roughly the shape of a normal distribution. If it does, use the method of Exercise 11 to estimate the mean and the standard deviation of the distribution and compare these results with the exact values previously obtained.

15. Plot the cumulative "less than" percentage distribution of whichever
     data you grouped among those of Exercises 12, 13, or 14 on pages
     26 and 27 on arithmetic probability paper, and judge whether the
     distribution of the data has roughly the shape of a normal distribution.
     If it does, use the method of Exercise 11 to estimate the mean and
     the standard deviation of the distribution and compare the results
     with the exact values previously obtained.

8.10 Applications of the Normal Distribution

Let us now consider some concrete examples dealing with random variables having at least approximately normal distributions. Suppose, for
instance, that the burning time of an experimental rocket is a random
variable which has roughly a normal distribution with $\mu = 5.12$ seconds
and $\sigma = 0.06$ seconds. *With this information we can determine all sorts of
probabilities, or percentages, about the burning times of these rockets.* For
instance, let us ask "What percentage of the burning times of these rockets
will exceed 5.25 seconds?" The answer to this question is given by the
area of the shaded region of Figure 8.18, namely, that to the right of

$$ z = \frac{5.25 - 5.12}{0.06} = \frac{0.13}{0.06} = 2.17 $$

Since the entry in Table I which corresponds to $z = 2.17$ is 0.4850, we find
that $0.5000 - 0.4850 = 0.0150$ *or 1.5 per cent of the burning times will
exceed 5.25 seconds.*

Continuing with the same example, let us also find the probability
that the burning time of one of these rockets will be anywhere from 5.00

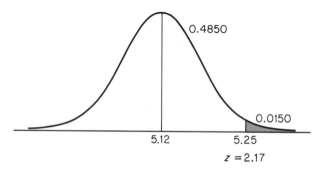

Figure 8.18. Normal distribution.

to 5.20 seconds. This probability is given by the shaded area of Figure 8.19, namely, that between

$$z = \frac{5.00 - 5.12}{0.06} = -2 \quad \text{and} \quad z = \frac{5.20 - 5.12}{0.06} = 1.33$$

and since the corresponding entries in Table I are 0.4772 for $z = 2$ and 0.4082 for $z = 1.33$, we find that the answer is $0.4772 + 0.4082 = 0.8854$.

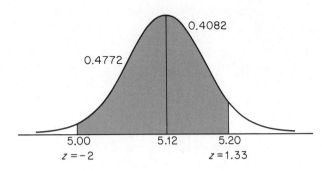

Figure 8.19. Normal distribution.

To consider a somewhat different example, suppose that the actual amount of instant coffee which a filling machine puts into a food processors's 6-ounce cans varies from can to can, and it can be looked upon as a random variable having a normal distribution with a standard deviation of 0.04 ounces. If only 2 per cent of the cans are to contain less than 6 ounces of coffee, *what will have to be the average amount of coffee which the filling machine puts into these cans?* This example differs from the preceding ones insofar as we are given $\sigma = 0.04$, a normal curve area (the one shaded in Figure 8.20), and we are asked to find μ. Since the value of z, for which the entry in Table I comes closest to $0.5000 - 0.0200 = 0.4800$, is 2.05, we have

$$-2.05 = \frac{6.00 - \mu}{0.04}$$

and, solving for μ, we get

$$6.00 - \mu = -2.05(0.04) = -0.082$$

and then

$$\mu = 6.00 + 0.08 = 6.08 \text{ ounces}$$

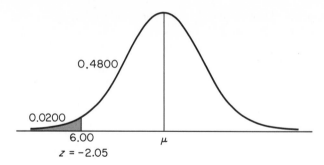

Figure 8.20. Normal distribution.

to the nearest hundredth of an ounce. This may not be very satisfactory so far as the processor of the coffee is concerned, and in Exercise 6 on page 217 the reader will be asked to show that if the variability of the machine is reduced so that $\sigma = 0.025$ ounces, this will lower the required average amount of coffee per can to $\mu = 6.05$ ounces, yet keep about 98 per cent of the cans above six ounces.

Although, strictly speaking, the normal distribution applies to continuous random variables, it is often used to *approximate* distributions of discrete random variables. This yields quite satisfactory results in many situations, but we have to be careful to make the *continuity correction* illustrated in the following example. Suppose that a baker knows that the daily demand for pecan pies is a random variable having the mean $\mu = 43.3$ and the standard deviation $\sigma = 4.6$. *What we would like to know is the probability that the demand for pecan pies will exceed 50 on any given day.* Assuming that it is reasonable to approximate this demand distribution with a normal curve, we shall thus have to look for the area of the shaded region of Figure 8.21, namely, that to the right of 50.5. The reason why we must look for the area to the right of 50.5 and *not* that to the right

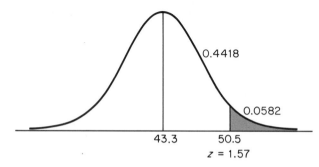

Figure 8.21. Normal distribution.

of 50 is that the number of pies the baker can sell is presumably a whole number. Hence, if we want to approximate the given demand distribution with a normal curve, we have to "spread" the values of this random variable over a continuous scale, and we can do this by representing each whole number k by the interval from $k - \dfrac{1}{2}$ to $k + \dfrac{1}{2}$. In particular, 20 is thus represented by the interval from 19.5 to 20.5, 25 is represented by the interval from 24.5 to 25.5, 50 is represented by the interval from 49.5 to 50.5, and the probability of getting a demand greater than 50 is given by the area under the curve to the right of 50.5. We thus get

$$z = \frac{50.5 - 43.3}{4.6} = 1.57$$

and it follows from Table I that the area of the shaded region of Figure 8.21 is $0.5000 - 0.4418 = 0.0582$. Thus, the probability is almost 0.06 that the demand for the baker's pecan pies will exceed 50.

8.11 The Normal Approximation of the Binomial Distribution

The normal distribution is sometimes introduced as a continuous distribution which provides a very close approximation to the binomial distribution when n, the number of trials, is very large and p, the probability of a success on an individual trial, is close to 0.50. Figure 8.22 contains the histograms of binomial distributions having $p = 0.50$ and $n = 2, 5, 10,$ and 25, and it can be seen that with increasing n these distributions approach the symmetrical bell-shaped pattern of the normal distribution. In fact, a normal curve with the mean $\mu = np$ and the standard deviation $\sigma = \sqrt{np(1 - p)}$ can often be used to approximate a binomial distribution even when n is "not too large" and p differs from 0.50, but is not too close to either 0 or 1. A good rule of thumb is to use this approximation only when np as well as $n(1 - p)$ are both greater than 5.

To illustrate this normal curve approximation of the binomial distribution, let us first consider the probability of getting 4 heads in 12 tosses of a balanced coin. Substituting $n = 12$, $x = 4$, $p = \dfrac{1}{2}$, and $\dbinom{12}{4} = 495$ into the formula on page 174, we get

$$495 \left(\frac{1}{2}\right)^4 \left(1 - \frac{1}{2}\right)^8 = \frac{495}{4,096}$$

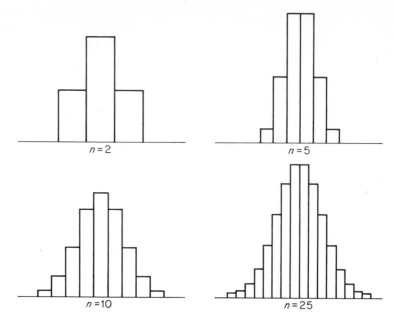

Figure 8.22. Binomial distribution with $p = 0.50$.

or approximately 0.12. To find the normal curve approximation to this probability, we shall again have to use the *continuity correction* mentioned on page 212, namely, represent 4 (heads) by the interval from 3.5 to 4.5 (see Figure 8.23). Since $\mu = 12(1/2) = 6$ and $\sigma = \sqrt{12(1/2)(1/2)} = 1.732$, it follows that the values between which we want to determine the area under the normal curve are, in standard units,

$$z = \frac{3.5 - 6}{1.732} = -1.44 \quad \text{and} \quad z = \frac{4.5 - 6}{1.732} = -0.87$$

The corresponding entries in Table I are 0.4251 and 0.3078, and the required probability is $0.4251 - 0.3078 = 0.1173$. Note that the difference between this value and the one obtained with the formula for the binomial distribution is negligible for most practical purposes.

The normal curve approximation of the binomial distribution is particularly useful in problems where we would otherwise have to use the formula for the binomial distribution repeatedly to obtain the values of many different terms. Suppose, for example, that we wanted to know the probability of getting at least 12 replies to mail-questionnaires sent to 100 persons, when the probability that any one of them will reply is 0.18.

In other words, we want to know the probability of getting at least 12 successes in 100 trials when the probability of a success on an individual trial is 0.18. If we tried to solve this problem by using the formula for the binomial distribution, we would have to find the sum of the probabilities corresponding to 12, 13, 14, ..., and 100 successes (or those corresponding to 0, 1, 2, ..., and 11). Evidently, this would involve a tremendous amount of work. On the other hand, using the normal curve approximation we have only to find the shaded area of Figure 8.24 namely, the area

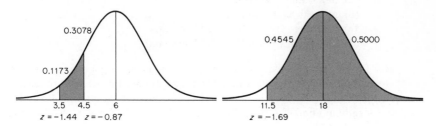

Figure 8.23. Distribution of number of heads for $n = 12$.

Figure 8.24. Distribution of number of replies.

to the right of 11.5. Note that we are again using the continuity correction according to which 12 is represented by the interval from 11.5 to 12.5, 13 is represented by the interval from 12.5 to 13.5, and so on.

Since $\mu = 100(0.18) = 18$ and $\sigma = \sqrt{100(0.18)(0.82)} = 3.84$, we find that in standard units 11.5 becomes

$$z = \frac{11.5 - 18}{3.84} = -1.69$$

and that the desired probability is $0.4545 + 0.5000 = 0.9545$. This means that we can expect to get at least 12 replies to 100 of these questionnaires *about 95 per cent of the time*, provided that 0.18 is the correct figure for the probability that any one person will reply. It is interesting to note that, rounded to two decimals, the *actual* value of this probability (obtained from an appropriate table) is 0.96.

The fact that the normal distribution is the only continuous distribution we have studied here (except for those of Exercises 1 through 3 on pages 206 and 207) may have given the erroneous impression that it is the only continuous distribution that really matters in the study of statistics. *Although it is true that the normal distribution plays an important role in many statistical problems, it is also true that its indiscriminate use can yield very misleading results.* As we shall see in Chapter 10 and later chapters, there are many other continuous distributions which play important roles in statistical inference.

EXERCISES

1. With reference to the burning-time example of page 210, what is the probability that this kind of rocket will burn
    (a) less than 5.10 seconds;
    (b) at least 5.30 seconds;
    (c) anywhere from 5.05 to 5.15 seconds;
    (d) anywhere from 5.13 to 5.14 seconds?

2. If the speed of cars recorded at a certain checkpoint is a random variable having (approximately) a normal distribution with the mean $\mu = 48.2$ miles per hour and the standard deviation $\sigma = 4.5$ miles per hour, find the probability that the speed of a car passing this checkpoint will
    (a) exceed the posted speed limit of 55 miles per hour;
    (b) be under 40 miles per hour;
    (c) be anywhere from 45 to 50 miles per hour.
    Since the speed of the cars is presumably measured on a continuous scale, no continuity correction is required.

3. The grapefruits grown in a large orchard have a mean weight of 19.3 ounces with a standard deviation of 2.2 ounces. Assuming that the distribution of the weight of these grapefruits has roughly the shape of a normal distribution, find
    (a) what percentage of the grapefruits weigh less than 18 ounces;
    (b) what percentage of the grapefruits weigh at least 20 ounces;
    (c) what percentage of the grapefruits weigh anywhere from 18.5 to 20.5 ounces;
    (d) the weight below which we will find the lightest 15 per cent of the grapefruits;
    (e) the weight above which we will find the heaviest 25 per cent of the grapefruits.
    Since the weights are presumably measured on a continuous scale, no continuity correction is required.

4. A manufacturer must buy coil springs which will stand at least a load of 20 pounds. If Supplier A can provide springs that can stand an average load of 24.5 pounds with a standard deviation of 2.1 pounds and Supplier B can provide springs that can stand an average load of 23.3 pounds with a standard deviation of 1.6 pounds, which of the two suppliers would provide the manufacturer with the smaller percentage of unsatisfactory springs? Assume that the distributions

of the loads can be approximated with a normal distribution; also, since the weights are presumably measured on a continuous scale, no continuity correction is required.

5. The average time required to perform Job A is 85 minutes with a standard deviation of 16 minutes, and the average time required to perform Job B is 110 minutes with a standard deviation of 11 minutes. Assuming normal distributions, what proportion of the time will Job A take longer than the average Job B, and what proportion of the time will Job B take less time than the average Job A? Since time is presumably measured on a continuous scale, no continuity correction is required.

6. With reference to the filling-machine example on page 211, verify the claim that if the variability of the machine is reduced so that $\sigma = 0.025$ ounces, this will lower the required average amount of coffee per can to $\mu = 6.05$ ounces, yet keep about 98 per cent of the cans above six ounces.

7. In a very large class in statistics, the final examination grades have a mean of 66.3 and a standard deviation of 14.5. Assuming that the distribution of the grades can be treated as if it had the shape of a normal distribution, find
    (a) what percentage of the grades should exceed 84;
    (b) what percentage of the grades should be 50 or less;
    (c) the lowest A, if the highest 10 per cent of the grades are to be regarded as A's;
    (d) the highest grade a student can get yet fail the test, if the lowest 25 per cent of the grades are to be regarded as failing grades.

8. The distribution of the I.Q.'s of the 4,000 employees of a large company has a mean of 104, a standard deviation of 14, and its shape is roughly that of a normal distribution. If it is known that a certain job requires a minimum I.Q. of 95 and bores those with an I.Q. over 110, how many of the company's employees should be suitable for the job on the basis of I.Q. alone. (Use the continuity correction.)

9. The head of the complaint department of a department store knows from experience that the number of complaints he receives per 8-hour day is a random variable with the mean $\mu = 33.4$ and the standard deviation $\sigma = 5.5$. Assuming that the distribution of the number of complaints has roughly the shape of a normal distribution, find the probabilities that he will receive
    (a) more than 40 complaints;
    (b) at least 40 complaints;
    (c) at most 30 complaints;
    (d) less than 30 complaints.

10. Find the probability of getting 6 heads in 14 tosses of a balanced coin
    (a) using the normal curve approximation;
    (b) referring to Table VI.

11. A multiple-choice test consists of 27 questions, each with four possible
    answers. If a student answers each question by drawing a card from
    an ordinary deck of 52 playing cards and checking the first, second,
    third, or fourth answer depending on whether he draws a club,
    diamond, heart, or spade, find with the normal curve approximation
    (a) the probability that he will get exactly 5 correct answers;
    (b) the probability that he will get exactly 10 correct answers;
    (c) the probability that he will get fewer than 5 correct answers;
    (d) the probability that he will get at least 8 correct answers.

12. Find the probability that a student will get at least 12 correct answers
    in a true-false test, if he decides how to mark each of the 20 questions
    by flipping a balanced coin. Use the normal curve approximation.

13. A television station claims that its Tuesday night movie regularly
    has 28 per cent of the total viewing audience in its area. If this claim
    is correct, what is the probability that among 400 viewers reached by
    phone on a Tuesday night, fewer than 100 are watching the station's
    movie?

14. If 40 per cent of the customers of a service station use their credit
    cards, what is the probability that among 500 customers more than
    320 *pay cash?*

15. If 25 per cent of all patients with high blood pressure have undesirable
    side effects from a certain kind of medicine, what is the probability
    that among 160 patients with high blood pressure who are treated
    with this medicine more than 45 have undesirable side effects?

BIBLIOGRAPHY

The following are two widely used tables of binomial probabilities.

Tables of the Binomial Probability Distribution, National Bureau of Standards
   Applied Mathematics Series No. 6. Washington, D.C.: U.S. Government
   Printing Office, 1950.

ROMIG, H. G., *50–100 Binomial Tables*. New York: John Wiley & Sons, Inc.,
   1953.

9

Sampling and Sampling
Distributions

9.1 Random Sampling

In Chapter 3 we made the following distinction between populations and samples: a population consists of all conceivably or hypothetically possible instances (or observations) of a given phenomenon, while a sample is simply a part (subset) of a population. Now let us distinguish further between populations that are *finite* and those that are *infinite*. A population is said to be *finite* if it consists of a finite, or fixed number, of elements (items, objects, measurements, or observations). In statistics, we are concerned mainly with populations whose elements are numbers, such as the finite population which consists of the I.Q.'s of all the students enrolled at U.C.L.A. or the finite population which consists of the relative humidity readings taken at all official weather stations throughout the United States at noon on January 1, 1972.

In contrast to finite populations, a population is said to be *infinite* if there is (at least hypothetically) no limit to the number of elements it can contain. For instance, the population consisting of the results of all hypothetically possible rolls of a pair of dice is an infinite population for there is no limit to the number of times they can be rolled, and so is the one consisting of all the hypothetically possible multiplications that could be performed by a digital computer. Also, we may want to look at the weights of 5 mice as a sample from the hypothetically infinite population consisting of the weights of all past, present, and future mice, and we may want to look at the number of mistakes three secretaries made in copying a technical report as a sample from the hypothetically infinite population consisting of the number of mistakes these secretaries (and, perhaps, other secretaries) might make while copying similar reports.

The purpose of most statistical investigations is to generalize from samples about both finite and infinite populations, and there are certain rules that must be observed to avoid getting results which are obviously poor, irrelevant, or invalid. Suppose, for instance, that we want to determine how much money the average person spends on his vacation. It seems unlikely that we would arrive at anything even remotely accurate, if we based our conclusions only on information supplied by Deluxe Class passengers on a two-week ocean cruise. Similarly, we can hardly expect to obtain reasonable generalizations about personal income in the United States from data reporting only the incomes of doctors, and we can hardly expect to infer much about wholesale prices of farm products in general on the basis of figures pertaining only to wholesale prices of fresh asparagus. These examples are, of course, extreme, but they serve to emphasize the point that sound generalizations (that is, sound inferences) do not come easily.

The whole problem of when and under what conditions samples permit reasonable generalizations is not easily answered. In most of the theory we shall develop in this book it will be assumed that we are dealing with a particular kind of sample called a *random sample*. (Other kinds of sampling procedures will be discussed briefly in Chapter 17). To illustrate the notion of a *random sample from a finite population*, let us consider first a finite population consisting of 5 elements which we shall label *a, b, c, d,* and *e*. (These might be the incomes of 5 professors, the weights of 5 students, the prices of 5 kinds of tires, and so on.) To begin with, let us see how many different samples of, say, size 3 can be taken from this finite population. To answer this question, we have only to refer to the rule on page 97, according to which there are $\binom{n}{r}$ ways in which r objects can be selected from a set of n objects. For our example, we have $n = 5$, $r = 3$, and there are therefore $\binom{5}{3} = 10$ different samples; one of these contains the elements *a, b, c,* another contains the elements *a, c, e,* a third contains the elements *b, d, e,* and so forth.

If we select one of the 10 possible samples in such a way that each has the same probability of being chosen, we say that we have a *simple random sample,* or more briefly, a *random sample.* One way in which this might be done is by writing each combination on a slip of paper, mixing the slips thoroughly, and then drawing one without looking. It would seem reasonable to say that with this method of selection each sample has a probability of 1/10 of being drawn. *More generally, a sample of size n taken from a population of size N is referred to as random if each of the $\binom{N}{n}$ possible* *samples has the same chance, namely, a probability of* $1/\binom{N}{n}$*, of being chosen.* This clearly conveys the idea that the selection of a random sample must, in some way, be left to chance; in fact, it is common practice to use various kinds of gambling devices as sampling aids.

In most realistic problems it is impossible, or at least impractical, to proceed as in our example; for instance, if a random sample of size 3 had to be drawn from a finite population of size 100, we would require

$$\binom{100}{3} = \frac{100 \cdot 99 \cdot 98}{3!} = 161,700$$

slips of paper to list all possible samples. Fortunately, such an elaborate and tedious procedure is not necessary, since we can achieve the identical results by choosing the sample values one at a time, making sure that in each successive drawing each of the remaining elements of the population has the same chance of being selected. To obtain a random sample of size 3 from a population of size 100, we might thus list each element on a slip of paper, mix the 100 slips of paper thoroughly, and then draw 3 in succession. Similarly, if we wanted to investigate the attitudes of the 220 members of a county's Medical Association toward a proposed insurance plan, we could select a random sample of, say, 10 by writing each name on a slip of paper, mixing them thoroughly, and drawing 10 without looking.

Even this relatively easy procedure can be simplified further in practice, where the best way of taking a random sample is with the use of a table of *random numbers*. Tables such as the one on pages 489 through 492 consist of pages on which the digits 0, 1, 2, 3, . . ., and 9 are recorded in much the same fashion as they might appear if they had been generated by means of a chance or gambling device giving each digit a probability of $\frac{1}{10}$ of appearing in any given place of the table. Actually, we could construct such a table ourselves by using a spinner like that shown in Figure 9.1, but in reality this is done by means of electronic computers. One of many commercially-published tables of random numbers is listed in the Bibliography at the end of this chapter.

Working with prepared tables of random numbers, it is quite simple to select a random sample from a finite population. Referring again to the problem of selecting 10 of the 220 members of a Medical Association, we might number the doctors 001, 002, 003, . . ., and 220, arbitrarily pick a page in a table of random numbers, three columns and a row from which to start, and then move down the page reading off 3-digit numbers, skipping those which do not apply. For instance, if we arbitrarily use the

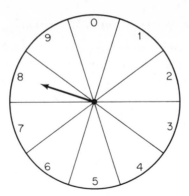

Figure 9.1. Spinner.

table on page 489 and the 6th, 7th, and 8th columns starting with the 5th row, our sample will consist of the Association's members whose numbers are

018   154   216   136   063   081   178   184   049   055

Note that in this selection we ignored numbers exceeding 220 and we also ignored (or would have ignored) each number after it has occurred for the first time. Incidentally, if we had wanted to be "even more random" in this example, we could have left the selection of the page, columns, and row to chance by using some gambling device or, perhaps, another page of random numbers.

Having defined random sampling from finite populations and having indicated how random samples can be obtained with the use of gambling devices or random numbers, we should add that in actual practice this is often easier said than done. For example, if we wanted to estimate (on the basis of a sample) the average outside diameter of a lot of 100,000 ball bearings packed in large cases, it would hardly be practical to number the bearings 000,001, 000,002, 000,003, ..., 100,000, choose 6-digit numbers from a table of random numbers, and locate the corresponding bearings as their numbers appeared. Similarly, it would be virtually impossible to sample trees in the Rocky Mountains by assigning a number to each tree or to take a sample of housewives in Chicago by assigning each one a number and then proceeding with the use of random numbers. In situations like these there generally is little choice but to proceed according to the dictionary definition of the word "random," namely, "haphazardly without definite aim or purpose," or perhaps improve the situation by using one of the special sample designs treated in Chapter 17. In the first of these alternatives we might well keep our fingers crossed that the samples we get will be such that statistical theory otherwise reserved for random samples can nevertheless be applied. This is true, particularly, in situations where we have little control over the selection of our data—for example, in medical studies where we often have to be satisfied with whatever cases happen to be available.

To this point we have been discussing only random sampling from finite populations. The concept of a random sample from an *infinite* population is somewhat more difficult to explain. To give a simple illustration, let us consider 10 flips of a balanced coin as a sample from the (hypothetically) infinite population which consists of all possible flips of the coin. We shall consider these 10 flips as a random sample provided the probability of getting heads is the same for each flip and if, furthermore, the 10 flips are independent. Also, we would be sampling from an infinite population if we sampled *with replacement* from a finite population, and our sample would be *random* if, for each draw, each element of the population has the same probability of being selected, and successive draws are independent. In general, *for a sample from an infinite population to be random, it is required that the selection of each sample value be controlled by the same probabilities, and that successive selections be independent.* All this is assumed when we refer to a set of measurements or observations (say, the weights of 12 cows, the lifetimes of 25 batteries, or the reaction times of 10 persons to a visual stimulus) as a random sample from an infinite population.

EXERCISES

1. How many different samples of size 2 can be selected
   (a) from a finite population of size 6;
   (b) from a finite population of size 12;
   (c) from a finite population of size 30?

2. How many different samples of size 3 can be selected
   (a) from a finite population of size 5;
   (b) from a finite population of size 15;
   (c) from a finite population of size 50?

3. What is the probability of each possible sample if a random sample of size 4 is to be taken
   (a) from a finite population of size 10;
   (b) from a finite population of size 25?

4. What is the probability of each possible sample if a random sample of size 5 is to be taken
   (a) from a finite population of size 12;
   (b) from a finite population of size 40?

5. Referring to the illustration on page 220, list the 10 possible samples of size 3 which can be drawn from the finite population consisting of the elements a, b, c, d, and e. If each of these samples is assigned the probability $\frac{1}{10}$, find

    (a) the probability that any specific element (say, the element b) will be contained in such a sample;

    (b) the probability that any specific pair of elements (say, the elements b and c) will be contained in such a sample.

6. List all possible choices of four of the following European cities: Amsterdam, Brussels, Copenhagen, Dresden, Edinburgh, and Florence. If a person randomly chooses four of these cities to visit on a trip, find

    (a) the probability that any one of these six cities (say, Brussels) will be included;

    (b) the probability that any two of these cities (say, Dresden and Florence) will be included;

    (c) the probability that any three of these cities (say, Amsterdam, Brussels, and Copenhagen) will be included.

7. On page 221 we suggested that a random sample can be chosen from a finite population by choosing the sample values one at a time and making sure in each case that all of the remaining elements have the same chance of being selected. Show that if a random sample is thus selected,

    (a) each possible sample of size 3 obtained from a finite population of size 5 has the probability $\frac{1}{10}$ (as on page 220);

    (b) each possible sample of size 3 obtained from a finite population of size 100 has the probability $\frac{1}{161,700}$, which agrees with the result obtained on page 221 that there are 161,700 possibilities.

8. Making use of the fact that among the $\binom{N}{n}$ possible random samples of size n which can be drawn from a finite population of size N there are $\binom{N-1}{n-1}$ which contain a specific element, show that the probability for any specific element to be contained in a random sample of size n from this finite population is $\frac{n}{N}$.

9. Suppose that a newspaper reporter wants to interview 12 of the 600 persons attending a political convention (who are listed in the program

in alphabetical order). Which ones (by number) would he interview, if he numbered the persons attending the convention from 001 through 600 and then chose a random sample by using the first three columns of the table on page 489 beginning with the first row.

10. The employees of a company have badges numbered serially from 1 through 643. Use the 16th, 17th, and 18th columns of the table on page 490 starting with the 11th row to select a random sample of 6 of the company's employees to serve on a committee.

11. Use random numbers to select a restaurant from among those listed in the yellow pages of your telephone directory (or that of a neighboring city.)

12. Suppose you have been in an accident and need three estimates of the damage done to your car. Use random numbers to select three automobile repair shops from among those listed in the yellow pages of a telephone directory.

13. Which requirement of random sampling is violated in part (b) of Exercise 3 on page 6. Explain your answer.

14. Which requirement of random sampling is violated in part (f) of Exercise 3 on page 6. Explain your answer.

9.2 Sampling Distributions

Let us now introduce the concept of the *sampling distribution of a statistic*, which is probably the most basic concept of statistical inference. As we shall see, this concept ties in closely with the idea of *chance variation*, or *chance fluctuations*, which we mentioned earlier to emphasize the need for measuring the variability of data. In this chapter we shall concentrate mainly on the sample mean and its sampling distribution, but in some of the exercises on page 236 and in later chapters we shall consider also the sampling distributions of other statistics.

There are two ways of approaching the study of sampling distributions. One, based on appropriate mathematical theory, leads to what is called a *theoretical sampling distribution;* the other, based on repeated samples from the same population, leads to what is called an *experimental sampling distribution.* The latter will prove to be very useful in our work as it will provide *experimental verification* of some important theorems, which cannot be derived formally at the level of this book.

Let us first give an example of a *theoretical sampling distribution of the mean* by considering the means of random samples of size $n = 2$ from the finite population of size $N = 5$, whose elements are the numbers 3, 5, 7,

9, and 11 (written on slips of paper). Using the formulas for μ and σ of Chapters 3 and 4, we find that the mean of this population is

$$\mu = \frac{3 + 5 + 7 + 9 + 11}{5} = 7$$

and that its standard deviation is

$$\sigma = \sqrt{\frac{(3 - 7)^2 + (5 - 7)^2 + (7 - 7)^2 + (9 - 7)^2 + (11 - 7)^2}{5}}$$
$$= \sqrt{8}$$

Then, if we take a random sample of size $n = 2$ from this population, the possible pairs of values we can get are 3 and 5, 3 and 7, 3 and 9, 3 and 11, 5 and 7, 5 and 9, 5 and 11, 7 and 9, 7 and 11, and 9 and 11. The means of these 10 samples are, respectively, 4, 5, 6, 7, 6, 7, 8, 8, 9, and 10, and (by virtue of the assumption of randomness which gives each of the samples the probability 1/10) we arrive at the following *theoretical sampling distribution of the mean:*

$\bar{x}$	Probability
4	$\dfrac{1}{10}$
5	$\dfrac{1}{10}$
6	$\dfrac{2}{10}$
7	$\dfrac{2}{10}$
8	$\dfrac{2}{10}$
9	$\dfrac{1}{10}$
10	$\dfrac{1}{10}$

A histogram of this sampling distribution is shown in Figure 9.2, and it gives us some idea about the chance fluctuations of means of random samples of size 2 from the given population. For instance, it tells us that the probability is $\dfrac{6}{10}$ that a sample mean will *not* differ from the popu-

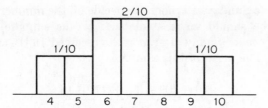

Figure 9.2. Theoretical sampling distribution of the mean.

lation mean $\mu = 7$ by more than 1, and that the probability is $\dfrac{8}{10}$ that a sample mean will *not* differ from the population mean $\mu = 7$ by more than 2; the first case corresponds to $\bar{x} = 6$, 7, or 8, and the second case corresponds to $\bar{x} = 5$, 6, 7, 8, or 9.

Further useful information about this sampling distribution of the mean can be obtained by calculating its mean $\mu_{\bar{x}}$ and its standard deviation $\sigma_{\bar{x}}$, where we are using the subscript $\bar{x}$ to distinguish these parameters from those of the original population. Using the formulas for the mean and the variance of a probability distribution on pages 187 and 189, we get

$$\mu_{\bar{x}} = 4 \cdot \frac{1}{10} + 5 \cdot \frac{1}{10} + 6 \cdot \frac{2}{10} + 7 \cdot \frac{2}{10} + 8 \cdot \frac{2}{10} + 9 \cdot \frac{1}{10} + 10 \cdot \frac{1}{10}$$
$$= 7$$

and

$$\sigma_{\bar{x}}^2 = (4 - 7)^2 \cdot \frac{1}{10} + (5 - 7)^2 \cdot \frac{1}{10} + (6 - 7)^2 \cdot \frac{2}{10} + (7 - 7)^2 \cdot \frac{2}{10}$$
$$+ (8 - 7)^2 \cdot \frac{2}{10} + (9 - 7)^2 \cdot \frac{1}{10} + (10 - 7)^2 \cdot \frac{1}{10}$$
$$= 3$$

so that $\sigma_{\bar{x}} = \sqrt{3}$. Note that the mean of this sampling distribution *equals* the mean of the population and that the standard deviation of the sampling distribution is *smaller* than that of the population, that is, $\sqrt{3}$ compared to $\sqrt{8}$. These are important relationships, and they will be discussed in detail later on.

To give an example of an *experimental sampling distribution of the mean*, let us suppose that an operator of tow trucks for an automobile association wants to determine the average number of service calls he can expect on a Sunday afternoon (to decide on the number of tow trucks and drivers he should make available). Let us suppose, furthermore, that on five recent Sunday afternoons he received 16, 15, 14, 12, and 18 service calls. The mean of this sample is

$$\bar{x} = \frac{16 + 15 + 14 + 12 + 18}{5} = 15$$

and in the absence of any other information he could use this figure as an estimate of μ, the true average number of service calls he will get on a Sunday afternoon. However, at the same time he must recognize the fact that if he had used data for five other Sunday afternoons, the mean would probably have been some number other than 15. Indeed, if he took several samples, each consisting of the number of service calls he receives on five different Sunday afternoons, he might well get such divergent values as 14.6, 18.8, 16.2, 12.6, 17.4, ..., for the corresponding means.

To get some idea how the means of such samples might vary *purely due to chance*, let us suppose that the number of service calls the tow truck operator gets on a Sunday afternoon is a random variable having a Poisson distribution with the mean $\lambda = 16$ (see page 181), and let us perform an experiment which consists of *simulating* the drawing of 50 samples, each

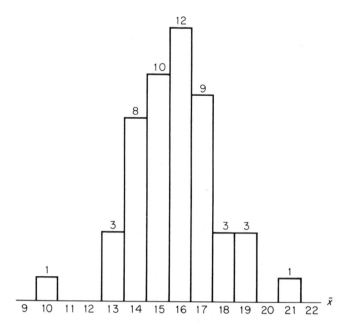

Figure 9.3. Experimental sampling distribution of the mean.

of size $n = 5$. How this can be done will be explained in the Technical Note on page 238; for the moment, though, let us merely use the results which are shown in the following table:

Sample	Number of Service Calls	Sample	Number of Service Calls
1	16, 15, 14, 12, 18	26	10, 18, 19, 13, 20
2	20, 18, 16, 19, 14	27	14, 19, 16, 13, 21
3	21, 18, 17, 26, 25	28	18, 14, 23, 23, 14
4	16, 10, 9, 19, 15	29	17, 16, 11, 17, 11
5	17, 16, 20, 17, 7	30	16, 13, 10, 14, 20
6	20, 8, 17, 16, 13	31	13, 17, 22, 19, 18
7	22, 21, 16, 15, 13	32	15, 13, 16, 14, 21
8	11, 23, 12, 20, 14	33	16, 15, 8, 12, 23
9	17, 22, 21, 16, 20	34	16, 15, 11, 20, 13
10	18, 13, 15, 11, 12	35	17, 17, 16, 21, 14
11	15, 11, 14, 14, 18	36	18, 20, 14, 26, 18
12	22, 15, 13, 19, 11	37	13, 16, 17, 11, 6
13	20, 17, 11, 19, 15	38	17, 19, 15, 19, 16
14	15, 16, 16, 15, 17	39	11, 13, 18, 23, 18
15	15, 16, 17, 17, 16	40	12, 25, 21, 18, 8
16	13, 15, 15, 13, 18	41	21, 12, 14, 17, 16
17	12, 11, 19, 17, 16	42	19, 10, 15, 16, 18
18	15, 26, 19, 20, 15	43	20, 10, 15, 15, 19
19	15, 11, 21, 8, 17	44	16, 10, 26, 14, 20
20	12, 21, 10, 15, 16	45	18, 12, 13, 19, 9
21	10, 11, 9, 11, 11	46	15, 14, 21, 17, 11
22	12, 12, 24, 11, 5	47	12, 13, 13, 14, 12
23	16, 18, 14, 9, 11	48	18, 8, 21, 14, 15
24	17, 9, 18, 16, 9	49	20, 17, 16, 18, 19
25	11, 17, 19, 20, 17	50	19, 17, 16, 13, 15

Each of these samples contains 5 values—the simulated number of service calls which the tow truck operator received on 5 Sunday afternoons. Calculating the means of these 50 samples, we get

15.0	17.4	21.4	13.8	15.4	14.8	17.4	16.0	19.2	13.8
14.4	16.0	16.4	15.8	16.2	14.8	15.0	19.0	14.4	14.8
10.4	12.8	13.6	13.8	16.8	16.0	16.6	18.4	14.4	14.6
17.8	15.8	14.8	15.0	17.0	19.2	12.6	17.2	16.6	16.8
16.0	15.6	15.8	17.2	14.2	15.6	12.8	15.2	18.0	16.0

and their distribution is given by the histogram of Figure 9.3.

As in the case of the theoretical sampling distribution of the mean which we constructed on page 226, a look at the experimental sampling distribution of $\bar{x}$ tells us a great deal about the way in which sample means tend to scatter among themselves due to chance. For instance, we find that the smallest mean is 10.4 and the largest is 21.4; furthermore, 31 out of 50 (or 62 per cent) of the means fell between 14.5 and 17.5, and 48 out of 50 (or 96 per cent) of the means fell between 12.5 and 19.5. These results will be more meaningful if we recall that they were obtained by simulating random sampling from a population which has the mean $\mu = 16$. Thus, we can restate our description of the experimental sampling distribution by saying that 62 per cent of the sample means were "off" (differed from the mean of the population) by less than 1.5, and that 96 per cent of the sample means were "off" by less than 3.5.

The two examples we have given in this section are really only *teaching aids* designed to introduce the concept of a sampling distribution. In actual practice, we ordinarily have only one mean, not 50, and we seldom can afford to go to the extreme of enumerating all possible samples from a finite population and, thus, construct a theoretical sampling distribution. In most practical situations we have no choice but to use special theorems, and so far as the sampling distribution of the mean is concerned, there are two theorems which generally provide all the information we need.

The first of these theorems expresses the fact which we discovered in connection with the example on page 227, namely, that $\mu_{\bar{x}} = \mu$ (the mean of the sampling distribution of $\bar{x}$ equals the mean of the population); it also provides formulas for the standard deviation $\sigma_{\bar{x}}$ of the sampling distribution of the mean. Formally,

> *For random samples of size n from a population having the mean μ and the standard deviation σ, the (theoretical) sampling distribution of $\bar{x}$ has the mean $\mu_{\bar{x}} = \mu$, and its standard deviation is given by*
>
> $$\sigma_{\bar{x}} = \frac{\sigma}{\sqrt{n}} \quad \text{or} \quad \sigma_{\bar{x}} = \frac{\sigma}{\sqrt{n}} \cdot \sqrt{\frac{N - n}{N - 1}}$$
>
> *depending on whether the population is infinite or finite of size N.*

It is customary to refer to $\sigma_{\bar{x}}$, the standard deviation of the sampling distribution of the mean, as the *standard error of the mean*. Its role in statistics is fundamental, as it measures the extent to which means fluctuate, or vary, due to chance. Regarding the two formulas for $\sigma_{\bar{x}}$, note that the first consists of the quotient $\sigma/\sqrt{n}$, while in the second this quotient is multiplied by the *"finite population correction factor"* $\sqrt{\dfrac{N-n}{N-1}}$. The quotient $\sigma/\sqrt{n}$ exhibits the following important information: (1) *If σ is large and there is considerable variation in the population from which the sample is obtained, we can expect proportionately large fluctuations in the distribution of the means;* (2) *The larger the sample size n, the smaller is the variation among the means and the closer we can expect a sample mean to be to the mean of the population.* Both of these arguments should seem plausible on intuitive grounds; after all, if there is considerable variability in the values from which the sample is obtained, this is apt to reflect also in the sample. So far as the second point is concerned, it stands to reason that if we have more information, we should also get better, that is, more reliable, estimates.

The factor $\sqrt{\dfrac{N-n}{N-1}}$ in the second formula for $\sigma_{\bar{x}}$ is generally omitted unless the sample constitutes a substantial portion (5 per cent or more) of the population. For instance, when $n = 100$ and $N = 10,000$ (and the sample constitutes but 1 per cent of the population),

$$\sqrt{\frac{N-n}{N-1}} = \sqrt{\frac{10,000 - 100}{10,000 - 1}} = 0.995$$

and this is so close to 1 that the correction factor can be omitted for most practical purposes.

To verify the second formula for $\sigma_{\bar{x}}$, let us return to the example where we constructed the theoretical sampling distribution of the mean for random samples of size $n = 2$ from the finite population which consists of the numbers 3, 5, 7, 9, and 11. As we showed on page 226, $\sigma = \sqrt{8}$ for this population, and if we now substitute this value together with $n = 2$ and $N = 5$ into the second formula for $\sigma_{\bar{x}}$, we get

$$\sigma_{\bar{x}} = \frac{\sqrt{8}}{\sqrt{2}} \cdot \sqrt{\frac{5-2}{5-1}} = \sqrt{3}$$

This agrees with the result we obtained on page 227, where we showed directly that the standard deviation of the given theoretical sampling distribution of the mean is $\sqrt{3}$. In the second example we took 50 random samples of size $n = 5$ from a population having the Poisson distribution with the mean $\mu = 16$ and, hence, the standard deviation $\sigma = \sqrt{16} = 4$ (see Exercises 12 on page 194). Thus, substitution into the first of the two formulas for $\sigma_{\bar{x}}$ yields

$$\sigma_{\bar{x}} = \frac{4}{\sqrt{5}} = 1.79$$

and it will be of interest to see how close the standard deviation of the 50 sample means of our experiment comes to this theoretical value. Actually calculating the mean and the standard deviation of the distribution of the 50 means shown in Figure 9.3, we get respectively, 15.76 and 1.89, which are *quite close* to $\mu = 16$ and $\sigma_{\bar{x}} = 1.79$. In other words, *the results of our experiment tend to support the theorem (on page 230) about the mean and the standard deviation of the sampling distribution of the mean.*

9.3 The Central Limit Theorem

To show how the theorem on page 230 is actually applied in practice, we require another theorem, called the *Central Limit Theorem*, which provides us with information about the over-all shape of a sampling distribution of the mean. However, even without it we can get some idea by applying *Chebyshev's Theorem* (see page 191), according to which we can now assert that *the probability of getting a sample mean which differs from the population mean by at least k standard deviations (namely, by at least $k \cdot \sigma_{\bar{x}}$) is at most $1/k^2$.* For instance, if we take a random sample of size $n = 64$ from an infinite population with $\sigma = 20$, then the probability of getting a sample mean which differs from the population mean by at least $2 \cdot \sigma_{\bar{x}} = 2 \cdot \dfrac{20}{\sqrt{64}} = 5$ is at most $1/2^2 = 0.25$. Similarly, if we take a random sample of size $n = 100$ from an infinite population with $\sigma = 4$, then the probability of getting a sample mean which differs from the mean of the population by at least $5 \cdot \dfrac{4}{\sqrt{100}} = 2$ is at most $1/5^2 = 0.04$. Note that we can thus make probability statements concerning the difference between the mean of a random sample and the mean of the population without having to go through the tedious process of constructing the complete theoretical sampling distribution (as we did in the example on page 226).

The *Central Limit Theorem* is of fundamental importance in statistics, as it justifies the use of normal curve methods in a great variety of problems. In words, the theorem can be stated as follows:

*If n is large, the (theoretical) sampling distribution of the mean
can be approximated closely with a normal distribution.*

In other words, if n is large, the sampling distribution of the statistic

$$z = \frac{\bar{x} - \mu}{\sigma_{\bar{x}}}$$

can be approximated closely with the *standard* normal distribution. It is
difficult to state precisely how large n must be so that this theorem applies; unless the distribution of the population has a very unusual shape,
however, the approximation will be good even if n is relatively small—
certainly, if n is 30 or more. Note that the distribution of Figure 9.3 is
fairly symmetrical and bell-shaped even though the sample size on which
each mean was based is only $n = 5$.

To illustrate the use of the Central Limit Theorem, let us find
the probability that the mean of a random sample of size 64, taken
from a population which has the standard deviation $\sigma = 20$, will differ
from the mean of the population by 5 or more. Since $\sigma_{\bar{x}} = \dfrac{20}{\sqrt{64}} = 2.5$,
we find that, in standard units, $\bar{x} = \mu + 5$ becomes

$$z = \frac{\bar{x} - \mu}{\sigma_{\bar{x}}} = \frac{(\mu + 5) - \mu}{2.5} = 2$$

and $\bar{x} = \mu - 5$ becomes

$$z = \frac{\bar{x} - \mu}{\sigma_{\bar{x}}} = \frac{(\mu - 5) - \mu}{2.5} = -2$$

and since the entry in Table I which corresponds to $z = 2$ is 0.4772, we
arrive at the result that the desired probability (also given by the total
area of the shaded regions of Figure 9.4) is $2(0.5000 - 0.4772) = 0.0456$.
It is of interest to note that when we applied Chebyshev's Theorem to
this example on page 232, we were able to show only that this probability
is "at most 0.25."

A sample of size $n = 5$ is ordinarily too small to apply the Central
Limit Theorem unless the distribution of the population, itself, follows
closely the pattern of a normal distribution. Nevertheless, let us check
briefly how the experimental sampling distribution of Figure 9.3, the one
based on the 50 sample means, agrees with what we might expect according to the theorem. On page 230 we pointed out that 31 of the 50 means
(or 62 per cent) fell between 14.5 and 17.5, namely, within 1.5 of the

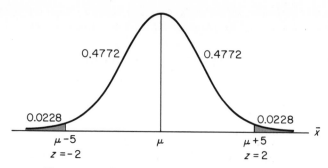

Figure 9.4. Sampling distribution of the mean.

population mean $\mu = 16$. Since the population standard deviation equaled 4, we have $\sigma_{\bar{x}} = \dfrac{4}{\sqrt{5}} = 1.79$, and the probability of getting a sample mean between 14.5 and 17.5 is given by the normal curve area between

$$z = \frac{14.5 - 16}{1.79} = -0.84 \quad \text{and} \quad z = \frac{17.5 - 16}{1.79} = 0.84$$

It follows from Table I that the desired probability (given also by the area of the shaded region of Figure 9.5) is $0.2995 + 0.2995 = 0.5990$, or in other words that we can expect about 60 per cent of the sample means to fall on the interval from 14.5 to 17.5. *This agrees quite well with the actual 62 per cent which we obtained in our experiment.* In Exercise 15, below, the reader will be asked to verify that the Central Limit Theorem would correspondingly lead us to expect that 95 per cent of the means should fall on the interval from 12.5 and 19.5, which is very close to the 96 per cent we actually obtained.

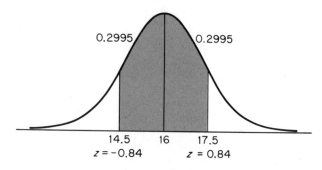

Figure 9.5. Sampling distribution of the mean.

The main purpose of this section and the last one has been to introduce the concept of a sampling distribution, and we chose for this purpose the sampling distribution of the mean. We could continue our study of sampling distributions by calculating the medians, the ranges, the standard deviations, or other statistics of the 50 samples on page 229 and group them into corresponding (experimental) sampling distributions (see Exercises 5 and 7 below). Then we could check how closely the standard deviations of these sampling distributions agree with the corresponding *standard errors* (obtained according to appropriate statistical theory), or otherwise compare the shape or other features of these sampling distributions with theoretical expectations.

EXERCISES

1. Random samples of size 2 are taken from the finite population which consists of the numbers 4, 5, 6, 7, 8, and 9.
    (a) Show that the mean and the standard deviation of this population are $\mu = 6.5$ and $\sigma = \sqrt{\dfrac{35}{12}}$.
    (b) List the 15 possible random samples of size 2 that can be taken from this finite population and calculate their respective means.
    (c) Using the results of part (b) and assigning each possible sample a probability of $\dfrac{1}{15}$, construct a sampling distribution of the mean for random samples of size 2 from the given finite population; that is, construct a table showing the probabilities associated with all the possible values of the mean.
    (d) Calculate the mean and the standard deviation of the probability distribution obtained in part (c) and verify the results with the use of the theorem on page 230.

2. Repeat parts (b), (c), and (d) of Exercise 1 for random samples of size $n = 3$ from the given finite population.

3. The finite population of Exercise 1 can be converted into an *infinite* population if we sample *with replacement*, that is, if we obtain the random sample of size 2 by first drawing one value and replacing it before drawing the second value.
    (a) List the 36 possible ordered samples of size 2 that can be drawn with replacement from the given population and calculate their respective means.
    (b) Using the results of part (a) and assigning each sample a probability of $\dfrac{1}{36}$, construct the sampling distribution of the mean for random samples of size 2 obtained with replacement from the given population.

(c) Calculate the mean and the standard deviation of the probability distribution obtained in part (b) and verify the results with the use of the theorem on page 230.

4. Convert the 50 samples on page 229 into 25 samples of size $n = 10$ by combining Samples 1 and 26, Samples 2 and 27, ..., and Samples 25 and 50. Calculate the mean of each of these samples of size 10 and determine their mean and their standard deviation. Compare this mean and this standard deviation with the corresponding values expected in accordance with the theorem on page 230.

5. Find the *medians* of the 50 samples on page 229, group them as we grouped the corresponding means, and construct a histogram like that of Figure 9.3. Comparing the standard deviation of this *experimental sampling distribution of the median* with that of the corresponding *experimental sampling distribution of the mean* (which was 1.89), what can we say about the relative "reliability" of the median and the mean in estimating the mean of the given population?

6. For random samples of size n from a population having the shape of a normal distribution, the *standard error of the median* (namely, the standard deviation of the sampling distribution of the median) is given by $\sqrt{\frac{\pi}{2}} \cdot \frac{\sigma}{\sqrt{n}}$ or approximately $1.25 \cdot \frac{\sigma}{\sqrt{n}}$. How close is the value obtained for the standard deviation of the medians of Exercise 5 to the value expected according to this formula with $\sigma = 4$ and $n = 5$?

7. Calculating the standard deviations of the 50 samples on page 229, we obtain the following *experimental sampling distribution of s:*

Sample Standard Deviation	Frequency
0.5–1.5	4
1.5–2.5	6
2.5–3.5	12
3.5–4.5	16
4.5–5.5	8
5.5–6.5	2
6.5–7.5	2

(Since none of the values are exactly 1.5, 2.5, ..., or 6.5, these overlapping class boundaries do not matter.)

(a) Calculate the mean and the standard deviation of this sampling distribution.

(b) For *large* samples, the formula $\frac{\sigma}{\sqrt{2n}}$ is sometimes used for the *standard error of a sample standard deviation.* Substituting $\sigma = 4$ and $n = 5$ into this formula, calculate the value of this standard error for the given example and compare it with the value of the standard deviation obtained for the experimental sampling distribution in part (a). (Note that since $n = 5$ is not exactly a "large" sample, we should not be surprised if the results are not too close.)

8. When we sample from an infinite population, what happens to the standard error of the mean (and, hence, to the expected size of the errors to which we are exposed when we use $\bar{x}$ to estimate μ) if
    (a) the sample size is increased from 80 to 320;
    (b) the sample size is increased from 100 to 225;
    (c) the sample size is increased from 75 to 675;
    (d) the sample size is decreased from 750 to 30.

9. What is the value of the *finite population correction factor* in the formula for $\sigma_{\bar{x}}$ when
    (a) $n = 5$ and $N = 200$;
    (b) $n = 10$ and $N = 400$;
    (c) $n = 100$ and $N = 5{,}000$?

10. Show that if the mean of a random sample of size n is used to estimate the mean of an infinite population with the standard deviation σ, there is a *fifty-fifty chance* that the error (that is, the difference between $\bar{x}$ and μ) is less than $0.67 \cdot \dfrac{\sigma}{\sqrt{n}}$. It has been the custom to refer to this quantity, or more precisely the quantity $0.6745 \cdot \dfrac{\sigma}{\sqrt{n}}$ as the *probable error of the mean;* nowadays, this term is used mainly in military applications.

11. If random samples of size $n = 60$ are taken from an infinite population with $\sigma = 24.5$, calculate the size of the *probable error* (see Exercise 10) and explain its significance.

12. On page 232 we used Chebyshev's Theorem to show that if a random sample of size 100 is taken from an infinite population with $\sigma = 4$, the probability that the mean of the sample will differ from the mean of the population by 2 or more is at most 0.04. What value would we get for this probability if we used the Central Limit Theorem? (*Hint:* if Table I were extended beyond $z = 3.09$, the entry corresponding to $z = 5.00$ would be 0.4999997.)

13. The mean of a random sample of size $n = 49$ is used to estimate the mean of a very large population (consisting of the lifetimes of certain batteries) which has a standard deviation of $\sigma = 35$ hours. What can

we say about the probability that the sample mean will be "off" either way by less than 12.5 hours
  (a) using Chebyshev's Theorem;
  (b) using the Central Limit Theorem.

14. The mean of a random sample of size $n = 225$ is used to estimate the mean of a very large population (consisting of the weights of certain animals) which has a standard deviation of 10 ounces. Using the Central Limit Theorem, what can we say about the probability that the value of $\bar{x}$ will be "off" either way by
  (a) less than 0.20 ounces;
  (b) less than 0.60 ounces;
  (c) at least 0.50 ounces;
  (d) at least 1.00 ounces?

15. Verify the statement made on page 234 that for the given example the Central Limit Theorem would lead us to expect that 95 per cent of the means should fall on the interval from 12.5 to 19.5.

16. If the distribution of the weights of all men traveling by air between Dallas and El Paso has a mean of 163 pounds and a standard deviation of 18 pounds, what is the probability that the combined gross weight of 36 men traveling on a plane between these two cities is more than 6,000 pounds?

9.4 Technical Note (Simulation)

Although we introduced random numbers originally to select random samples from finite populations, they are used for many other purposes. They serve to *simulate* almost any kind of gambling device; in fact, they can be used to simulate any situation involving an element of uncertainty or chance. For example, we can play the game of "Heads or Tails" without ever flipping a coin by letting the digits 0, 2, 4, 6, and 8 represent *heads* while the digits 1, 3, 5, 7, and 9 represent *tails*. Thus, using the 5th column of the table on page 490 starting at the top, we get 1, 2, 2, 6, 6, 7, 0, 4, 5, ..., and we interpret this as *tail, head, head, head, head, tail, head, head, tail,* ....

In recent years, techniques based on random numbers have been applied to a great variety of problems in the physical, social, and biological sciences. Referred to under the name of *Monte Carlo Methods*, they have been used to *simulate* such things as the spread of cholera epidemics, the collision of photons with electrons, the scattering of neutrons in a nuclear reactor, traffic congestion on freeways, air turbulence and its effect on airplane wings, to mention but a few. In this way, methods based on random numbers (and often high-speed computers) make it possible to simulate experiments which either cannot be performed in the laboratory (such as the spreading of an epidemic) or which would otherwise require prohibitively expensive equipment. Monte Carlo methods have also found wide application in business research, where they are used for solving inventory problems, or questions arising in connection with the allocation of resources, advertising, competition, and over-all planning and organization.

In this section we shall concern ourselves mainly with the problem of *simulating sampling experiments;* that is, we shall use random numbers to simulate the observation or measurement of values of random variables. Although we shall limit ourselves here to discrete random variables, random numbers can be used also to simulate the observation of values of continuous random variables. How this is done is discussed in the author's textbook on mathematical statistics, which is referred to in the Bibliography on page 298.

To illustrate the simulation of a discrete random variable, let us consider a very simple experiment which consists of repeated flips of 3 balanced coins. There are many different ways in which this experiment can be simulated: one possibility is to proceed as in the first paragraph of this section, letting 0, 2, 4, 6, and 8 represent *heads* while 1, 3, 5, 7, and 9 represent *tails*, and using three random digits to represent the results obtained with three coins. Thus, if we used the first three columns of page 491 starting at the top, we would obtain 244, 574, 776, 683, 644, 882, 984, . . ., and we would interpret this as 3, 1, 1, 2, 3, 3, 2, . . ., heads on successive flips of three coins.

Whenever possible, it is preferable to get values of random variables directly from random numbers, that is, without having to worry about the interpretation of each individual digit. Thus, remembering from Chapter 8 that the probabilities of getting 0, 1, 2, or 3 heads with 3 balanced coins (or in 3 flips of one balanced coin) are, respectively, 1/8, 3/8, 3/8, and 1/8, we might use the following scheme:*

Number of Heads	Random Numbers
0	000–124
1	125–499
2	500–874
3	875–999

* Observe that we are using three-digit random numbers rather than, say, two-digit numbers, because 1,000 is divisible by 8 while 100 is not; in other words, we can allocate 1/8 of the 1,000 three-digit numbers from 000 to 999 to "0 heads" whereas we could not have done this with the 100 two-digit numbers from 00 to 99 without leaving some of them out.

Note that with this scheme 125 of the three-digit random numbers from 000 to 999 (or one-eighth) represent 0 heads, 375 of the random numbers (or three-eights) represent 1 head, and so on. If we got the random numbers 095, 632, 715, 309, and 897, for example, we would interpret this as representing 0, 2, 2, 1, and 3 heads.

To give another example—one which does not directly pertain to a game of chance—let us refer to the probability function on page 153 and simulate the occurrence of hurricanes in the given county for twenty years. Since the probabilities were rounded to two decimals, we shall use the two-digit random numbers from 00 through 99 and let the first 9 (the ones from 00 through 08) represent the occurrence of 0 hurricanes. The next 22 random numbers (the ones from 09 through 30) are used to represent the occurrence of 1 hurricane; the next 26 after that (the ones from 31 through 56) are used to represent the occurrence of 2 hurricanes; and so on. We thus get

Number of Hurricanes	Probability	Random Numbers
0	0.09	00–08
1	0.22	09–30
2	0.26	31–56
3	0.21	57–77
4	0.13	78–90
5	0.06	91–96
6	0.02	97–98
7	0.01	99

and if we arbitrarily use the 21st and 22nd columns of the table on page 491 starting with the 16th row, we get 18, 92, 78, 23, 35, 55, 22, 83, 50, 97, 49, 78, 30, 19, 30, 42, 39, 88, 16, and 77. This corresponds to the occurrence of 1, 5, 4, 1, 2, 2, 1, 4, 2, 6, 2, 4, 1, 1, 1, 2, 2, 4, 1, and 3 hurricanes, respectively, for twenty years.

Finally, let us explain how we obtained the 50 samples on page 229. Looking up the values of the Poisson distribution with $\lambda = 16$ in an appropriate table, we obtained the figures shown in the middle column of the table which follows, and then we distributed the random numbers from 000 through 999 so that the number assigned to each value of the random variable is proportional to the corresponding probability. Thus, one random number is assigned to 5 service calls for which the probability is 0.001, three random numbers are assigned to 6 service calls for which the probability is 0.03, six random numbers are assigned to 7 service calls for which the probability is 0.006, ..., and the simulation was performed by means of the following scheme:

Number of Service Calls	Probability	Random Numbers
5	0.001	000
6	0.003	001–003
7	0.006	004–009
8	0.012	010–021
9	0.021	022–042
10	0.034	043–076
11	0.050	077–126
12	0.066	127–192
13	0.082	193–274
14	0.093	275–367
15	0.099	368–466
16	0.099	467–565
17	0.093	566–658
18	0.083	659–741
19	0.070	742–811
20	0.056	812–867
21	0.043	868–910
22	0.031	911–941
23	0.022	942–963
24	0.014	964–977
25	0.009	978–986
26	0.006	987–992
27	0.003	993–995
28	0.002	996–997
29	0.001	998
30	0.001	999

Thus, if we got the random numbers 183, 709, 051, 255, and 412, for example, this would correspond to five Sundays on which there were, respectively, 12, 18, 10, 13, and 15 service calls.

EXERCISES

1. In Chapter 5 we illustrated the Law of Large Numbers by repeatedly flipping a coin, determining the accumulated number of heads after each fifth flip, and then plotting the results as in Figure 5.3. To duplicate this, simulate 100 flips of a coin with the use of random numbers (as suggested in the first paragraph of this section), determine the accumulated number of heads after each fifth flip, and plot the results as in Figure 5.3.

2. Using one-digit random numbers (omitting 7, 8, 9, and 0), simulate 120 rolls of a balanced die. Also determine the accumulated proportion of threes after each tenth "roll of the die," and plot the results as in Figure 5.3 to illustrate the Law of Large Numbers.

3. Using four random digits to represent the results obtained when tossing four balanced coins (0, 2, 4, 6, and 8 represents *heads* while 1, 3, 5, 7, and 9 represents *tails*), simulate an experiment which consists of 160 tosses of four coins. Compare the observed number of times that 0, 1, 2, 3, and 4 heads occurred with the corresponding *expected frequencies* which are 10, 40, 60, 40, and 10.

4. Repeat Exercise 3 using the following scheme:

Number of Heads	Random Numbers
0	0000–0624
1	0625–3124
2	3125–6874
3	6875–9374
4	9375–9999

5. Referring to part (b) of Exercise 1 on page 235, label the 15 possible samples 01, 02, 03, ..., 14, and 15, and use random numbers to simulate an experiment in which 100 random samples of size 2 are taken from the given population. Compare the distribution of the means of these samples with the corresponding theoretical sampling distribution obtained in part (c), and its mean and standard deviation with the values expected in accordance with the theorem on page 230.

6. In Exercise 12 on page 156 we gave the probability distribution for the number of dresses a woman entering the given shop will buy.
   (a) Construct a table showing how the random numbers 00 through 99 might be assigned to the five values of the random variable.
   (b) Use random numbers and the scheme of part (a) to simulate the number of dresses that will be bought by 50 women who enter the store. How does the mean of these values compare with the expected number which the reader was asked to calculate in that exercise.

7. On page 175 we gave the binomial distribution for the number of drivers, among 5, who will use their seat belts on short trips. Construct a table showing how the random numbers 000 through 999 might be assigned to the six values of this random variable.

8. On page 181 we gave the Poisson distribution for the number of policy
    holders filing claims against the given insurance company each year.
    Construct a table showing how the random numbers 000 through 998
    might be assigned to the eleven values of this random variable.

9. On page 241 we gave the scheme which was used to obtain the 50
    samples on page 229. Use this scheme to get 40 simulated random
    samples of size 4 from the given population (which has the mean
    $\mu = 16$ and the standard deviation $\sigma = 4$), calculate their mean and
    their standard deviation, and compare the results with those expected
    in accordance with the theorem on page 230.

10. The owner of a bakery knows that the daily demand for a highly-
    perishable cheese cake is as shown in the following table:

Daily Demand (Number of Cheese Cakes)	Probability
0	0.05
1	0.15
2	0.25
3	0.25
4	0.20
5	0.10

    (a) Distribute the two-digit numbers from 00 through 99 among
        these six possibilities so that the corresponding random numbers
        can be used to simulate the daily demand.
    (b) Use the results of part (a) to simulate the demand for the cheese
        cake on 30 consecutive business days.
    (c) If the baker makes a profit of $2.00 on each cake that he sells,
        but loses $1.00 on each cake that goes to waste (namely, each
        cake that cannot be sold on the day it is baked), find the baker's
        profit or loss for each of the 30 days of part (b), assuming that
        each day he bakes 3 of these cakes. Also find his *average profit*
        per day.
    (d) Repeat part (c) assuming that each day he bakes 4 rather than
        3 of these cakes. Which of the two appears to be more
        profitable?
    (e) Referring to the original probabilities, calculate the baker's
        expected daily profit if he bakes 3 cakes each day and also if he
        bakes 4. (*Hint:* if he bakes 3 and sells 0 he loses $3.00, and this
        happens with a probability of 0.05; if he bakes 3 and sells 1 he
        breaks even, and this happens with a probability of 0.15; if he
        bakes 3 and sells 2 he makes a profit of $3.00, and this happens
        with a probability of 0.25; and if he bakes 3 and sells them all he

makes a profit of $6.00, and this happens with a probability of $0.25 + 0.20 + 0.10 = 0.55$.)

BIBLIOGRAPHY

Among the many published tables of random numbers, the following is one of the most widely used

RAND Corporation, *A Million Random Digits with 100,000 Normal Deviates.* New York: The Free Press, 1955.

PART 3

Statistical Inference

10

Inferences Concerning Means

10.1 Introduction

In recent years attempts have been made to treat all problems of statistical inference (that is, all problems concerning generalizations based on samples) within the framework of a unified theory. Although this theory, called *decision theory*, has many conceptual and theoretical advantages, its application poses problems which are difficult to overcome. To understand these problems, one must appreciate the fact that no matter how objectively an experiment or an investigation may be planned, it is impossible to eliminate all elements of subjectivity. It is at least partially a subjective decision whether to base an experiment (say, the determination of an index of diffraction) on 3 measurements, on 5 measurements, or on 10 or more. Also, subjective factors invariably enter the design of equipment, the hiring of personnel, and even the precise formulation of a problem one wants to investigate. An element of subjectivity enters even when we define such terms as "good" or "best" in connection with the choice between different decision criteria (say, when deciding between a sample mean and a sample median in a problem of estimation), or when looking for the straight line which "best" fits a set of paired data. Above all, subjective judgments are practically unavoidable when one is asked to put "cash values" on the risks to which one is exposed. In contrast to the examples which we used in our discussion of *decision making* in Chapter 7, it is generally impossible in statistics to be completely objective in specifying rewards for being right (or close) and penalties for being wrong (or not close enough). After all, if a scientist is asked to judge the safety of a piece of equipment, how can he put a cash value on the consequences of a possible error on his part, if such an error may result in the loss of human lives?

The general approach we shall use in this book to problems of statistical inference may be called the *classical approach*, insofar as it does not *formally* take into account the various subjective factors we have mentioned. In other words, the subjective elements will not appear as part of the formulas, themselves; rather, they will appear in the choice among formulas to be used in a given situation, in decisions concerning the size of a sample, in specifying the probabilities with which we are willing to incur certain risks, in specifying the maximum errors we consider acceptable, and so forth. The *Bayesian approach*, which can account for some of these subjective factors, is introduced briefly in Sections 10.4 and 12.3.

10.2 Problems of Estimation

According to some dictionaries, an estimate is a valuation based on opinion or roughly made from imperfect or incomplete data. Although this definition may apply to a parent's opinionated estimate of the ability of his child, or a politician's wishful thinking based on incomplete returns from his own precinct, this is *not* how the term "estimate" is used in statistics. In statistics we allow estimates based on opinions or incomplete information only if such opinions are based on sound judgment or experience and if the samples are scientifically selected.

Statistical methods of estimation find applications almost anywhere, in science, in business, as well as in everyday life. *In science*, a biologist may wish to estimate what proportion of a certain kind of insect is born physically defective, a psychologist may wish to estimate the average (mean) time it takes an adult to react to a given stimulus, and an engineer may wish to estimate how much variability there is in the strength of a new alloy. *In business*, a finance company may wish to estimate what proportion of its customers plan to buy a new car within the next year, a contractor may wish to estimate the average monthly rent paid for two-room apartments in a suburb where he is planning to erect some new units, and a manufacturer of television tubes may wish to estimate how much variation there is in the lifetimes of his product. Finally, *in everyday life*, we may want to estimate what percentage of car accidents are due to faulty brakes, we may be interested in estimating the average time it takes to iron a pair of pajamas, and we may wish to know how much variation one can expect in a child's performance in school. Note that in each case we gave three examples: one dealing with the estimation of a percentage or proportion, one dealing with the estimation of a mean, and one dealing with an appropriate measure of variation. These (and particularly the first two) are the parameters with which we are concerned in most problems of estimation.

Referring again to the science examples, the biologist may estimate the proportion of physically defective insects as 0.08, the psychologist may estimate the average time it takes to react to the given stimulus as 0.32 seconds, and the engineer may estimate the standard deviation of measurements of the strength of different specimens of the new alloy as 240 pounds per square inch. Estimates like these are called *point estimates*, since each one consists of a single number, namely, a single point on the real number scale. Although this may be the most common way of expressing an estimate, point estimates have the serious shortcoming that they do not tell us anything about their relative merits; that is, they do not tell us how close we can expect them to be to the quantities they are supposed to estimate. In other words, *point estimates do not tell us anything about the intrinsic reliability or precision of the method of estimation which is being used.* For instance, if an advertisement claims on the basis of "scientific" evidence that 80 per cent of all doctors prefer Brand X cigarets, this would *not* be very meaningful if the claim were based on interviews with only 5 doctors, among whom 4 happen to prefer Brand X. However, the claim would become more and more meaningful if it were based on interviews with 100 doctors, 400 doctors, or perhaps even 1,000 doctors. This illustrates why point estimates should always be accompanied by some information which makes it possible to judge their merits. How this is done will be explained in the next section.

In most of the methods discussed in this chapter we shall assume that our estimates are to be based only on *direct* observations or measurements. If this kind of information is to be supplemented with collateral information based on *indirect data* or a person's *subjective judgment*, it may be necessary to use some form of *Bayesian inference;* perhaps, the method described in Section 10.4.

10.3 The Estimation of Means

To illustrate some of the problems we face in the estimation of means, let us consider a study in which space scientists want to estimate the average (mean) increase in the pulse rate of astronauts performing a certain task in outer space. Simulating weightlessness, they obtain the following data (increases in pulse rate in beats per minute) for 32 persons who perform the given task:

26	20	17	22	23	21	25	33
32	25	30	27	28	24	12	21
20	14	29	23	22	36	25	21
23	26	24	19	27	24	30	26

The mean of this sample is $\bar{x} = 24.2$ beats per minute, and in the absence of any other information this figure may well have to serve as an estimate of μ, the true average increase in the pulse rate of astronauts performing the given task in outer space.

This is quite alright, but to comply with the suggestion that point estimates should always be accompanied by information which makes it possible to judge their merit, we might add that the size of the sample on which the estimate is based is $n = 32$ and that the sample standard deviation is 5.15 beats per minute. Unfortunately, this kind of information is meaningful only to someone who has some knowledge of statistics; to make it more meaningful to the layman, let us go back briefly to the discussion of the preceding chapter, in particular to that dealing with the sampling distribution of the mean. Of course, we know that sample means (of data describing the same phenomenon) will fluctuate from sample to sample, but we also know that the mean and the standard deviation of the sampling distribution which describes these fluctuations are μ and $\sigma/\sqrt{n}$. Here μ and σ are the mean and the standard deviation of the (supposedly infinite) population from which the sample was obtained. Also making use of the Central Limit Theorem (see page 233) according to which this sampling distribution can be approximated closely with a normal curve, we can now assert with a probability of $1 - \alpha$ that $\bar{x}$ will differ from μ by less than $z_{\alpha/2}$ standard deviations. As defined in Exercise 7 on page 208, $z_{\alpha/2}$ denotes the value for which the area *to its right* under the standard normal distribution is $\alpha/2$, and, hence, the area under the curve between $-z_{\alpha/2}$ and $z_{\alpha/2}$ is $1 - \alpha$ (as is illustrated in Figure 10.1). In other words, *we can assert with a probability of $1 - \alpha$ that*

$\bar{x}$ will differ from μ by less than $z_{\alpha/2} \dfrac{\sigma}{\sqrt{n}}$, and since $\bar{x} - \mu$ is the error we make

when we use $\bar{x}$ as an estimate of μ, we can assert with a probability of $1 - \alpha$

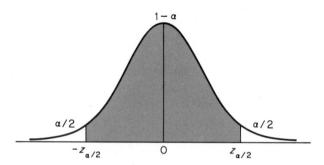

Figure 10.1. Normal distribution.

that the size of this error will be less than $z_{\alpha/2} \dfrac{\sigma}{\sqrt{n}}$. Using the numerical results obtained in Exercise 7 on page 208, namely, that $z_{0.025} = 1.96$, $z_{0.010} = 2.33$, and $z_{0.005} = 2.58$, we find that we can assert with a probability of 0.95 that the error will be less than $1.96 \dfrac{\sigma}{\sqrt{n}}$, with a probability of 0.98 that the error will be less than $2.33 \dfrac{\sigma}{\sqrt{n}}$, and with a probability of 0.99 that the error will be less than $2.58 \dfrac{\sigma}{\sqrt{n}}$.

The result we have obtained involves one complication: to judge the size of the error we might make when using $\bar{x}$ as an estimate of μ, we must know σ, the standard deviation of the population. Since this is seldom the case in practical situations, we have no choice but to replace σ with an estimate, usually the sample standard deviation s. Generally speaking, this is reasonable provided the sample size is sufficiently large, and by "sufficiently large" we mean $n = 30$ or more.

Returning now to our numerical example, we can assert with a probability of 0.95 that if we estimate the *true* average increase in the pulse rate (of astronauts performing the given task) as 24.2 beats per minute, the error of our estimate is less than

$$1.96 \frac{s}{\sqrt{n}} = 1.96 \frac{5.15}{\sqrt{32}} = 1.78$$

beats per minute. Of course, the error of our estimate is either less than 1.78 or it is not (and we really do not know which), but *if we had to bet,* 95 to 5 (or 19 to 1) would be fair odds that the error *is* less than 1.78 beats per minute. Similarly, 98 to 2 (or 49 to 1) would be fair odds that the error is less than

$$2.33 \frac{s}{\sqrt{n}} = 2.33 \frac{5.15}{\sqrt{32}} = 2.12$$

beats per minute, and 99 to 1 would be fair odds that it is less than

$$2.58 \frac{s}{\sqrt{n}} = 2.58 \frac{5.15}{\sqrt{32}} = 2.35$$

beats per minute. Note that these odds are based on the *success-ratio* of the method we have used, namely, the per cent of the time it can be expected to work. When σ is unknown and n is less than 30, the method we have just discussed cannot be used, but there exists a modification which applies also to small samples (n less than 30), provided we can assume that we are sampling from a population having roughly the shape of a normal distribution. We shall discuss this method later, on page 256.

An interesting feature of the formula for the maximum error is that it can also be used to determine the sample size that is required to attain a desired degree of precision. Suppose we want to use the mean of a random sample to estimate the mean of a population and we want to be able to assert with a probability of $1 - \alpha$ that the error of this estimate will be less than some quantity E. In accordance with the theory on page 251, we can thus write

$$▲ \qquad E = z_{\alpha/2}\frac{\sigma}{\sqrt{n}} \qquad ▲$$

and upon solving this equation for n we get

$$▲ \qquad n = \left[\frac{z_{\alpha/2}\cdot\sigma}{E}\right]^2 \qquad ▲$$

Note that this formula cannot be used unless we know (or can approximate) the standard deviation of the population whose mean we want to estimate.

To illustrate this technique, suppose we want to estimate (on the basis of a sample) the average I.Q. of all applicants to medical colleges in a given year and that we want this estimate to be off by at most 1.0 with a probability of 0.99. Suppose also that (on the basis of experience with similar data) it is reasonable to let σ be 6.4. Substituting these values together with $z_{0.005} = 2.58$ into the formula for n, we obtain

$$n = \left[\frac{2.58(6.4)}{1.0}\right]^2$$

which is 273 *rounded up* to the nearest whole number. Thus, a sample of size $n = 273$ will suffice for the stated purpose; in other words, if we base our estimate on a random sample of size $n = 273$, we can assert with a probability of 0.99 that our estimate (namely, the sample mean) will be within 1.0 of the true mean.

The error we make when using a sample mean to estimate the mean of a population is given by the difference $\bar{x} - \mu$, and the fact that the *magnitude* of this error is less than $z_{\alpha/2}\frac{\sigma}{\sqrt{n}}$ can be expressed by means of the inequality

$$-z_{\alpha/2}\frac{\sigma}{\sqrt{n}} < \bar{x} - \mu < z_{\alpha/2}\frac{\sigma}{\sqrt{n}}$$

(In case the reader is not familiar with inequality signs, let us point out that $a < b$ means "a is less than b," while $a > b$ means that "a is greater than b." Also, $a \le b$ means "a is less than or equal to b," while $a \ge b$ means "a is greater than or equal to b.") Applying some simple algebra, we can rewrite the above inequality as

▲ $$\bar{x} - z_{\alpha/2}\frac{\sigma}{\sqrt{n}} < \mu < \bar{x} + z_{\alpha/2}\frac{\sigma}{\sqrt{n}}$$ ▲

and we can now assert with a probability of $1 - \alpha$ that this inequality is satisfied for any given sample, namely, that the interval from $\bar{x} - z_{\alpha/2}\dfrac{\sigma}{\sqrt{n}}$ to $\bar{x} + z_{\alpha/2}\dfrac{\sigma}{\sqrt{n}}$ actually contains the mean we are trying to estimate. An interval like this is called a *confidence interval*, its endpoints are called *confidence limits*, and the probability $1 - \alpha$ with which we can assert that such an interval will "do its job," namely, that it will contain the quantity we are trying to estimate, is called the *degree of confidence*. The values most commonly used for the degree of confidence are $1 - \alpha = 0.95, 0.98,$ or 0.99, and, as we pointed out earlier, the corresponding values of $z_{\alpha/2}$ are 1.96, 2.33, and 2.58.

When σ is unknown and n is 30 or more, we proceed as before and estimate σ with the sample standard deviation s. The resulting $1 - \alpha$ *large-sample confidence interval* for μ becomes

▲ $$\bar{x} - z_{\alpha/2}\frac{s}{\sqrt{n}} < \mu < \bar{x} + z_{\alpha/2}\frac{s}{\sqrt{n}}$$ ▲

If we apply this technique to the pulse-rate example on page 249, where we had $n = 32$, $\bar{x} = 24.2$, and $s = 5.15$, we obtain the following 0.95 confidence interval for the true average increase in the pulse rate of astronauts performing the given task in outer space:

$$24.2 - 1.96\frac{5.15}{\sqrt{32}} < \mu < 24.2 + 1.96\frac{5.15}{\sqrt{32}}$$

$$22.4 < \mu < 26.0$$

Had we wanted to calculate a 0.98 confidence interval for this example, we would have obtained

$$22.1 < \mu < 26.3$$

and this illustrates the interesting fact that *the surer we want to be in connection with a confidence interval, the less we have to be sure of.* In other words, if we increase the degree of certainty (the degree of confidence), the confidence interval becomes wider and thus tells us less about the quantity we want to estimate.

When we estimate the mean of a population with the use of a confidence interval, we refer to this kind of estimate as an *interval estimate.* In contrast to a point estimate, an interval estimate requires no further elaboration about its relative merits; this is taken care of indirectly by the degree of confidence and its actual width.

So far we have assumed that the sample size was large enough to treat the sampling distribution of the mean as if it were a normal distribution, and to replace σ with s in the formula for the standard error. In order to develop corresponding theory which applies also to *small samples*, let us now assume that the population from which we are sampling can be approximated closely with a normal curve. We can then base our methods on the statistic

▲
$$t = \frac{\bar{x} - \mu}{s/\sqrt{n}}$$
 ▲

whose sampling distribution is called the *t distribution.* (More specifically, it is called the *Student-t distribution,* as it was first investigated by W. S. Gosset, who published his writings under the pen name of "Student.") The shape of this distribution is very much like that of the normal curve; it is symmetrical with zero mean, but there is a slightly higher probability of getting values falling into the two tails (see Figure 10.2). Actually, the shape of the t distribution depends on the size of the sample or, better, on the quantity $n - 1$, which in this connection is called the *number of degrees of freedom.**

For the standard normal distribution, we defined $z_{\alpha/2}$ in such a way that the area *to its right* under the normal curve equals $\alpha/2$ and, hence, the area under the curve between $-z_{\alpha/2}$ and $z_{\alpha/2}$ equals $1 - \alpha$. As is shown in Figure 10.3, the corresponding values for the t distribution are $-t_{\alpha/2}$ and $t_{\alpha/2}$; the main difference is that these values depend on $n - 1$ (the number of degrees of freedom) and, hence, must be looked up in each case in a special table. Table II at the end of the book contains (among others) the values of $t_{0.025}$, $t_{0.010}$, and $t_{0.005}$, with the number of degrees of freedom

---

* In other applications of the t distribution, for example, on page 281, the number of degrees of freedom may be given by a different expression.

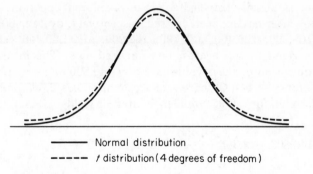

——— Normal distribution

- - - - - *t* distribution (4 degrees of freedom)

Figure 10.2. Standard normal distribution and example of *t* distribution.

going from 1 to 29. Observe that as the number of degrees of freedom increases, $t_{0.025}$ approaches 1.96, $t_{0.010}$ approaches 2.33, and $t_{0.005}$ approaches 2.58, the corresponding values for the normal distribution.

Duplicating the argument on page 253, we now find that a $1 - \alpha$ *small-sample confidence interval for μ is given by*

$$\bar{x} - t_{\alpha/2} \frac{s}{\sqrt{n}} < \mu < \bar{x} + t_{\alpha/2} \frac{s}{\sqrt{n}}$$

The only difference between this confidence-interval and the second one on page 253 is that $t_{\alpha/2}$ takes the place of $z_{\alpha/2}$; of course, the formula for t does not contain the population standard deviation σ, so that we do not have to make the substitution (s for σ) which restricted the confidence interval on page 253 to large samples.

To illustrate the calculation of a small-sample confidence interval for μ, let us consider an experiment conducted by a highway department to test the durability of a new paint for white center lines. Tests strips were

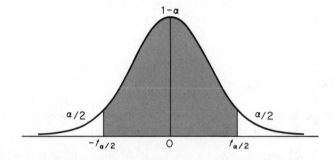

Figure 10.3. *t* distribution.

painted across heavily-traveled roads in 8 different locations and they deteriorated after having been crossed, respectively, by 142,600, 167,800, 136,500, 108,300, 126,400, 133,700, 162,000, and 149,400 cars. These figures are given to the nearest one hundred cars. The mean and the standard deviation of these values are $\bar{x}$ = 140,800 and s = 19,200, and since $t_{0.025}$ for $8 - 1 = 7$ degrees of freedom equals 2.365, substitution into the formula for a 0.95 confidence interval yields

$$140,800 - 2.365\,\frac{19,200}{\sqrt{8}} < \mu < 140,800 + 2.365\,\frac{19,200}{\sqrt{8}}$$

or

$$124,700 < \mu < 156,900$$

This is an interval estimate of the total amount of traffic which this paint can takes before it deteriorates.

The method which we used on page 251, to indicate the possible size of the error we make when using a sample mean to estimate the mean of a population, can easily be adapted to small samples (provided the population from which we are sampling has roughly the shape of a normal distribution). All we have to do is substitute s for σ and $t_{\alpha/2}$ for $z_{\alpha/2}$ in the formula for the maximum error E. To give an example, suppose that in 4 test runs an experimental engine consumed on the average 13.8 gallons per minute with a standard deviation of 1.6 gallons. Since $t_{0.005}$ equals 5.841 for $4 - 1 = 3$ degrees of freedom, substitution into the formula for E yields

$$E = t_{0.005}\cdot\frac{s}{\sqrt{n}} = 5.841\cdot\frac{1.6}{\sqrt{4}} = 4.7$$

Thus, if we use the mean of 13.8 gallons per minute as an estimate of the *true* average gasoline consumption of the engine, we can assert with a probability of 0.99 that the error of this estimate is at most 4.7. This also illustrates the fact that *although we can make logically correct inferences on the basis of very small samples, our results may well involve considerable errors and our confidence intervals are apt to be very wide.*

10.4 A Bayesian Estimate*

In recent years there has been mounting interest in methods of inference which look upon parameters (for example, a population mean μ) as ran-

---

* The material in this section is somewhat more advanced, and it may be omitted without loss of continuity.

dom variables. The whole idea is not really new, but these *Bayesian methods*, as they are called, have received a considerable impetus and much wider applicability through the concept of personal, or subjective, probability. In fact, this is why supporters of the personal concept of probability refer to themselves as *Bayesians*, or *Bayesian statisticians*.

In this section we shall present a Bayesian method of estimating the mean μ of a population. Proponents of the subjective, or personal, point of view in probability look upon μ as a random variable whose distribution is indicative of *how strongly a person feels about the various values which μ can take on*. In other words, they suggest that in any problem where we want to estimate the mean of a population, a person will feel most strongly about some particular value of μ, and that this enthusiasm will diminish for values of μ which are further and further away from the one he likes the most. Like any distribution, this kind of *subjective prior distribution* for the possible values of μ has a mean which we shall denote μ_0 and a standard deviation which we shall denote σ_0.

In the Bayesian estimation of a mean, these *prior feelings* about the possible values of μ are combined with *direct sample evidence* consisting of a random sample of size n, its mean $\bar{x}$, and its standard deviation s (which may have to serve as an estimate of σ, the standard deviation of the population). This can be accomplished by means of the formula

▲
$$\text{Estimate} = \frac{\dfrac{n}{\sigma^2} \cdot \bar{x} + \dfrac{1}{\sigma_0^2} \cdot \mu_0}{\dfrac{n}{\sigma^2} + \dfrac{1}{\sigma_0^2}}$$
▲

which may be used under fairly general conditions. (Usually, it is based on the assumption that the distribution of the population from which we are sampling as well as the subjective prior distribution of μ have roughly the shape of normal distributions.)

Before we apply this formula, let us briefly examine some of its most important features. To begin with, it should be observed that it is a *weighted mean* (see Section 3.3) of $\bar{x}$ and μ_0, whose respective weights are $\dfrac{n}{\sigma^2}$ and $\dfrac{1}{\sigma_0^2}$, namely, the *reciprocals* of the variances of their distributions. When no direct information is available and $n = 0$, the weight of $\bar{x}$ equals 0 and the estimate is based entirely on the subjective prior information. However, when more and more direct evidence becomes available (that is, when n becomes larger and larger), the weight shifts more and more toward the direct sample evidence, namely, the sample mean $\bar{x}$. There are two other points worth mentioning: When the subjective feelings about the possible values of μ are rather indefinite, that is, when σ_0 is relatively large, the estimate will be based to a greater extent on $\bar{x}$; on the other hand, when there is a great deal of variability in the population from which we are sampling, that is, when σ is relatively large, the estimate will be based to a greater extent on μ_0.

To illustrate this subjective approach, suppose that someone is planning to open a new bowling alley, and that a business consultant feels most strongly that he should net on the average $2,600 a month; also, the subjective prior distribution which he attaches to the various possible values of μ has a standard deviation of $130. In other words, $\mu_0 = \$2,600$ and $\sigma_0 = \$130$. Now, if during the first 9 months the operation of the bowling alley nets, say, $2,810, $2,690, $2,350, $2,400, $2,320, $2,250, $2,430, $2,600, and $2,670, the problem is *how to modify the original estimate of $\mu_0 = \$2,600$ in the light of this information.* Having $n = 9$, $\bar{x} = \$2,502$, and $s = \$195$ (as can easily be verified), substitution into the formula on page 257 yields

$$\text{Estimate} = \frac{\dfrac{9}{195^2} \cdot 2,502 + \dfrac{1}{130^2} \cdot 2,600}{\dfrac{9}{195^2} + \dfrac{1}{130^2}} = \$2,522$$

and this accounts for both, the consultant's feelings as well as the direct sample evidence. It is doubtful, perhaps, whether we should substitute s for σ when n is only 9, but as the problem has been stated, we have no choice.

This introduction to the Bayesian approach has been very brief, but it should have served to bring out the following two points: (1) *In Bayesian statistics the parameter about which an inference is to be made is looked upon as a random variable having a distribution of its own, and* (2) *this kind of inference permits the use of direct as well as collateral information.* To clarify the last point, let us add that in the bowling-alley example the subjective prior distribution of the consultant may have been based on a subjective evaluation of various factors (say, business conditions in general) or perhaps on *indirect* objective information about the "performance" of other bowling alleys.

EXERCISES

1. To estimate the average time it takes to assemble a certain computer component, the efficiency expert of an electronics firm timed 40 technicians in the performance of this task, getting a mean of 14.63

minutes and a standard deviation of 2.45 minutes. What can one say with a probability of 0.95 about the possible size of the error, if the efficiency expert uses 14.63 minutes as an estimate of the true average time it takes to assemble the given computer component?

2. Use the data of Exercise 1 to construct a 0.99 confidence interval for the true average time it takes to assemble the given computer component.

3. A study of the annual growth of a certain kind of cactus showed that (under controlled conditions) 60 of them grew on the average 42.8 mm per year with a standard deviation of 5.4 mm. Construct a 0.95 confidence interval for the true average growth of this kind of cactus (under the given controls).

4. Referring to Exercise 3, what can one assert with a probability of 0.98 about the possible size of the error, if the sample mean of 42.8 mm is used as an estimate of the true average annual growth of the given kind of cactus?

5. In a study of automobile collision insurance costs, a random sample of 80 body repair costs on a particular kind of damage had a mean of $432.56 and a standard deviation of $67.31.
   (a) Construct a 0.95 confidence interval for the true average cost of this kind of body repair.
   (b) What can be said with a probability of 0.99 about the possible size of the error, if the sample mean of $432.56 is used as an estimate of the true average cost of this kind of body repair?

6. In a reading achievement test, 120 fifth graders of a certain (very large) school district had a mean score of 83.4 with a standard deviation of 14.8.
   (a) Assuming that the 120 students were chosen at random, construct a 0.98 confidence interval for the mean score which all the fifth graders in this school district would get if they took the test.
   (b) Assuming that the 120 students were chosen at random, what can be said with a probability of 0.95 about the possible size of the error, if the sample mean of 83.4 is used as an estimate of the average performance on this test of all fifth graders in the school district?

7. A random sample of 50 cans of pear halves has a mean weight of 29.0 ounces and a standard deviation of 0.8 ounces. If this mean of 29.0 ounces is used as an estimate of the true mean weight of all the cans of pear halves from which this sample was obtained, with what probability can we assert that this estimate is "off" by at most 0.2 ounces? (*Hint:* substitute z for $z_{\alpha/2}$ in the formula for E on page 252, solve for z, and find the area under the standard normal distribution between $-z$ and z.)

8. Taking a random sample from its very extensive files, a power company finds that the amounts owed in 200 delinquent accounts have a mean of \$21.44 and a standard deviation of \$6.19. If this mean of \$21.44 is used as an estimate of the true average amount owed by all of the power company's delinquent accounts, with what probability can we assert that this estimate is "off" by at most \$0.50?

9. In a study of television viewing habits, it is desired to estimate the average number of hours a person over 65 spends watching per week. Assuming that it is reasonable to use a standard deviation of 3.4 hours, how large a sample would be required if one wants to be able to assert with a probability of 0.95 that the sample mean will be "off" by a quarter hour or less?

10. An efficiency expert wants to determine the average number of minutes a customer spends in a certain candy store. Assuming that it is reasonable to use a standard deviation of 1.6 minutes, how large a sample would he have to take to be able to assert with a probability of 0.99 that his estimate, the sample mean, will be "off" by at most 0.5 minutes?

11. Suppose we want to estimate the average score adults living in a large rural area would get in a current events test, and we want to be "98 per cent certain" that our estimate, the mean of a suitable sample, will be "off" by at most 2.5. How large a sample will we need, assuming that it is known from similar studies that the standard deviation of such scores is 10?

12. If a sample constitutes an appreciable portion of a finite population (say, 5 per cent or more), the various formulas of Section 10.3 must be modified by basing them on the second standard-error formula of the theorem on page 230 instead of the first. For instance, the formula for E becomes

$$E = z_{\alpha/2} \frac{\sigma}{\sqrt{n}} \sqrt{\frac{N-n}{N-1}}$$

Repeat Exercise 8 assuming that there are 800 delinquent accounts in the power company's files.

13. Rework part (b) of Exercise 6, given that there are only 300 fifth graders in the school district. (*Hint:* use the method of Exercise 12.)

14. Use the finite population correction factor $\sqrt{\dfrac{N-n}{N-1}}$ to modify the confidence interval formula on page 253 and, thus, make it applicable to problems in which a sample constitutes a substantial portion of a finite population. Rework part (a) of Exercise 6, given that there are only 300 fifth graders in the school district.

15. If the population from which we sample has roughly the shape of a normal distribution, the confidence interval formula on page 253 applies also when n is small. Since $\sigma = 4$ in the number-of-service-calls example on page 229, calculate a 0.95 confidence interval of the form

$$\bar{x} \pm 1.96 \cdot \frac{4}{\sqrt{5}}$$

for each of the 50 samples and check what proportion actually contains the population mean $\mu = 16$. How does this proportion compare with theoretical expectations?

16. In establishing the authenticity of an ancient coin, its weight is often of critical importance. If four experts independently weighed a Phoenician tetradrachm and obtained a mean of 14.30 grams and a standard deviation of 0.03 grams, construct a 0.95 confidence interval for the true weight of this coin.

17. In an air pollution study, the following amounts of suspended benzene-soluble organic matter (in micrograms per cubic meter) were obtained at an experiment station for eight different samples of air: 2.2, 1.8, 3.1, 2.0, 2.4, 2.0, 2.1, and 1.2. Construct a 0.99 confidence interval for the corresponding true mean.

18. In setting the type for a book, a compositor makes, respectively, 10, 11, 14, 8, 12, and 17 mistakes in six galleys. What can we assert with a probability of 0.95 about the possible size of our error, if we use the mean of this sample as an estimate of the average number of mistakes this compositor makes per galley?

19. In 5 determinations of the melting point of tin, a chemist obtained a mean of 232.27 degrees centigrade with a standard deviation of 0.64 degrees.
    (a) Construct a 0.95 confidence interval for the actual melting point of tin.
    (b) If the chemist uses his mean of 232.27 degrees as *the* melting point of tin, what can he assert with a probability of 0.98 about the possible size of his error?

20. In an experiment, the average amount of time it took 12 fuses to blow with a 20 per cent overload was 12.43 minutes with a standard deviation of 2.6 minutes.
    (a) Construct a 0.99 confidence interval for the true average time it takes such fuses to blow with a 20 per cent overload.
    (b) If the sample mean of 12.43 minutes is used as an estimate of the true average time it takes such fuses to blow with a 20 per cent overload, what can we say with a probability of 0.95 about the possible size of the error?

21. In a survey conducted in a retirement community, it was found that ten "senior citizens" visited a physician on the average 6.5 times per year with a standard deviation of 1.8.

  (a) Construct a 0.98 confidence interval for the corresponding true mean.

  (b) What can we assert with a probability of 0.99 about the possible size of our error, if we estimate the true average number of times such a person visits a physician per year as 6.5?

22. A distributor of soft-drink vending machines knows that in a super-market one of his machines will sell on the average 835 drinks per week. Of course, this *mean* will vary somewhat from market to market, and this variation is measured by a standard deviation of 12.4. So far as a machine placed in a particular market is concerned, the number of drinks sold will vary from week to week, and this variation is measured by a standard deviation of 43.6.

  (a) If this distributor plans to put one of his soft-drink vending machines into a brand new supermarket, what estimate would he use for the number of drinks he can expect to sell per week?

  (b) How would he modify his estimate if during 10 weeks the machine sells on the average 592 drinks per week?

  (c) How would he modify his original estimate if during 50 weeks the machine sells on the average 917 drinks per week?

23. A college professor is making up a final examination in history which is to be given to a large group of students. His feelings about the average grade they should get is expressed subjectively by a distribution which has the mean $\mu_0 = 64$ and the standard deviation $\sigma_0 = 1.4$. If, subsequently, the examination is tried on a random sample of 50 students whose grades have a mean of 73.9 and a standard deviation of 7.6, find a Bayesian estimate of the average grade all the students in the large group should get in this test.

10.5 Hypothesis Testing: Two Kinds of Errors

So far, the problems we have treated in this chapter have all been prob-lems of *estimation:* In each case we had to determine the actual value of the mean of a population, such as the "true" average increase in the pulse rate of astronauts performing a given task, the true average I.Q. of all applicants to medical colleges, the true durability of a paint, and so on. In the remainder of this chapter we shall study problems of a somewhat different nature; we shall have to decide, for example, whether it is rea-sonable to maintain that the average annual income of families living in a certain area is $8,400, whether one method of teaching foreign languages is more effective than another, whether a person living in a given city averages 1.4 traffic tickets per year, and so on. In each of these problems we must *test a hypothesis* concerning a mean; that is, we must decide whether to accept or reject a hypothesis (an assumption, or claim) concerning the mean of a population. In later chapters we shall also investigate tests concerning other parameters. For instance, we shall be asked to decide whether a new chemical will actually remove 85 per cent of all spots, or whether the variability of a child's reaction to a visual stimulus is within certain limits. It is important to keep in mind, however, that although such tests may concern a "true" percentage or a "true" standard deviation rather than a "true" mean, the general approach is very much the same.

To give an example that is typical of the kind of situation we face when testing a statistical hypothesis, suppose that a developer who is planning to build a new shopping center in a certain location wants to check the claim that the average annual income of families living within five miles of the location is $8,400. Thus, he asks his staff to interview 200 families, selected at random within the given area, and decides beforehand that *he will accept the claim if the mean annual income of these 200 families falls between $8,300 and $8,500; otherwise, he will reject the claim* (and take whatever action is thus called for in the planning of the new shopping center).

This provides a clear-cut criterion for deciding whether the average annual income of all families in the given area is $8,400, but unfortunately, it is not infallible. Since the decision is based on a sample, it could happen *purely by chance* that the sample mean will exceed $8,500 or be less than $8,300 even though the true mean *is* $8,400. *If this happened, the developer might well make uncalled-for changes in his plans for the shopping center and thus, unnecessarily, waste time and money.* Of course, he could reduce the chances of this happening by changing the decision criterion, say, by choosing wider limits and accepting the hypothesis $\mu = \$8,400$ so long as the sample mean falls between $8,250 and $8,550. Unfortunately, though, this would have the undesirable effect of *increasing another risk*, namely, the risk of obtaining a sample mean between $8,250 and $8,550 when the true average annual income of the families in the given area is actually *not* $8,400.

Before adopting the original decision criterion, it would thus be wise to investigate the chances that it will lead to a wrong decision, and to this end let us assume that it is known that $\sigma = \$760$ for income distributions like the one with which the developer is concerned. Let us first investigate the probability of getting a sample mean less than $8,300 or greater than $8,500 even though the true mean is $8,400. In other words, we will have to determine the probability of getting a sample mean less than $8,300 or greater than $8,500 when a random sample of size $n = 200$

is taken from a population whose mean and standard deviation are $\mu = \$8,400$ and $\sigma = \$760$. This probability is represented by the area of the shaded region of Figure 10.4, and it can easily be found by approximating the sampling distribution of the mean with a normal distribution.

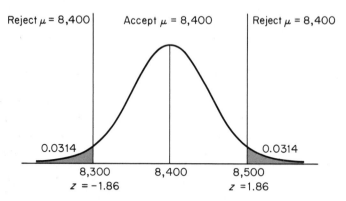

Figure 10.4. Test criterion.

Assuming that the population from which the sample is taken is so large that it can be treated as infinite, the first of the two formulas for $\sigma_{\bar{x}}$ on page 230 yields

$$\frac{\sigma}{\sqrt{n}} = \frac{760}{\sqrt{200}} = 53.7$$

Consequently, the dividing lines of the criterion, in standard units, are

$$z = \frac{8,300 - 8,400}{53.7} = -1.86 \quad \text{and} \quad z = \frac{8,500 - 8,400}{53.7} = 1.86$$

and it follows from Table I that the area in each "tail" of the sampling distribution of Figure 10.4 is $0.5000 - 0.4686 = 0.0314$. Hence, the probability of getting a value in either tail of this sampling distribution, namely, *the probability of erroneously rejecting the hypothesis* $\mu = \$8,400$ is $0.0314 + 0.0314 = 0.0628$. Whether this is an acceptable risk is for the developer to decide; it would have to depend on the "cash value" he puts on the consequences of making such an error.

Let us now look at the other kind of situation, where the test fails to detect the fact that μ is not equal to \$8,400. Suppose, for instance, that $\mu = \$8,550$, in which case the probability of *not* detecting the fact that $\mu \neq \$8,400$ is given by the area of the shaded region of Figure 10.5, namely, the area under the curve between 8,300 and 8,500. The mean of the sampling distribution is now $\mu = 8{,}550$, its standard deviation is as before $\sigma_{\bar{x}} = 53.7$, and the dividing lines of the criterion, in standard units, are

$$z = \frac{8{,}300 - 8{,}550}{53.7} = -4.66 \quad \text{and} \quad z = \frac{8.500 - 8.550}{53.7} = -0.93$$

Since the area under the curve to the left of $z = -4.66$ is negligible, it follows from Table I that the area of the shaded region of Figure 10.5 is

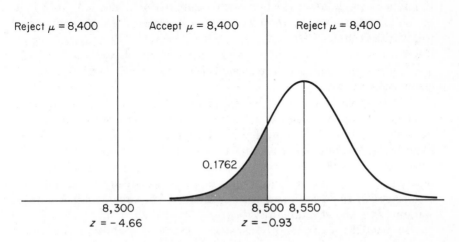

Reject $\mu = 8{,}400$ Accept $\mu = 8{,}400$ Reject $\mu = 8{,}400$

0.1762

8,300   8,500 8,550

$z = -4.66$ $z = -0.93$

Figure 10.5. Test criterion.

$0.5000 - 0.3238 = 0.1762$. Again, it is up to the developer to decide whether this represents an acceptable risk.

To summarize what we have discussed, let us refer to the hypothesis that the average annual income of all the families in the given area is $8,400 (the hypothesis $\mu = \$8{,}400$) as hypothesis H. Evidently, this hypothesis is either true or false, it is either accepted or rejected, and the developer is faced with the situation described in the following table:

	H is True	H is False
Accept H	Correct decision	Type II error
Reject H	Type I error	Correct decision

If the hypothesis is *true and accepted* or *false and rejected,* he is in either case making a correct decision; if the hypothesis is *true but rejected,* he is committing an error called a *Type I error* (the error of rejecting a true hypothesis); if the hypothesis is *false but accepted,* he is committing an error called a *Type II error* (the error of accepting a false hypothesis). Thus, we showed in our example that the probability that he will commit a Type I error is about 0.06, and when $\mu = \$8,550$ the probability that he will commit a Type II error is about 0.18. Whether an error is a Type I error or a Type II error really depends on how we formulate whatever hypothesis we want to test. For example, if H is the hypothesis that the Republican candidate will win a certain gubernatorial election, we would be committing a Type I error if we *erroneously* predicted that his opponent will win the election. However, this same error would be a Type II error if we formulated H as the hypothesis that the Republican candidate will lose. In the first case we erroneously reject H and in the second case we erroneously accept it.

The scheme outlined above is reminiscent of what we did in Chapter 7. Analogous to the decision which the research manager had to make in the example on page 158, we now have to decide whether to accept or reject hypothesis H. The main difficulty in carrying this analogy much further is that in actual practice we can seldom put "cash values" on the various outcomes, as we did in Chapter 7. As we have already indicated, the seriousness of the consequences will generally determine whether a decision presents an acceptable risk.

On page 264 we arbitrarily chose $\mu = \$8,550$ as the alternative value for which we calculated the probability of committing a Type II error. Had the alternative value been very close to $8,400 (say, $8,390 or $8,410), the consequences of accepting the hypothesis $\mu = \$8,400$ would have been trivial; on the other hand, it may be of interest to know the probability of erroneously accepting the hypothesis $\mu = \$8,400$ when actually $\mu = \$8,200$ or $8,600. Duplicating the method which we used on page 265, namely, approximating the sampling distribution of the mean with a normal curve, the reader will be asked to verify in Exercise 6 on page 274 the values shown in the table on the next page. Since a Type II error is committed when hypothesis H is accepted and μ does *not* equal $8,400 in our example, the entries in the right-hand column of the table are identical with those of the middle column, except for the value which corresponds to $\mu = \$8,400$. When $\mu = \$8,400$ and, hence, hypothesis H is *true,* the probability of accepting H is the probability of *not* committing a Type I error, namely, $1 - 0.0628 = 0.9372$, or approximately 0.94. (This probability of committing a Type I error was found earlier on page 264.)

If we plot the probabilities of accepting H as we did in Figure 10.6 and

Value of μ	Probability of Type II Error	Probability of Accepting H
8,150	0.003	0.003
8,200	0.03	0.03
8,250	0.18	0.18
8,300	0.50	0.50
8,350	0.82	0.82
8,400	–	0.94
8,450	0.82	0.82
8,500	0.50	0.50
8,550	0.18	0.18
8,600	0.03	0.03
8,650	0.003	0.003

fit a smooth curve, we obtain the *operating characteristic curve* of the given test criterion, or simply its *OC-curve*. An operating characteristic curve provides a good over-all picture of the advantages and disadvantages of a test criterion. For all values of the parameter except the one assumed under hypothesis H it gives the probability of committing a Type II error; for the value assumed under hypothesis H (in our case $\mu = \$8,400$), it gives the probability of *not* committing a Type I error. A visual inspection of an operating characteristic curve thus enables us to decide whether the risks connected with a particular decision criterion are acceptable, too high, or too low; at least, it aids in the decision. Of course, it must be remembered that the OC-curve of Figure 10.6 applies only to the special

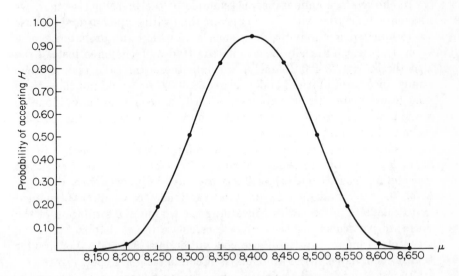

Figure 10.6. Operating characteristic curve.

test where the hypothesis $\mu = 8,400$ is accepted if the mean of a random sample of size 200 falls between 8,300 and 8,500, while otherwise it is rejected, and σ is known to be 760. Changes in the test criterion or the sample size will automatically change the shape of the operating characteristic curve (see Exercises 7 and 8 on page 274); in fact, an OC-curve can often be made to assume a particular desired shape by a suitable choice of the sample size and (or) the dividing lines of the test criterion.

If we had plotted the probabilities of *rejecting H* instead of those of *accepting H*, we would have obtained the graph of the *power function* of the test criterion rather than its OC-curve. The concept of an OC-curve is used more widely in applications, especially in industrial applications, while the concept of a power function is used more widely in matters that are of theoretical interest. Either curve aids in the evaluation of the risks to which one is exposed by a test criterion, and the general idea of an OC-curve or a power function is, of course, not limited to the particular test concerning the average annual income of families in the given area. Hypothesis H could have been the hypothesis that one fertilizer is better than another, the hypothesis that a new antibiotic is effective, the hypothesis that one teaching method is more effective than another—to mention but a few of countless possibilities.

10.6 Hypothesis Testing: Null Hypotheses and Significance Tests

In the average-annual-income example of the preceding section, we had more difficulties with Type II errors than with Type I errors, because we formulated our hypothesis in such a way that the probability of a Type I error could easily be calculated. Had we formulated instead the hypothesis $\mu \neq \$8,400$ (namely, the hypothesis that the true average annual income of all the families is *not* $8,400), we would not have been able to calculate this probability; at least, we could not have done so without actually specifying the amount by which the true average annual income differs from $8,400.

In choosing the hypothesis which we referred to as hypothesis H in our example, we followed the general rule of *formulating hypotheses in such a way that we know what to expect if they are true*. As it was known that $\sigma = \$760$, the hypothesis $\mu = \$8,400$ (and the theory of Chapter 9) enabled us to calculate probabilities concerning the sampling distribution of the mean; in particular, we were able to calculate the probability that the mean of a random sample of size 200 will be less than $8,300 or greater than $8,500, namely, the probability of committing a Type I error.

In order to follow the rule given in italics in the preceding paragraph, we often have to assume (hypothesize) the exact opposite of what we may want to prove. If we wanted to show that one psychological test is more reliable than another, we would have to formulate the hypothesis that there is *no difference* in their reliability; if we wanted to show that a new medicine is more effective than an old one, we would have to formulate the hypothesis that there is *no difference*, namely, that the two medicines are equally effective; and if we wanted to show that one kind of ore has a higher uranium content than another, we would have to formulate the hypothesis that their uranium contents are *the same*. Since we assumed, respectively, that there is no difference in reliability, no difference in effectiveness, and no difference in uranium content, we refer to each of these hypotheses as a *null hypothesis* and denote it H_0. This explains how the term "null hypothesis" arose, although nowadays it is used for any hypothesis set up primarily to see whether it can be rejected. Note that the idea of setting up a null hypothesis is not uncommon even in non-statistical thinking. It is precisely what is done in criminal court proceedings, where an accused is assumed to be innocent unless his guilt is proven beyond a reasonable doubt. The assumption that the accused is not guilty is a null hypothesis.

Although we avoid one kind of difficulty by formulating hypotheses so that the probability of a Type I error can be calculated, this will not help insofar as Type II errors are concerned. The only time there is no difficulty in finding the probabilities of either kind of error is when we test a *specific* hypothesis against a *specific* alternative. This would have been the case in our example if we had been interested *only* in testing the claim that the average annual salary of all the families is \$8,400 against the conflicting claim that it is \$8,550.

A possible escape from the difficulties with Type II errors is to avoid this kind of error altogether. To illustrate how this might be done, suppose we want to test a sociologist's claim that in a given city persons over 65 average more traffic tickets per year than the average of 1.4 for all age groups combined. To test this claim we decide to check the records of 50 randomly-selected persons over 65 in the given city, and base our decision on the following criterion.

> *Reject the null hypothesis $\mu = 1.4$ (namely, the hypothesis that the average for persons over 65 is the same as that for all age groups combined) if the sample mean exceeds 1.6; otherwise, we reserve judgment.*

Note that with this criterion there is no need to calculate the probability of committing a Type II error; we never really accept the null hypothesis and, hence, we cannot possibly make the mistake of accepting a false null hypothesis.

The procedure we have just outlined is referred to as a *test of significance*. If the difference between what we expect and what we get is so large that it cannot reasonably be attributed to chance, we reject the null hypothesis on which our expectation is based. If the difference between what we expect and what we get is so small that it can well be attributed to chance, we say that the result is *not (statistically) significant*. We then reserve judgment or accept the null hypothesis depending on whether a definite decision, a definite action one way or the other, is required. In the above example, the claim that persons over 65 average more traffic tickets per year is confirmed if the ones in the sample average more than 1.6; in that case it is felt that *the difference between the sample mean and μ = 1.4 is too large to be attributed to chance*. In fact, in Exercise 9 on page 274 the reader will be asked to show that when $\mu = 1.4$ and $\sigma = 0.60$, the probability that the mean of a random sample of size 50 will exceed 1.6 is about 0.01. If the sample mean is 1.6 or less, we simply say that *the null hypothesis cannot be rejected* and, hence, that *the investigation does not confirm the sociologist's claim*. Note that we do not say that the sociologist is *wrong*—we merely say that *we were unable to show that he is right*. This is a negative sort of result and it may well call for a more thorough check of the sociologist's claim, perhaps, one based on a larger sample.

Referring again to the average-annual-income example on page 263, we could convert the criterion into that of a significance test by writing it as follows:

> *Reject the null hypothesis μ = $8,400 if the mean of a random sample of size 200 is less than or equal to $8,300 or greater than or equal to $8,500; reserve judgment if the mean falls between $8,300 and $8,500.*

So far as the rejection of the null hypothesis is concerned, the criterion has remained unchanged, and hence, the probability of a Type I error is still 0.0628. However, so far as its acceptance is concerned, we are now playing it safe by reserving judgment. Of course, this raises the question whether it is actually feasible to reserve judgment in this kind of example—if a decision one way or the other is required (say, to decide on what kind of stores to put into the shopping center), the developer would have to use a criterion like the original one on page 263.

Reserving judgment in a test of significance is similar to what happens in court proceedings where the prosecution does not have sufficient evidence to get a conviction, but where in one's mind it would be going too far to say that the defendant is definitely not responsible for the crime. Thus, whether we can afford to reserve judgment in any given situation will have to depend on the nature of the circumstances and the objectives of the problem.

Since the general problem of testing hypotheses and constructing statistical decision criteria often presents some difficulties to beginners, it will help to proceed systematically as outlined in the following five steps:

(a)   *We formulate a (null) hypothesis in such a way that the probability of a Type I error can be calculated.*

(b)   *We formulate an alternative hypothesis so that the rejection of the null hypothesis is equivalent to the acceptance of the alternative hypothesis.*

In the average-annual-income example the null hypothesis was $\mu = \$8,400$, and even though we did not say so specifically, the alternative hypothesis was $\mu \neq \$8,400$. (Presumably, the developer of the new shopping center is interested in knowing whether the true average might be higher and also whether it might be lower.) We refer to this kind of alternative as a *two-sided alternative,* since we shall want to reject the null hypothesis if the true value of μ is *on either side* of \$8,400. An illustration of where we use a *one-sided alternative* is provided by the traffic-ticket example. Here the null hypothesis is $\mu = 1.4$ tickets per year and the alternative hypothesis is $\mu > 1.4$, for this corresponds to the sociologist's claim. Had we been interested in checking the claim that persons over 65 actually average *fewer* traffic tickets than all age groups combined, we would have used the alternative hypothesis $\mu < 1.4$, and if we had been interested in checking whether their performance is *different* (better or worse) than that of all age groups combined, we would have used the alternative hypothesis $\mu \neq 1.4$. Beginners often find it difficult to decide upon an appropriate one-sided or two-sided alternative, and we want to stress the fact that there is no general rule—how one proceeds in a given example depends entirely on the nature of the problem.

Having formulated a null hypothesis and a suitable alternative, we then proceed with the following steps:

(c)   *We specify the probability with which we are willing to risk a Type I error; if possible, desired, or necessary, we may also make some specifications about the probabilities of Type II errors for given alternative values of the parameter with which we are concerned.*

(d)   *Using appropriate statistical theory we construct the test criterion.*

It is here that we depart from the average-annual-income example, where we first specified the criterion and then calculated the probabilities of committing Type I and Type II errors. In general, we reverse this procedure, specifying the probability of committing a Type I error (and, perhaps, Type II errors) and then choosing an appropriate test criterion.

The probability of committing a Type I error is generally referred to as the *level of significance* at which the test is being conducted, and it is denoted by the Greek letter α (*alpha*). Thus, if we write $\alpha = 0.05$ in connection with a decision criterion, this means that we risk a Type I error with a probability of 0.05. Of course, the decision whether to use $\alpha = 0.05$, $\alpha = 0.01$, or some other value, will have to depend on whatever consequences there may be to committing a Type I error in a particular problem. Generally speaking, the more serious the consequences of an error of this kind, the smaller the risk of committing it one is willing to take. Note that with reference to the courtroom procedures mentioned earlier, the level of significance expresses formally what we mean by the phrase "beyond a reasonable doubt."

So far as Step (d) is concerned, in the average-annual-income example we based the test criterion on the normal curve approximation of the sampling distribution of the mean. In general, it will have to depend on the *statistic* upon which we want to base the decision and on its sampling distribution. Looking back at the two examples of this chapter, we find that the construction of a test criterion depends also on the alternative hypothesis we happen to choose. In the average-annual-income example, the two-sided alternative $\mu \neq \$8,400$ led to a *two-sided test criterion* (or *two-tail test*), in which the null hypothesis is rejected for values of $\bar{x}$ that are either too large or too small; in the traffic-ticket example, the one-sided alternative $\mu > 1.4$ led to a *one-sided test criterion* (or *one-tail test*), in which the null hypothesis is rejected only for values of $\bar{x}$ that are too large. In general, a test is said to be *one-sided* or *two-sided* (*one-tailed* or *two-tailed*), depending on whether the null hypothesis is rejected only for values of the test statistic falling into one of the two tails of its sampling distribution, or for values falling into either tail of this distribution.

(e) *Finally, we specify whether the alternative to rejecting the hypothesis formulated in (a) is to accept it or to reserve judgment.*

As we saw in our examples, this will have to depend on the nature of the problem, possible consequences and risks, and whether a decision one way or the other must be reached. Quite often we accept a null hypothesis with the tacit hope that we are not exposing ourselves to excessively high risks of committing *serious* Type II errors.

The purpose of this discussion has been to present some of the basic problems connected with the testing of statistical hypotheses. Although the methods we have mentioned are *objective*, that is, two research workers analyzing the same data under the same conditions should arrive at identical results, their use does entail some arbitrary, or subjective, considerations. For instance, in the traffic-ticket example if was partly a subjective decision to use a sample of size 50 and to "draw the line" at 1.6. More generally, the choice of α, the probability of a Type I error, and perhaps also the probability of a Type II error, often denoted β (*beta*), must depend to some extent on the consequences of making wrong decisions. Although it may be very difficult to put exact "cash values" on all such eventualities, they must nevertheless be considered, at least indirectly or tacitly, in choosing suitable test criteria.

Before we go into the various special tests treated in the remainder of this chapter, let us point out that the preceding discussion is not limited to tests concerning means. The concepts we have introduced apply equally well to tests of hypotheses concerning proportions, population standard deviations, the randomness of samples, relationships among several variables, and so on.

EXERCISES

1. Suppose that a team of doctors is examining applicants to see whether they are physically fit to become astronauts. What type of error would they commit if they erroneously accepted the hypothesis that an applicant is physically fit to become an astronaut? What type of error would they commit if they erroneously rejected the hypothesis that an applicant is physically fit to become an astronaut?

2. Suppose we want to test the hypothesis that an antipollution device for cars is effective. Explain under what conditions we would be committing a Type I error and under what conditions we would be committing a Type II error.

3. An airline wants to test the hypothesis that at least 60 per cent of its passengers enjoy being served a cocktail on transcontinental flights. What type of error is committed if this hypothesis is erroneously rejected? What type of error is committed if this hypothesis is erroneously accepted?

4. Whether an error is a Type I error or a Type II error depends on how we formulate the hypothesis we want to test. To illustrate this, rephrase the hypothesis of Exercise 3 so that the Type I error becomes a Type II error, and vice versa.

5. A professor of education is concerned with the effectiveness of a programmed teaching technique.

(a) What hypothesis is he testing, if he is committing a Type I error when he erroneously concludes that the programmed teaching technique is effective?

(b) What hypothesis is he testing, if he is committing a Type II error when he erroneously concludes that the programmed teaching technique is effective?

6. Verify the values given in the table on page 267 by duplicating the method used in the text to calculate the probability of a Type II error for $\mu = \$8,550$.

7. Suppose that in the average-annual-income example the criterion is changed so that the hypothesis $\mu = \$8,400$ is accepted if the mean of a random sample of size 200 falls between \$8,280 and \$8,520, while otherwise the hypothesis is rejected.

(a) Calculate the probability of committing a Type I error with this new criterion.

(b) Calculate the probabilities of committing Type II errors with this new criterion for $\mu = \$8,150, \$8,200, \$8,250, \$8,300, \$8,350, \$8,450, \$8,500, \$8,550, \$8,600,$ and \$8,650.

(c) Draw a graph of the OC-curve of this criterion.

8. Suppose that in the average-annual-income example the sample size is increased from 200 to 250, while the criterion remains as stated on page 263.

(a) Calculate the probability of erroneously rejecting the hypothesis $\mu = \$8,400$ with this procedure.

(b) Calculate the probabilities of erroneously accepting the hypothesis $\mu = \$8,400$ with this procedure, when actually $\mu = \$8,200, \$8,250, \$8,300, \$8,350, \$8,450, \$8,500, \$8,550,$ and \$8,600.

(c) Draw a graph of the OC-curve of this criterion.

9. Verify for the traffic-ticket example on page 270 that the probability of committing a Type I error is 0.01 (when $\sigma = 0.60$).

10. The average drying time of a manufacturer's paint is 20 minutes. Investigating the effectiveness of a modification in the chemical composition of his paint, the manufacturer wants to test the null hypothesis $\mu = 20$ minutes against a suitable alternative, where μ is the average drying time of the new paint.

(a) What alternative hypothesis should the manufacturer use if he does not want to make the modification in the chemical composition of the paint unless it is definitely proven superior?

(b) What alternative hypothesis should the manufacturer use if the new process is actually cheaper and he wants to make the modification unless it increases the drying time of the paint?

11. A city police department is considering replacing the tires on its cars with radial tires. If μ_1 is the average number of miles they get out of

their old tires and μ_2 is the average number of miles they will get out of the new tires, the null hypothesis they shall want to test is $\mu_1 = \mu_2$.

(a) What alternative hypothesis should they use if they do not want to buy the radial tires unless they are definitely proven to give a better mileage? In other words, the burden of proof is put on the radial tires and the old tires are to be kept unless the null hypothesis can be rejected.

(b) What alternative hypothesis should they use if they are anxious to get the new tires (which have some other nice features) unless they actually give a poorer mileage than the old tires? Note that now the burden of proof is on the old tires, which will be kept only if the null hypothesis can be rejected.

(c) What alternative hypothesis would they have to use so that the rejection of the null hypothesis can lead either to keeping the old tires or to buying the new ones?

12. The proportion of uncollectible charge accounts of a certain department store is $p = 0.008$, and to see whether this figure can be reduced, the store's credit manager tries a more threatening kind of collection letter on a random sample of the store's delinquent accounts.

(a) Against what alternative should he test the null hypothesis $p = 0.008$ if he is a very careful man and does not want to introduce the more threatening collection letters unless they really work?

(b) Against what alternative should he test the null hypothesis $p = 0.008$ if he wants to use the new letter unless it actually makes things worse?

10.7 Tests Concerning Means

In Sections 10.5 and 10.6 we used tests concerning means to illustrate the basic principles of hypothesis testing; now we shall consider more generally the problem of testing the null hypothesis that the mean of a population equals some specified value μ_0 against an appropriate alternative. Usually, the alternative hypothesis is of the form $\mu < \mu_0$, $\mu > \mu_0$, or $\mu \neq \mu_0$; that is, it specifies that the population mean is less than, greater than, or not equal to the value assumed under the null hypothesis. Thus, if we wanted to test the hypothesis that the mean distance required to stop a car going 20 miles per hour is 25 feet against the alternative hypothesis that it takes more than 25 feet, we would begin by writing

$$\textit{Null Hypothesis:} \qquad \mu = 25 \ \text{feet}$$
$$\textit{Alternative Hypothesis:} \quad \mu > 25 \ \text{feet}$$

To abbreviate our notation, we often use the symbols H_0 and H_A (or H_1) to denote the null hypothesis and the alternative.

Next, we specify the level of significance as $\alpha = 0.01$, $\alpha = 0.05$, or some other value as the case may be, and after that we depart slightly from the procedure used in the examples of the preceding section. In both the average-annual-income and traffic-ticket examples, we formulated the test criterion in terms of possible values of $\bar{x}$; now we shall base it on the statistic

$$z = \frac{\bar{x} - \mu_0}{\sigma/\sqrt{n}}$$

which simply means that we work in *standard units*. The reason for this is easy to explain: *using standard units, we can formulate criteria which are applicable to a great variety of problems, and not just one.* Note that in the above formula for z the population standard deviation is assumed to be known while μ_0 is the value of the population mean assumed under the null hypothesis.

If we approximate the sampling distribution of the mean, as before, with a normal distribution, we can now use the test criteria shown in Figure 10.7; depending on the choice of the alternative hypothesis, the dividing line (or lines) of the criterion is $-z_\alpha$ or z_α for the *one-sided* alternatives, and $-z_{\alpha/2}$ and $z_{\alpha/2}$ for the *two-sided* alternative. As we explained on page 208, z_α and $z_{\alpha/2}$ are values for which the area *to their right* under the standard normal distribution is α and $\alpha/2$, respectively. Symbolically, we can formulate these criteria as follows:

Alternative Hypothesis	Reject the Null Hypothesis if	Accept the Null Hypothesis or Reserve Judgment if
$\mu < \mu_0$	$z < -z_\alpha$	$z \geq -z_\alpha$
$\mu > \mu_0$	$z > z_\alpha$	$z \leq z_\alpha$
$\mu \neq \mu_0$	$z < -z_{\alpha/2}$ or $z > z_{\alpha/2}$	$-z_{\alpha/2} \leq z \leq z_{\alpha/2}$

The most commonly used values of the level of significance are 0.05 and 0.01. If $\alpha = 0.05$, the dividing lines of the criteria are -1.64 or 1.64 for

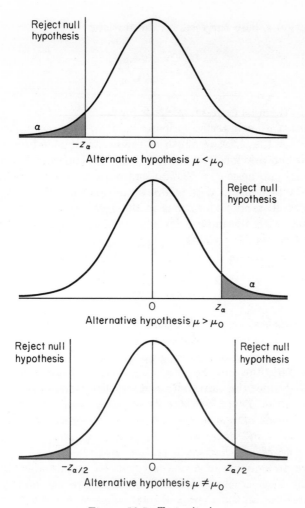

Figure 10.7. Test criteria.

the one-sided alternatives, and -1.96 and 1.96 for the two-sided alternative; if $\alpha = 0.01$, the dividing lines of the criteria are -2.33 or 2.33 for the one-sided alternatives, and -2.58 and 2.58 for the two-sided alternative. All these values were obtained earlier when we discussed confidence intervals for μ.

A serious shortcoming of the tests just described is that they require knowledge of σ, the standard deviation of the population from which the sample is obtained. Since σ is unknown in most practical applications, we usually have no choice but to replace it with an estimate in the formula for z. If n is large (30 or more), we usually substitute for σ the sample standard deviation s, base the tests on the statistic

$$\blacktriangle \qquad z = \frac{\bar{x} - \mu_0}{s/\sqrt{n}} \qquad \blacktriangle$$

and refer to them as *large-sample tests* for μ.

To illustrate such tests, suppose than an oceanographer wants to check whether the average depth of the ocean in a certain area is 42.0 fathoms, as had previously been recorded. More specifically, he wants to test the null hypothesis $\mu = 42.0$ fathoms against the alternative hypothesis $\mu \neq 42.0$ fathoms at the level of significance $\alpha = 0.05$. Thus, taking soundings at 40 random locations in the given area, he gets a mean of 43.7 fathoms with a standard deviation of 5.1, so that substitution into the above formula for z yields

$$z = \frac{43.7 - 42.0}{5.1/\sqrt{40}} = 2.11$$

Since this exceeds $z_{0.025} = 1.96$, the null hypothesis will have to be rejected, and it seems that the original figure was a bit too low. It is of interest to note, however, that if the oceanographer had chosen the level of significance $\alpha = 0.01$, the null hypothesis $\mu = 42.0$ fathoms could *not* have been rejected, since the value obtained for z lies between $-z_{0.005} = -2.58$ and $z_{0.005} = 2.58$. *This illustrates the very important point that the level of significance should always be specified before any statistical tests are actually performed. This will spare us the temptation of later choosing a level of significance which may happen to suit our purpose.*

To give an example of a one-tail test, let us consider the following study made by a trucking firm. To check a tire manufacturer's claim that his tires last on the average 24,000 miles, they put 100 of these tires on company trucks, getting a mean of 23,720 miles and a standard deviation of 1,840 miles. Their null hypothesis is $\mu = 24,000$ miles and their alternative hypothesis is $\mu < 24,000$, since they are interested mainly in determining whether the tires might not be as good as claimed. Let us suppose also that their level of significance is $\alpha = 0.01$. Substituting the given values of $\bar{x}$, μ_0, n, and s into the formula for z, they obtain

$$z = \frac{23,720 - 24,000}{1,840/\sqrt{100}} = -1.52$$

and since this is *not* less than $z_{0.01} = -2.33$, they find that *the null hypothesis cannot be rejected.* Even though the tires did not perform quite as well as claimed, the difference between 23,720 and 24,000 is sufficiently small here to be attributed to chance.

If n is small (less than 30), we can proceed as on page 254, where we based small-sample confidence intervals for μ on the t distribution. Assuming again that the population from which we are sampling has roughly the shape of a normal distribution, we now base our decision on the statistic

$$\blacktriangle \qquad\qquad t = \frac{\bar{x} - \mu_0}{s/\sqrt{n}} \qquad\qquad \blacktriangle$$

where μ_0 is the value of the population mean specified under the null hypothesis. Observe that this formula for t is identical with the one for z on page 278, where we substituted s for σ and, hence, restricted the method based on normal curve theory to large samples.

To perform tests based on the t statistic, we have only to refer to Figure 10.7 or the table on page 276 with z replaced by t, and z_α and $z_{\alpha/2}$ replaced by t_α and $t_{\alpha/2}$, respectively. As we explained on page 254, t_α and $t_{\alpha/2}$ are values for which the area *to their right* under the t distribution is equal to α and $\alpha/2$, respectively. All these dividing lines of the test criteria may be obtained from Table II, *with the number of degrees of freedom equaling $n - 1$.*

To illustrate this kind of *small-sample test for μ*, suppose that we want to investigate a vacuum cleaner manufacturer's claim that the noise level of his latest model is at most 75 decibels, and that we have at our disposal readings of the sound intensity produced by 12 of his machines. The null hypothesis is $\mu = 75$, the alternative hypothesis is $\mu > 75$, and let us say that the level of significance is to be $\alpha = 0.05$. If, furthermore, the mean of the 12 readings is 77.3 and their standard deviation is 3.6, substitution of these values together with $n = 12$ and $\mu_0 = 75$ into the formula for t yields

$$t = \frac{77.3 - 75}{3.6/\sqrt{12}} = 2.21$$

Since this exceeds 1.796, the value of $t_{0.050}$ for $12 - 1 = 11$ degrees of freedom, we find that *the null hypothesis will have to be rejected*. In other words, the vacuum cleaners *are* noisier than claimed.

10.8 Differences Between Means

There are many applied problems in which we must decide whether an observed difference between two sample means can be attributed to chance or whether it is an indication of the fact that the samples came from populations with unequal means. We may want to decide, for instance, whether there is really a difference in the performance of two kinds of batteries, if a sample of one kind lasted on the average for 65.3 hours while a sample of another kind lasted on the average for 67.9 hours. Similarly, we may want to decide on the basis of samples whether men can perform a given task faster than women, whether the students in one school have a higher average I.Q. than those of another, whether one ceramic product is more brittle than another, and so on.

The method we shall employ to test whether an observed difference between two sample means can be attributed to chance is based on the following theory. *If $\bar{x}_1$ and $\bar{x}_2$ are the means of two large independent random samples of size n_1 and n_2, the sampling distribution of the statistic $\bar{x}_1 - \bar{x}_2$ can be approximated closely with a normal curve having the mean $\mu_1 - \mu_2$ and the standard deviation*

$$\sqrt{\frac{\sigma_1^2}{n_1} + \frac{\sigma_2^2}{n_2}}$$

where μ_1 and μ_2 are the means of the populations from which the two samples were obtained and σ_1 and σ_2 are their standard deviations. The above formula is also referred to as the standard error of the difference between two means.

By "independent" samples we mean that the selection of one sample is in no way affected by the selection of the other. Thus, the theory does not apply to "before and after" kinds of comparisons, nor does it apply, say, to the comparison of the I.Q.'s of husbands and wives. A special method for handling this kind of problem is given in Exercise 22 on page 286.

One difficulty in applying the above theory is that in most practical situations σ_1 and σ_2 are unknown. However, if we limit ourselves to *large samples* (neither n_1 nor n_2 should be less than 30), it is reasonable to substitute the sample standard deviations s_1 and s_2 for σ_1 and σ_2, and base the test of the null hypothesis $\mu_1 = \mu_2$ on the statistic

$$z = \frac{\bar{x}_1 - \bar{x}_2}{\sqrt{\dfrac{s_1^2}{n_1} + \dfrac{s_2^2}{n_2}}}$$

Note that this formula for z is obtained by subtracting from $\bar{x}_1 - \bar{x}_2$ the mean of its sampling distribution, which under the null hypothesis is $\mu_1 - \mu_2 = 0$, and then dividing by the standard deviation of its sampling distribution with s_1 and s_2 substituted for σ_1 and σ_2.

Depending on whether the alternative hypothesis is $\mu_1 < \mu_2$, $\mu_1 > \mu_2$, or $\mu_1 \neq \mu_2$, the criteria on which we base the actual tests are again the ones shown in Figure 10.7 and also in the table on page 276 with μ_1 and μ_2 substituted for μ and μ_0. To illustrate this large-sample technique, suppose that we want to check whether the average weekly food expenditures of families with two children is the same in two cities. The data which we have at our disposal show that 80 families in one city averaged $42.61 with a standard deviation of $6.85, while 100 families in the other city averaged $39.12 with a standard deviation of $5.92. To test the null hypothesis $\mu_1 = \mu_2$ (where μ_1 and μ_2 are the respective true average weekly food expenditures) against the alternative hypothesis $\mu_1 \neq \mu_2$, say, at the level of significance $\alpha = 0.05$, we have only to substitute all the given values into the formula for z, getting

$$ z = \frac{42.61 - 39.12}{\sqrt{\dfrac{(6.85)^2}{80} + \dfrac{(5.92)^2}{100}}} = 3.61 $$

Since this exceeds $z_{0.025} = 1.96$, *the null hypothesis will have to be rejected;* in other words, the sample data confirm that there *is* a difference between the average weekly food expenditures of families with two children in the two cities.

The Student-t distribution also provides criteria for *small-sample tests* concerning differences between two means. To use these criteria, we have to assume that the two samples come from populations which can be approximated closely with normal distributions and which, furthermore, have *equal variances*. Specifically, we base the test of the null hypothesis $\mu_1 = \mu_2$ on the statistic

$$ \blacktriangle \qquad t = \frac{\bar{x}_1 - \bar{x}_2}{\sqrt{\dfrac{\Sigma\,(x_1 - \bar{x}_1)^2 + \Sigma\,(x_2 - \bar{x}_2)^2}{n_1 + n_2 - 2} \cdot \left(\dfrac{1}{n_1} + \dfrac{1}{n_2}\right)}} \qquad \blacktriangle $$

where $\Sigma\,(x_1 - \bar{x}_1)^2$ is the sum of the squared deviations from the mean for the first sample while $\Sigma\,(x_2 - \bar{x}_2)^2$ is the sum of the squared deviations from the mean for the second sample. Note that since, by definition, $\Sigma\,(x_1 - \bar{x}_1)^2 = (n_1 - 1)s_1^2$ and $\Sigma\,(x_2 - \bar{x}_2)^2 = (n_2 - 1)s_2^2$, the above formula for t can be simplified somewhat when the two sample variances (or standard deviations) have already been calculated from the data.

Under the given assumptions, it can be shown that the sampling distribution of this t statistic is the Student-t distribution with $n_1 + n_2 - 2$ degrees of freedom. Depending on whether the alternative hypothesis is

$\mu_1 < \mu_2$, $\mu_1 > \mu_2$, or $\mu_1 \neq \mu_2$, the criteria on which we base the *small-sample tests for the significance of the difference between two means* are again the ones shown in Figure 10.7 and also in the table on page 276, with t substituted throughout for z, and μ_1 and μ_2 substituted for μ and μ_0.

To illustrate this kind of test, suppose that someone wants to compare the fat content of two kinds of ice cream, Brand A and Brand B, on the basis of the following two random samples of size five:

> Brand A: 13.2, 14.0, 12.9, 13.5, and 13.4 per cent
> Brand B: 13.5, 13.4, 14.1, 14.5, and 14.0 per cent

The means of these two samples are 13.4 and 13.9, and it is desired to test the null hypothesis $\mu_1 = \mu_2$ (where the subscripts 1 and 2 refer to Brand A and Brand B) against the alternative hypothesis $\mu_1 \neq \mu_2$ at the level of significance $\alpha = 0.05$. To calculate t in accordance with the above formula, the following two quantities will first have to be determined:

$$\Sigma (x_1 - \bar{x}_1)^2 = (13.2 - 13.4)^2 + (14.0 - 13.4)^2 + (12.9 - 13.4)^2$$
$$+ (13.5 - 13.4)^2 + (13.4 - 13.4)^2$$
$$= 0.66$$

and

$$\Sigma (x_2 - \bar{x}_2)^2 = (13.5 - 13.9)^2 + (13.4 - 13.9)^2 + (14.1 - 13.9)^2$$
$$+ (14.5 - 13.9)^2 + (14.0 - 13.9)^2$$
$$= 0.82$$

Substitution of these values together with $n_1 = 5$, $n_2 = 5$, $\bar{x}_1 = 13.4$, and $\bar{x}_2 = 13.9$ into the formula for t, yields

$$t = \frac{13.4 - 13.9}{\sqrt{\dfrac{0.66 + 0.82}{5 + 5 - 2} \cdot \left(\dfrac{1}{5} + \dfrac{1}{5}\right)}} = -1.84$$

and since this value falls between -2.306 and 2.306, where 2.306 is the value of $t_{0.025}$ for $5 + 5 - 2 = 8$ degrees of freedom, it follows that *the null hypothesis cannot be rejected*. If the person conducting the investigation suspects nevertheless that the second kind of ice cream has a higher average fat content, he would be wise to repeat the experiment *with larger samples*. Note also that if the person had wanted to compare the fat content of three or four ice creams instead of two, he would have had to use a more general method, which we shall discuss in Chapter 13.

EXERCISES

1. A criminologist claims that a person convicted of counterfeiting spends on the average 19.4 months in jail. A law student who feels that this figure seems high, takes a random sample of 35 such cases from court files and gets a mean of 18.2 months spent in jail with a standard deviation of 3.6. What can he conclude from these figures at the level of significance $\alpha = 0.01$?

2. In a labor-management discussion it was brought up that it takes the average worker (employed in a very large plant) 41.3 minutes to get to work. Is this figure substantiated by a survey in which a sample of 60 workers took on the average 39.7 minutes to get to work with a standard deviation of 8.2 minutes? Use a level of significance of 0.05.

3. According to the norms established for a test designed to measure a person's ability to think logically, college sophomores should average 72.6 points with a standard deviation of 10.8. This test is given to a group of 40 college sophomores who have just completed a required course in mathematics which is supposed to sharpen their general powers of reasoning. What can one conclude about the effectiveness of the course if these students averaged 77.8 on the test? Use $\alpha =$ 0.01.

4. An ambulance service claims that it takes it on the average 8.9 minutes to reach its destination in emergency calls. To check on this claim, the agency which licenses ambulances services has them timed on 50 emergency calls, getting a mean of 9.3 minutes with a standard deviation of 1.6 minutes. What can they conclude at the level of significance $\alpha = 0.05$?

5. A process for making steel pipe is under control if the diameter of the pipe has a mean of $\mu = 3.0$ inches with a standard deviation of 0.015 inches. In order to test whether the process is under control (so far as μ is concerned), a random sample of size 30 is taken four times a day, and it is decided in each case on the basis of $\bar{x}$ whether to accept or reject the null hypothesis $\mu = 3.0$. Using a level of significance of $\alpha = 0.01$, should they adjust the process if they get a mean of 3.0078?

6. A real estate broker who is trying to sell a piece of property to a restaurant chain assures them that during the summer months *on the average* 4,000 cars pass by the property each day. Making their own investigation, the management of the restaurant chain obtains a mean of 3,694 cars and a standard deviation of 353 cars for 32 days. Using the level of significance $\alpha = 0.01$, what can they conclude about the real estate broker's claim?

7. A survey conducted by an insurance company showed that a random sample of 200 private passenger cars were driven on the average 12,840 miles a year with a standard deviation of 2,670 miles. Use this information to test the hypothesis that the average private passenger car is driven 12,500 miles a year against the alternative that the average is higher
    (a) at the level of significance $\alpha = 0.05$;
    (b) at the level of significance $\alpha = 0.01$.

8. A random sample of 6 steel beams has a mean compressive strength of 58,392 psi (pounds per square inch) with a standard deviation of 648 psi. Use this information and the level of significance $\alpha = 0.05$ to test whether the true average compressive strength of the steel from which this sample came is 58,000 psi.

9. A special diet given to eight overweight women helped them to lose 10, 15, 12, 18, 19, 11, 14, and 13 pounds within a period of three months. Use a level of significance of 0.01 to test the claim that on the average the diet will help an overweight woman to lose 10.5 pounds within three months. Formulate the alternative hypothesis so that the burden of proof will be on the diet.

10. A reading teacher wants to determine whether a certain student has an average reading speed of $\mu = 800$ words per minute or whether his average reading speed is less than 800 words per minute. What can he conclude at the level of significance $\alpha = 0.05$ if in six minutes the student read, respectively, 780, 800, 790, 750, 750, and 810 words?

11. In the past, a golfer has averaged 86 on a certain course with a standard deviation of 2.6. If, with a new set of clubs, he averages 81 over four rounds, can we really say that the new clubs have improved his game or could it be just chance? Use a level of significance of 0.05.

12. The yield of alfalfa from six test plots is 1.1, 2.1, 1.8, 1.0, 1.7, and 1.3 tons per acre. Test at the level of significance $\alpha = 0.05$ whether this supports the contention that the true average yield for this kind of alfalfa is 1.4 tons per acre.

13. A soft-drink vending machine is set to dispense 6 ounces per cup. If the machine is tested 8 times, yielding a mean cup fill of 5.8 ounces with a standard deviation of 0.16 ounces, test at the level of significance $\alpha = 0.05$ whether this apparent underfill is significant.

14. To compare high school seniors' knowledge of current events in two different school districts, samples of 60 seniors from each district were given a special test. If those of the first district obtained an average score of 78 with a standard deviation of 5, while those of the second district obtained an average score of 74 with a standard deviation of 4, test at the level of significance $\alpha = 0.05$ whether the difference between the two means is significant.

15. A company claims that its light bulbs are superior to those of a competitor on the basis of a study which showed that a sample of 40 of its bulbs had an average "lifetime" of 628 hours of continuous use with a standard deviation of 27 hours, while a sample of 30 bulbs made by the competitor had an average "lifetime" of 619 hours of continuous use with a standard deviation of 25 hours. Check, at the level of significance $\alpha = 0.05$, whether this claim is justified.

16. Sample surveys conducted in a certain large county in 1950 and again in 1965 showed that in 1950 the average height of 400 ten-year-old boys was 53.2 inches with a standard deviation of 2.4 inches, while in 1965 the average height of 500 ten-year-old boys was 53.9 inches with a standard deviation of 2.5 inches. Use the level of significance $\alpha = 0.05$ to test whether this increase in height is significant.

17. Suppose that in Exercise 16 we had been interested in knowing whether the true average increase in height is at least 0.5 inches. Thus, rework the exercise and test the hypothesis $\mu_2 - \mu_1 = 0.5$ against the alternative hypothesis $\mu_2 - \mu_1 > 0.5$ at the level of significance $\alpha = 0.05$. (*Hint:* substitute $\bar{x}_1 - \bar{x}_2 + 0.5$ for $\bar{x}_1 - \bar{x}_2$ in the numerator of the formula for z.)

18. In the comparison of two kinds of paint, a consumer-testing service found that four one-gallon cans of Brand A cover on the average 514 square feet with a standard deviation of 32 square feet, while four one-gallon cans of Brand B cover on the average 487 square feet with a standard deviation of 27 square feet. Use the level of significance $\alpha = 0.05$ to test whether the difference between the two sample means is significant.

19. Measurements performed on random samples of two kinds of cigarettes yielded the following results on their nicotine content (in milligrams):

$$\text{Brand A:} \quad 21.4, \quad 23.6, \quad 24.8, \quad 22.4, \quad 26.3$$
$$\text{Brand B:} \quad 22.4, \quad 27.7, \quad 23.5, \quad 29.1, \quad 25.8$$

Use the level of significance $\alpha = 0.01$ to check on the claim that Brand B has a higher average nicotine content than Brand A.

20. Coal produced by different mines may have different heat-producing capacities. Test at the level of significance $\alpha = 0.05$ whether the difference between the means of the following samples from two mines (data are in millions of calories per ton) is significant:

$$\text{Mine 1:} \quad 8,400, \quad 7,880, \quad 8,420, \quad 7,950, \quad 8,250$$
$$\text{Mine 2:} \quad 7,820, \quad 8,040, \quad 7,800, \quad 7,790, \quad 7,850$$

21. Six guinea pigs which were injected with 0.5 mg of a tranquilizer took 11, 13, 9, 14, 15, and 13 seconds to fall asleep, while six guinea pigs which were injected with 1.5 mg of the tranquilizer took 10, 5, 8, 9, 6, and 10 seconds to fall asleep. Use a level of significance of 0.05 to test the null hypothesis that the difference in dosage has no effect.

22. If we want to study the effectiveness of a diet on the basis of weights "before and after," or if we want to study whatever differences there may be between the I.Q.'s of husbands and wives, the methods of Section 10.8 cannot be used. The samples are not independent; in fact, in each case the data are *paired*. To handle problems of this kind, we work with the differences of the paired data (retaining their signs), and test whether these differences may be looked upon as a sample from a population for which $\mu = 0$. If the sample is small, we use the t test on page 279, and if the sample is large, we use the test on page 278. Apply this technique to the following data designed to test whether there is a difference in the number of miles obtained per gallon with two kinds of gasoline:

	Gasoline I	Gasoline II
Test car 1	14.1	14.2
Test car 2	15.2	15.5
Test car 3	12.5	12.5
Test car 4	13.0	13.6
Test car 5	15.5	15.8
Test car 6	14.4	14.3
Test car 7	12.8	13.0
Test car 8	13.5	13.7

Use the level of significance $\alpha = 0.05$.

23. In a study of the effectiveness of physical exercise in weight reduction, a group of 16 persons engaged in a prescribed program of physical exercise for one month showed the following results:

Weight Before (Pounds)	Weight After (Pounds)	Weight Before (Pounds)	Weight After (Pounds)
209	196	170	164
178	171	153	152
169	170	183	179
212	207	165	162
180	177	201	199
192	190	179	173
158	159	243	231
180	180	144	140

Use the level of significance $\alpha = 0.01$ to test the null hypothesis that the prescribed program of exercise is *not* effective (against a suitable alternative).

24. To determine the effectiveness of an industrial safety program, the following data were collected over a period of one year on the average weekly loss of man hours due to accidents in 12 plants "before and after" the program was put into operation:

$$
\begin{array}{llll}
37 \text{ and } 28, & 72 \text{ and } 59, & 26 \text{ and } 24, & 125 \text{ and } 120 \\
45 \text{ and } 46, & 54 \text{ and } 43, & 13 \text{ and } 15, & 79 \text{ and } 75 \\
12 \text{ and } 18, & 34 \text{ and } 29, & 39 \text{ and } 35, & 26 \text{ and } 24
\end{array}
$$

Use the level of significance $\alpha = 0.01$ to check whether the null hypothesis that the safety program is *not* effective can be rejected (against a suitable alternative).

BIBLIOGRAPHY

An informal introduction to interval estimation is given under the heading of "How to be precise though vague," in

MORONEY, M. J., *Facts from Figures.* London: Penguin Books, Inc., 1956.

A discussion of the theoretical foundation of the t distribution as well as other mathematical details omitted in this book may be found in most textbooks on mathematical statistics. Some of the theory of Bayesian estimation pertaining to Section 10.4 is discussed on pages 441–442 of

SCHLAIFER, R., *Probability and Statistics for Business Decisions*, New York: McGraw-Hill, Inc., 1959

and in Section 5.1 of

LINDLEY, D. V., *Introduction to Probability and Statistics from a Bayesian Viewpoint, Part 2.* Cambridge: Cambridge University Press, 1965.

11

Inferences Concerning

Standard Deviations

11.1 The Estimation of σ

Many of the methods described in Chapter 10 required knowledge of the population standard deviation σ, a quantity which, unfortunately, is often unknown. We got around this difficulty in the large-sample confidence interval for μ on page 253 and in the large-sample tests concerning μ on page 278 by substituting for σ the sample standard deviation s. Of course, this raises the question whether it is at all reasonable to make such a substitution, and to answer this question even in part, we shall have to investigate the general problem of estimating population standard deviations.

Although the sample standard deviation is by far the most popular estimate of the standard deviation of a population, there are other methods, other statistics, which are sometimes used to get quick estimates of σ. We already saw in Exercise 11 on page 209 that an estimate of σ can be obtained from the graph of a distribution on arithmetic probability paper, provided the distribution has roughly the shape of a normal curve. Other short-cut estimates based, respectively, on the sample range and on certain fractiles of a distribution are mentioned in Exercises 8 and 9 on pages 292 and 293.

In the remainder of this section we shall limit our discussion to problems which arise when we use s as an estimate of σ (or s^2 as an estimate of σ^2); in particular, we shall establish confidence intervals for σ based on sample standard deviations. The theory on which these confidence intervals are based assumes that the population from which we are sampling has roughly the shape of a normal distribution. It can then be shown that the statistic

▲
$$\chi^2 = \frac{(n-1)s^2}{\sigma^2}$$
▲

called "chi-square," has as its sampling distribution a well-known continuous distribution, the *chi-square distribution*. The mean of this distribution is $n-1$ and, as in connection with the t distribution, we refer to this
quantity as the *number of degrees of freedom*, or simply the *degrees of freedom*. An example of a chi-square distribution is shown in Figure 11.1; in

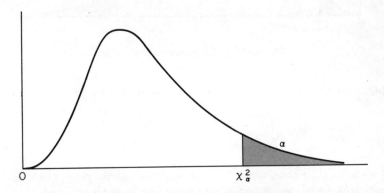

Figure 11.1. Chi-square distribution.

contrast to the normal and t distributions, its domain is restricted to the
nonnegative real numbers.

   Analogous to z_α and t_α, we define χ^2_α as the value for which the area *to
its right* under the chi-square distribution is equal to α. Thus, $\chi^2_{\alpha/2}$ is such
that the area *to its right* under the curve is $\alpha/2$, while $\chi^2_{1-\alpha/2}$ is such that
the area *to its left* under the curve is $\alpha/2$ (see also Figure 11.2). We made
this distinction because the chi-square distribution is *not* symmetrical,
and we shall need values corresponding to areas in either tail of the dis-
tribution. Among others, values of $\chi^2_{0.995}$, $\chi^2_{0.975}$, $\chi^2_{0.025}$, and $\chi^2_{0.005}$ are given
in Table III at the end of the book for 1, 2, 3, ..., and 30 degrees of
freedom.

   Referring to Figure 11.2, it is apparent that we can assert with a probability of $1-\alpha$ that a random variable having the chi-square distribution
will assume a value between $\chi^2_{1-\alpha/2}$ and $\chi^2_{\alpha/2}$. Making use of this fact in
connection with the χ^2 statistic given above, we can assert with a probability of $1-\alpha$ that

$$\chi^2_{1-\alpha/2} < \frac{(n-1)s^2}{\sigma^2} < \chi^2_{\alpha/2}$$

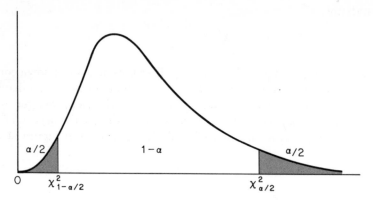

Figure 11.2. Chi-square distribution.

Applying some algebra, we can rewrite this inequality as

$$\blacktriangle \qquad \sqrt{\frac{(n-1)s^2}{\chi^2_{\alpha/2}}} < \sigma < \sqrt{\frac{(n-1)s^2}{\chi^2_{1-\alpha/2}}} \qquad \blacktriangle$$

and we have thus obtained a $1 - \alpha$ *confidence interval for* σ. Omitting the square root signs, we obtain a corresponding confidence interval for the population variance σ^2.

To illustrate the use of this confidence interval formula, let us return to the example on page 279, in which we were concerned with the noise level of a certain kind of vacuum cleaner. We had $s = 3.6$ decibels for the 12 readings, and since $\chi^2_{0.975}$ and $\chi^2_{0.025}$ for $12 - 1 = 11$ degrees of freedom equal 3.816 and 21.920, respectively, substitution into the formula yields*

$$\sqrt{\frac{11(3.6)^2}{21.920}} < \sigma < \sqrt{\frac{11(3.6)^2}{3.816}}$$
$$\sqrt{6.50} < \sigma < \sqrt{37.4}$$
$$2.55 < \sigma < 6.12$$

This is a 0.95 confidence interval for the population standard deviation, and it provides us with information about the *true variability* of the noise produced by the given kind of vacuum cleaner.

The confidence interval we have just described is often referred to as a *small-sample* confidence interval; it is used mainly when n is small and, of

* We used $\chi^2_{0.975}$ and $\chi^2_{0.025}$ so that the probability in each tail of the distribution is 0.025, and, hence, the degree of confidence is 0.95.

course, only when one can assume that the shape of the population is close to that of a normal curve. Otherwise, we make use of the fact that *for large samples* the sampling distribution of s can be approximated with a normal distribution having the mean σ and the standard deviation $\sigma/\sqrt{2n}$ (see Exercise 7 on page 236). In this connection, $\sigma/\sqrt{2n}$ is referred to as the *standard error* of s. We can now assert with a probability of $1 - \alpha$ that

$$s - z_{\alpha/2}\frac{\sigma}{\sqrt{2n}} < \sigma < s + z_{\alpha/2}\frac{\sigma}{\sqrt{2n}}$$

and simple algebra leads to the following $1 - \alpha$ *large-sample confidence interval for the population standard deviation* σ:

$$\blacktriangle \qquad \frac{s}{1 + \dfrac{z_{\alpha/2}}{\sqrt{2n}}} < \sigma < \frac{s}{1 - \dfrac{z_{\alpha/2}}{\sqrt{2n}}} \qquad \blacktriangle$$

Referring to the example on page 249, where we were interested in the increase of the pulse rate of astronauts performing a certain task, we substitute $n = 32$ and $s = 5.15$, and for $1 - \alpha = 0.95$ we obtain

$$\frac{5.15}{1 + \dfrac{1.96}{\sqrt{64}}} < \sigma < \frac{5.15}{1 - \dfrac{1.96}{\sqrt{64}}}$$

$$4.14 < \sigma < 6.82$$

We can thus assert with a probability of 0.95 that the interval from 4.14 to 6.82 beats per minute contains σ, the true standard deviation of the increase of the pulse rate of astronauts performing the given task.

EXERCISES

1. Referring to the data of Exercise 17 on page 261, construct a 0.95 confidence interval for σ, the true standard deviation of the amount of suspended benzene-soluble organic matter in the air at the location where the experiment was conducted.

2. Referring to Exercise 19 on page 261, construct a 0.99 confidence interval for σ, which is a measure of the inherent precision of the chemist's method of determining the melting point of tin.

3. Referring to Exercise 20 on page 261, construct a 0.95 confidence interval for σ, the true standard deviation of the blowing time of the given kind of fuse.

4. Referring to Exercise 21 on page 262, construct a 0.99 confidence interval for σ, which measures the true variability in the number of times a "senior citizen" visits a physician per year.

5. Referring to Exercise 1 on page 258, construct a 0.95 large-sample confidence interval for σ, the true standard deviation of the length of time it takes a technician to assemble the given computer component.

6. Referring to Exercise 5 on page 259, construct a 0.98 large-sample confidence interval for σ, which measures the true variability in the cost of the given body repairs.

7. Referring to Exercise 6 on page 259, construct a 0.99 large-sample confidence interval for σ, the true standard deviation of fifth graders in the given school district in the reading achievement test.

8. When dealing with very small samples, good estimates of the population standard deviation can often be obtained on the basis of the sample range (the largest sample value minus the smallest). Such quick estimates of σ are given by the sample range divided by the divisor d, which depends on the size of the sample; for samples from populations having roughly the shape of a normal distribution, its values are as shown in the following table:

n	2	3	4	5	6	7	8	9	10	11	12
d	1.13	1.69	2.06	2.33	2.53	2.70	2.85	2.97	3.08	3.17	3.26

For example, in the illustration on page 256, which dealt with the test strips painted across heavily-traveled roads in 8 locations, the sample range is $167,800 - 108,300 = 59,500$, and, hence, we estimate σ, which in this case measures the true variation in the durability of the paint as

$$\frac{59,500}{2.85} = 20,878 \text{ cars}$$

This differs somewhat from the value of s given on page 256, namely, 19,200, but not knowing the value of σ, we cannot say which of the two estimates is actually closer.

    (a) Referring to Exercise 17 on page 261, use this method to estimate σ, which measures the actual variability of the pollution of the air at the given location.

(b) Referring to Exercise 18 on page 261, use this method to estimate σ, the true standard deviation of the number of mistakes
the compositor makes per galley.

9. Quick estimates of the standard deviation of a distribution (and, hence,
of the population from which the data were obtained) can also be found
by means of the formula $\frac{1}{3}(F_{15/16} - F_{1/16})$, where the two fractiles $F_{15/16}$
and $F_{1/16}$ are as defined in Exercise 24 on page 58.
   (a) Use this method to estimate the standard deviation of the distribution of the scores of the 150 applicants on page 14, and
   compare it with the actual value obtained on page 72.
   (b) Use this method to estimate the standard deviation of the data
   of Exercise 10 on page 25, the one pertaining to the number of
   customers a restaurant served for lunch on 120 weekdays, and
   compare the result with the standard deviation of the data obtained in Exercise 17 on page 77.

11.2 Tests Concerning Standard Deviations

In this section we shall discuss two kinds of tests concerning population
standard deviations—the first concerns the null hypothesis that a population standard deviation equals a specified constant σ_0, and the second concerns the equality of the standard deviations of two populations. The
first kind of test is required whenever we want to check on the uniformity
of a product, process, or operation. For instance, we may want to test
whether a given kind of glass is sufficiently homogeneous for making
delicate optical equipment, whether a group of trainees is sufficiently uniform to be taught in one class, whether the variation in the potency of a
medicine is within permissible limits, and so on. The second kind of test
is often used in conjunction with the two-sample t test described in Section
10.8, since this test requires that the standard deviations of the two populations are equal. For instance, in the ice-cream-fat-content example on
page 282, the two samples had the standard deviations

$$s_1 = \sqrt{\frac{0.66}{4}} = 0.41 \quad \text{and} \quad s_2 = \sqrt{\frac{0.82}{4}} = 0.45$$

and although their difference is small, we should really have put this to a
test before we performed the t test for differences between means.

The test of the null hypothesis $\sigma = \sigma_0$, namely, the hypothesis that a
population standard deviation equals a specified constant, is based on the
same assumptions, the same statistic, and the same sampling theory as the
small-sample confidence interval of Section 11.1. Assuming that our sam-

ple comes from a population having roughly the shape of a normal distribution, we base our decision on the statistic

$$\blacktriangle \qquad \chi^2 = \frac{(n-1)s^2}{\sigma_0^2} \qquad \blacktriangle$$

where n and s^2 are the sample size and the sample variance, while σ_0 is the value of the population standard deviation assumed under the null hypothesis. As we pointed out on page 289, the sampling distribution of this statistic is the chi-square distribution with $n - 1$ degrees of freedom; hence, the criteria for testing the null hypothesis $\sigma = \sigma_0$ against the alternative hypothesis $\sigma < \sigma_0$, $\sigma > \sigma_0$, or $\sigma \neq \sigma_0$ are as shown in Figure 11.3. For the one-sided alternative $\sigma < \sigma_0$ we reject the null hypothesis for values of χ^2 falling into the left-hand tail of its sampling distribution; for the one-sided alternative $\sigma > \sigma_0$ we reject the null hypothesis for values of χ^2 falling into the right-hand tail of its sampling distribution; and for the two-sided alternative $\sigma \neq \sigma_0$ we reject the null hypothesis for values of χ^2 falling into either tail of its sampling distribution. The quantities χ^2_α, $\chi^2_{\alpha/2}$, and $\chi^2_{1-\alpha/2}$ are defined on page 289, and in practice they are obtained from Table III.

To illustrate this kind of test, suppose that it is of special interest to psychologists to know the variability of the time it takes drivers to react to a visual stimulus and that in a random sample the reaction times of 16 persons have a standard deviation of 0.0072 seconds. Suppose, furthermore, that it is desired to check whether the true variability is not greater than $\sigma = 0.005$ seconds. In order to test the null hypothesis $\sigma = 0.005$ against the alternative $\sigma > 0.005$, say, at the level of significance $\alpha = 0.05$, we first calculate the value of the chi-square statistic, getting

$$\chi^2 = \frac{(16-1)(0.0072)^2}{(0.005)^2} = 31.1$$

Since this exceeds 24.996, the value of $\chi^2_{0.05}$ for $16 - 1 = 15$ degrees of freedom, we find that *the null hypothesis will have to be rejected*, and we conclude that the actual variability of drivers' reaction time to the given visual stimulus is measured by a standard deviation *greater than 0.005*.

Given independent random samples from two populations, we usually base tests of the *equality* of the two population variances on the ratios s_1^2/s_2^2 or s_2^2/s_1^2, where s_1 and s_2 are the two sample standard deviations. Assuming that the populations from which the two samples were obtained have roughly the shape of normal distributions, it can be shown that the sampling distribution of such a ratio (appropriately called a *variance ratio*) is a continuous distribution called the *F distribution*. This distribution depends on the sample sizes n_1 and n_2, or better on the two

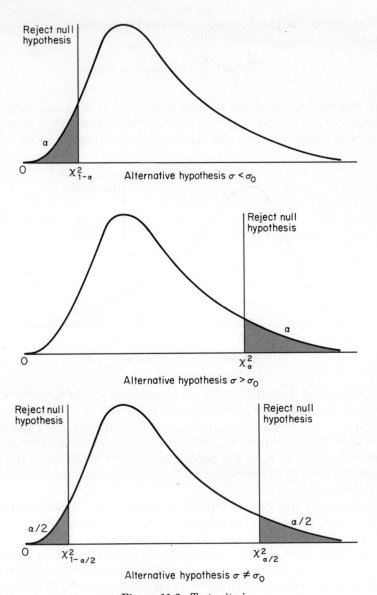

Figure 11.3. Test criteria.

parameters $n_1 - 1$ and $n_2 - 1$, referred to as the respective *degrees of freedom*. One difficulty connected with this distribution is that most tables give only values of $F_{0.05}$ (defined in the same way as $z_{0.05}$, $t_{0.05}$, and $\chi^2_{0.05}$) and $F_{0.01}$, so that we can work only with the right-hand tail of the distribution. It is for this reason that we base our decision concerning the equality of the two population standard deviations σ_1 and σ_2 on the statistic

▲ $$F = \frac{s_1^2}{s_2^2} \quad or \quad \frac{s_2^2}{s_1^2} \quad whichever \; is \; larger$$ ▲

With this statistic we reject the null hypothesis $\sigma_1 = \sigma_2$ and accept the alternative $\sigma_1 \neq \sigma_2$ when the observed value of F exceeds $F_{\alpha/2}$, where α is the level of significance (see also Figure 11.4). Note that by using a right-

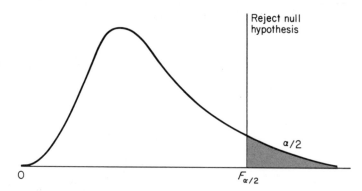

Figure 11.4. F distribution.

hand tail area of $\alpha/2$ instead of α, we compensate for the fact that we always use the larger of the two variance ratios. The necessary values of $F_{\alpha/2}$ for $\alpha = 0.02$ or 0.10, namely, $F_{0.01}$ and $F_{0.05}$, are given in Table IV at the end of the book, where the number of degrees of freedom for the numerator is $n_1 - 1$ or $n_2 - 1$ depending on whether we are using the ratio s_1^2/s_2^2 or the ratio s_2^2/s_1^2; correspondingly, the number of degrees of freedom for the denominator is $n_2 - 1$ or $n_1 - 1$.

To illustrate this kind of test, let us return to the second example of Section 10.8, where we used a t test to check on the significance of the difference between two means. As we pointed out on page 281, this test requires that the standard deviations of the populations from which the two samples were obtained are equal, and this is something which should really have been checked. Since we had 5 observations for each brand of ice cream and the two sample variances were $s_1^2 = \dfrac{0.66}{4} = 0.165$ and

$s_2^2 = \dfrac{0.82}{4} = 0.205$, we now get

$$F = \frac{s_2^2}{s_1^2} = \frac{0.205}{0.165} = 1.24$$

and since this is less than 16.0, the value of $F_{0.01}$ for 4 and 4 degrees of freedom, we find that *the null hypothesis $\sigma_1 = \sigma_2$ cannot be rejected* at the level of significance $\alpha = 0.02$. In fact, since the value which we obtained for F is quite close to 1, there would seem to be very little risk in concluding that the two samples came from populations with equal or nearly equal standard deviations.

EXERCISES

1. In a random sample, the time which 20 women took to complete the written test for their driver's licenses had a variance of 6.2 minutes. Test the null hypothesis $\sigma^2 = 8$ minutes against the alternative hypothesis $\sigma^2 < 8$ minutes at the level of significance $\alpha = 0.05$.

2. Use a level of significance of 0.01 to test the null hypothesis $\sigma = 0.01$ for the diameters of certain bolts, if in a random sample of size 10 the diameters of the bolts had the standard deviation $s = 0.008$.

3. Referring to Exercise 12 on page 284, test the null hypothesis that $\sigma = 0.3$ tons per acre. Use a level of significance of 0.05.

4. Referring to Exercise 13 on page 284, use the level of significance $\alpha = 0.05$ to test the null hypothesis $\sigma = 0.10$ ounces.

5. Large-sample tests concerning population standard deviations are usually based on the fact that, *for large n*, the sampling distribution of

$$\blacktriangle \qquad z = \sqrt{2\chi^2} - \sqrt{2(n-1)} \qquad \blacktriangle$$

is approximately the standard normal distribution, where χ^2 is the statistic given on page 294. One can thus test the null hypothesis $\sigma = \sigma_0$ against the alternatives $\sigma < \sigma_0, \sigma > \sigma_0$, or $\sigma \neq \sigma_0$ with the use of the criteria shown in Figure 10.7.
    (a) Referring to Exercise 4 on page 283, use this method to test at the level of significance $\alpha = 0.05$ whether $\sigma = 2.0$ minutes for the time it takes the ambulances to reach their destination in emergency calls.
    (b) Referring to Exercise 7 on page 284, use this method to test at the level of significance $\alpha = 0.01$ whether $\sigma = 2,400$ miles.

6. Referring to Exercise 18 on page 285, test at the level of significance $\alpha = 0.02$ whether it is reasonable to assume that the two samples come from populations having the same standard deviation.

7. Referring to Exercise 19 on page 285, test whether it is reasonable to assume that the populations from which the two samples came have the same standard deviation. Use $\alpha = 0.02$.

8. Referring to Exercise 20 on page 285, test at the level of significance $\alpha = 0.02$ whether it is reasonable to assume that the two samples come from populations having the same standard deviation.

9. Referring to Exercise 21 on page 286, test at the level of significance $\alpha = 0.02$ whether it is reasonable to assume that the two samples come from populations having the same standard deviation.

BIBLIOGRAPHY

Discussions of the chi-square and F distributions may be found in most textbooks on mathematical statistics; for example, in the author's *Mathematical Statistics*, 2nd ed., published by Prentice-Hall, Inc. in 1971, and in

HOEL, P. G., *Introduction to Mathematical Statistics*, 4th ed. New York: John Wiley & Sons, Inc., 1971.

12

Inferences Concerning

Proportions

12.1 Introduction

The work of this chapter will be very similar to that of Chapters 10 and 11, at least in principle. In *problems of estimation* we shall again construct confidence intervals and worry about the possible size of our error, and in the *testing of hypotheses* we shall again have to be careful in formulating null hypotheses and their alternatives, in deciding between one-tail tests and two-tail tests, in choosing the level of significance, and so on. The main difference is that the parameters with which we will now be concerned are "true" proportions (percentages, or probabilities) instead of population means. Hence, we shall base our methods on *counts* instead of measurements.

12.2 The Estimation of Proportions

The kind of information that is usually available for the estimation of a "true" proportion (percentage, or probability) is the *relative frequency* with which an event has occurred, namely, a *sample proportion*. If an event occurs x times out of n, the relative frequency of its occurrence is x/n, and we generally use this sample proportion to estimate the corresponding "true" proportion, which we shall denote p. For example, if 108 of 150 persons with high blood pressure report an immediate improvement upon the injection of a new drug, then

$$\frac{x}{n} = \frac{108}{150} = 0.72$$

and we can use this figure as an estimate of the true proportion of persons with high blood pressure who will immediately benefit from the new drug. Note that since a *percentage* is simply a proportion multiplied by 100 and a *probability* can be interpreted as a proportion "in the long run," we could also say that we are thus estimating that 72 per cent of all persons with high blood pressure will immediately benefit from the new drug, or that we are estimating the probability that a person with high blood pressure will immediately benefit from the new drug as 0.72. We have made this point to emphasize the fact that *the problem of estimating a "true" percentage or a "true" probability is basically the same as that of estimating a "true" proportion.*

Throughout this section it will be assumed that the situations with which we are dealing satisfy (at least approximately) the conditions underlying the *binomial distribution;* that is, our information will always tell us how many successes there are in a given number of *independent* trials, and it will be assumed that for each trial the probability of a success has the same value p. Thus, the sampling distribution of the *counts* on which our methods will be based is the binomial distribution, whose mean and standard deviation are given by the formulas

$$\mu = np \quad \text{and} \quad \sigma = \sqrt{np(1-p)}$$

as we indicated on pages 188 and 190.

The fact that the formula for σ involves the quantity p (which we want to estimate) leads to some difficulties, but we can avoid them, at least for the moment, by constructing confidence intervals for p with the use of Tables Va and Vb at the end of the book. (So far as the construction of these special tables is concerned, let us merely point out that the statisticians who made them used methods which are very similar to what we did on page 253 except that they referred to very detailed tables of binomial probabilities instead of areas under the normal curve.)

Tables Va and Vb provide 0.95 and 0.99 confidence intervals for proportions; they are easy to use and require no calculations. If a sample proportion x/n is less than or equal to 0.50, we begin by marking its value on the *bottom scale;* then we go up vertically until we reach the two contour lines (curves) which correspond to the size of the sample, and finally we read the confidence limits for p off the *left-hand scale.* If the sample proportion exceeds 0.50, we mark its value on the scale which is *at the top,* go down vertically until we reach the two contour lines (curves) which correspond to the size of the sample, and then we read the confidence limits for p off the *right-hand scale.*

To illustrate the use of Tables Va and Vb when the sample proportion is less than or equal to 0.50, suppose that in a random sample of 400 persons who live in a large Western city there are only 136 who felt that the city's public transportation system is adequate. This is $\dfrac{136}{400} = 0.34$ or 34 per cent. To estimate the corresponding true proportion of persons in that city who feel that the transportation system is adequate, we locate 0.34 on the bottom scale of Table Va, and proceeding as in Figure 12.1 we obtain the 0.95 confidence interval

$$0.29 < p < 0.39$$

Had we wanted to use a degree of confidence of 0.99, Table Vb would correspondingly have yielded the interval

$$0.28 < p < 0.40$$

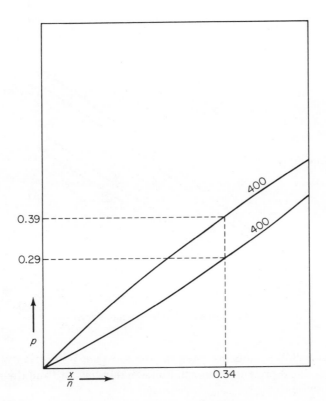

Figure 12.1. Confidence limits for p.

and this illustrates again that *an increase in the degree of confidence will lead to a wider interval and, hence, to less specific information about the quantity we are trying to estimate.*

To illustrate the use of Tables Va and Vb when the sample proportion exceeds 0.50, let us refer again to the problem on page 299 which concerned the persons with high blood pressure and the effect of a new drug. The sample proportion of persons with high blood pressure who immediately benefited from the new drug was $\frac{108}{150} = 0.72$, and the first thing we discover is that there are no curves corresponding to $n = 150$ in Tables Va and Vb. Reading between the lines, however, as in Figure 12.2, we

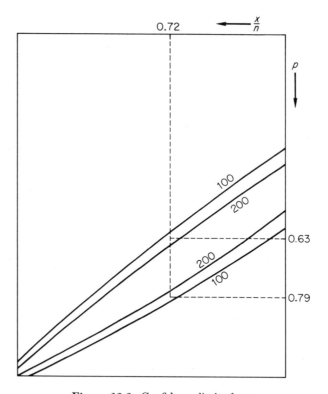

Figure 12.2. Confidence limits for p.

find that a 0.95 confidence interval for the true proportion of persons with high blood pressure who will immediately benefit from the new drug is given by

$$0.63 < p < 0.79$$

and that a corresponding 0.99 confidence interval, read off Table Vb instead of Table Va, is given by

$$0.61 < p < 0.81$$

Note that Tables Va and Vb give contour lines (curves) only for samples of size $n = 8, 10, 12, 16, 20, 24, 30, 40, 60, 100, 200, 400,$ and $1,000$, so that for all other values of n we have to proceed as in this last example.

When n is large and p is not close to either 0 or 1, large-sample confidence intervals for p can be based on the normal approximation of the binomial distribution. Using the method of Section 8.11, we can treat

$$z = \frac{x - np}{\sqrt{np(1 - p)}}$$

as a value of a random variable having the standard normal distribution and, duplicating the work on page 253, we obtain the following $1 - \alpha$ *large-sample confidence interval for p:*

$$\frac{x}{n} - z_{\alpha/2} \sqrt{\frac{p(1 - p)}{n}} < p < \frac{x}{n} + z_{\alpha/2} \sqrt{\frac{p(1 - p)}{n}}$$

This formula cannot be used *as is*, since the confidence limits, themselves, involve the parameter p we are trying to estimate. The quantity

$$\sqrt{\frac{p(1 - p)}{n}}$$

which appears in these confidence limits, is generally called the *standard error of a proportion;* it *is* the standard deviation of the sampling distribution of a sample proportion. We could manipulate the above inequality so that the middle term is p and the other two terms consist of expressions which can be calculated without knowledge of p (see Exercise 11 on page 310), but it is more common to use the $1 - \alpha$ *large-sample confidence interval*

$$\blacktriangle \qquad \frac{x}{n} - z_{\alpha/2} \sqrt{\frac{\frac{x}{n}\left(1 - \frac{x}{n}\right)}{n}} < p < \frac{x}{n} + z_{\alpha/2} \sqrt{\frac{\frac{x}{n}\left(1 - \frac{x}{n}\right)}{n}} \qquad \blacktriangle$$

which is obtained by substituting the sample proportion x/n for p in the standard error formula $\sqrt{\frac{p(1 - p)}{n}}$. Here $z_{\alpha/2}$ is as defined on page 250, and the values most often used are again $z_{0.025} = 1.96$, $z_{0.01} = 2.33$, and $z_{0.005} = 2.58$.

To illustrate this technique, let us return to the example on page 301, where only 136 of 400 persons felt that their city's public transportation system is adequate. Substituting $x = 136$, $n = 400$, and $z_{0.025} = 1.96$ (for a degree of confidence of 0.05) into the above formula, we get

$$\frac{136}{400} - 1.96 \sqrt{\frac{\frac{136}{400}\left(1 - \frac{136}{400}\right)}{400}} < p < \frac{136}{400} + 1.96 \sqrt{\frac{\frac{136}{400}\left(1 - \frac{136}{400}\right)}{400}}$$

or

$$0.294 < p < 0.386$$

and this is very close, indeed, to the confidence interval $0.29 < p < 0.39$ obtained previously with the use of Table Va. In fact, if we round to two decimals the results are the same.

The theory presented here can also be used to judge the possible size of the error one makes when using a sample proportion x/n as a point estimate of p. *In fact, we can assert with a probability of $1 - \alpha$ that the size of this error, that is, its magnitude, will be less than*

$$\blacktriangle \quad z_{\alpha/2} \sqrt{\frac{p(1 - p)}{n}} \quad \textit{or approximately} \quad z_{\alpha/2} \sqrt{\frac{\frac{x}{n}\left(1 - \frac{x}{n}\right)}{n}} \quad \blacktriangle$$

The first of these two formulas cannot be used in actual practice since p is always unknown in problems of this kind; the second formula should be used only when n is large.

To illustrate this technique, let us refer back to the first of the two examples of this section, namely, the one where 108 of 150 persons with high blood pressure felt an immediate improvement after an injection of a new drug. Now, if we use $\frac{108}{150} = 0.72$ as a *point estimate* of the true proportion of persons with high blood pressure who will feel an immediate improvement upon the administration of the drug, we can assert with a probability of 0.99 that this estimate is "off" either way by less than

$$2.58 \sqrt{\frac{(0.72)(1 - 0.72)}{150}} = 0.10$$

rounded to two decimals.

As in the estimation of means, we can use the expression for the maximum error to determine how large a sample is needed to attain a desired degree of precision. If we want to be able to assert with a probability of $1 - \alpha$ that our sample proportion will differ from the "true" proportion p by less than some quantity E, we can write

$$E = z_{\alpha/2} \sqrt{\frac{p(1 - p)}{n}}$$

and upon solving for n, we get

$$\blacktriangle \qquad n = p(1 - p)\left[\frac{z_{\alpha/2}}{E}\right]^2 \qquad \blacktriangle$$

Since this formula requires knowledge of p, the quantity we are trying to estimate, it cannot be used exactly as it stands. However, it can be shown that $p(1 - p)$ is at most equal to $\frac{1}{4}$, and that it assumes this maximum value only when $p = \frac{1}{2}$. *It follows that it is always safe to use the above formula with $p = \frac{1}{2}$, although the resulting sample size may be unnecessarily large.* In case we do have some preliminary information about the possible range of values on which p might fall, we can take this into account in determining n. For instance, if it is reasonable to suppose that a proportion we are trying to estimate lies on the interval from 0.60 to 0.80, we substitute into the above formula for n whichever value of p is closest to $\frac{1}{2}$; in this particular case we would substitute $p = 0.60$.

To illustrate this technique, suppose we want to estimate what proportion of the graduates of four-year colleges hold their first job (after graduation) for at least a year, and we want to be "95 per cent sure" that the error of our estimate will be less than 0.04. If we have no idea what the true proportion might be, we substitute $E = 0.04$, $p = \frac{1}{2}$, and $z_{0.025} = 1.96$ into the above formula for n, and we get

$$n = \tfrac{1}{2}(1 - \tfrac{1}{2})\left[\frac{1.96}{0.04}\right]^2$$
$$= 601$$

rounded up to the nearest integer. Hence, if we base our estimate on a random sample of size $n = 601$, we can assert with a probability of *at least* 0.95 that the sample proportion we get will not be "off" by more than 0.04. (We added the words "at least" because $n = 601$ may actually be larger than necessary because we substituted $\frac{1}{2}$ for p.)

Had we known in this example that the proportion we are trying to estimate is in the neighborhood of, say, 0.30, the formula for n would have yielded

$$n = (0.30)(1 - 0.30) \left[\frac{1.96}{0.04} \right]^2$$
$$= 505$$

rounded up to the nearest integer. This illustrates the fact that if we *do* have some information about the possible size of the proportion we are trying to estimate, this can appreciably reduce the size of the required sample.

12.3 A Bayesian Estimate*

In the preceding section we looked upon the "true" proportions we tried to estimate as *unknown constants;* in Bayesian estimation these parameters are looked upon as *random variables* having *prior distributions* reflecting either the strength of one's belief about the possible values they can assume, or other indirect information. As in Section 10.4, we are thus faced with the problem of *combining prior information with direct sample evidence.*

To illustrate how this might be done, consider a large company which routinely pays thousands of invoices submitted by its suppliers. Of course, it is of interest to know what proportion of these invoices might contain errors, and interviews of three of the company's executives reveal the following information: Mr. Martin "feels" that only 0.005 (or a half of one per cent) of the invoices contain errors; Mr. Green, who is generally regarded to be as reliable in his estimates as Mr. Martin, "feels" that 0.01 (or one per cent) of the invoices contain errors; and Mr. Jones, who is generally regarded to be twice as reliable in his estimates as either Mr. Martin or Mr. Green, "feels" that 0.02 (or two per cent) of the invoices contain errors. Assuming that $p = 0.005$, $p = 0.01$, and $p = 0.02$ are the only possibilities, we thus have the following *prior distribution* for the proportion of invoices containing errors:

Value of p	Prior Probability
0.005	0.25
0.010	0.25
0.020	0.50

* This section may be omitted without loss of continuity.

Note that the prior probabilities assigned to Mr. Martin's and Mr. Green's estimates are the same, while that assigned to Mr. Jones' estimate is twice as large.

Now suppose that a random sample of $n = 200$ of the invoices is carefully checked, and that it turns out that only $x = 1$ of them contains an error. The probabilities of this happening for $p = 0.005$, $p = 0.01$, and $p = 0.02$ are, respectively,

$$\binom{200}{1} (0.005)^1 (0.995)^{199} = 0.37$$

$$\binom{200}{1} (0.010)^1 (0.990)^{199} = 0.27$$

and

$$\binom{200}{1} (0.020)^1 (0.980)^{199} = 0.07$$

where we used the formula for the binomial distribution on page 174, and logarithms to perform the calculations. Combining these probabilities with the prior probabilities on page 306 by means of the formula for Bayes' rule, which we studied in Section 6.5, we find that the *posterior probability* of Mr. Martin's estimate of 0.005 being right is

$$\frac{(0.25)(0.37)}{(0.25)(0.37) + (0.25)(0.27) + (0.50)(0.07)} = 0.47$$

while the corresponding *posterior probabilities* of Mr. Green's and Mr. Jones' estimates of 0.01 and 0.02 are, respectively,

$$\frac{(0.25)(0.27)}{(0.25)(0.37) + (0.25)(0.27) + (0.50)(0.07)} = 0.35$$

and

$$\frac{(0.50)(0.07)}{(0.25)(0.37) + (0.25)(0.27) + (0.50)(0.07)} = 0.18$$

We have thus arrived at the following *posterior distribution* for the proportion of invoices containing errors:

Value of p	Posterior Probability
0.005	0.47
0.010	0.35
0.020	0.18

This distribution reflects the prior judgments as well as the direct sample evidence, and it should not come as a surprise that the highest posterior probability goes to $p = 0.005$. After all, the sample proportion actually equaled $\dfrac{x}{n} = \dfrac{1}{200} = 0.005$.

To continue this example, if we ask for a *point estimate* of the true proportion of invoices containing errors, we could use the *mean of the posterior distribution*, namely,

$$(0.005)(0.47) + (0.010)(0.35) + (0.020)(0.18) = 0.009$$

Note that in contrast to the *mean of the prior distribution*, which was

$$(0.005)(0.25) + (0.010)(0.25) + (0.020)(0.50) = 0.014$$

the mean of the posterior distribution reflects the prior judgments as well as the direct sample evidence.

In the example of this section, it was assumed that p had to be 0.005, 0.010, or 0.020, and this restriction was imposed mainly to simplify the calculations. The method of analysis would have been the same, however, if we had considered 10 different values of p, or even 100. In fact, there exist Bayesian techniques, somewhat similar to that of Section 10.4, in which the unknown proportion p can take on any value on the continuous interval from 0 to 1, and its prior distribution is correspondingly given by a continuous distribution.

EXERCISES

1. In a sample survey, 184 of 400 persons interviewed in a large city said that they opposed the construction of any more freeways. Use Table V to construct a 0.95 confidence interval for the corresponding true proportion.

2. In a random sample of 60 claims filed against a company writing collision insurance on cars, 36 exceeded $750. Use Table V to construct a 0.99 confidence interval for the true proportion of claims filed against this company that exceed $750.

3. In a random sample of 300 high school seniors in a large city, 198 said that they expect to continue their education. Use Table V to construct a 0.95 confidence interval for the true proportion of high school seniors in the given city who expect to continue their education.

4. In a sample survey, 200 persons with incomes of $10,000 or more were asked "Where would you be most likely to find out all there is to know

about some news in which you are very much interested?" If 110 replied "television," use Table V to construct a 0.99 confidence interval for the true proportion of persons in this income group who would respond in this way to the given question.

5. In the sample survey of Exercise 4, another 200 persons were asked where they are most likely to find advertising that can be trusted. If 46 replied "in newspapers," construct a 0.99 confidence interval for the corresponding true proportion
    (a) using Table V;
    (b) using the large-sample confidence interval formula on page 303.

6. In a random sample of 100 cans of mixed nuts (taken from a very large shipment), 16 contained no pecans. Construct a 0.95 confidence interval for the *probability* that there will be no pecans in a can which is randomly selected from this shipment
    (a) using Table V;
    (b) using the large-sample confidence interval formula on page 303.

7. In a random sample of 250 television viewers in a certain area, 95 had seen a certain controversial program. Construct a 0.99 confidence interval for the actual *percentage* of television viewers in that area who saw the program
    (a) using Table V;
    (b) using the large-sample confidence interval formula on page 303.

8. In a random sample of 500 women over the age of 21 interviewed in a large city, 315 said that they held full- or part-time jobs.
    (a) Use Table V to construct a 0.95 confidence interval for the true proportion of women over 21 in the given city who hold full- or part-time jobs.
    (b) If we use $\dfrac{315}{500} = 0.63$ as an estimate of the true proportion of women over 21 in the given city who hold full- or part-time jobs, what can we say with a probability of 0.95 about the possible size of our error?

9. In a random sample of 600 cars stopped at a roadblock on *Interstate 40*, there were 132 which did not meet certain federal safety standards.
    (a) Use the formula on page 303 to construct a 0.99 confidence interval for the true proportion of cars on that highway which do not meet the given standards.
    (b) If we use $\dfrac{132}{600} = 0.22$ as an estimate of the true proportion of cars on that highway which do not meet the given standards, what can we say with a probability of 0.98 about the possible size of our error?

10. In a study conducted by a trading stamp company, it was found that among 400 randomly selected male shoppers who were deliberately not offered trading stamps, 328 asked for them.
   (a) Use Table V to construct a 0.95 confidence interval for the corresponding true proportion.
   (b) Use the large-sample confidence interval formula on page 303 to construct a 0.98 confidence interval for the corresponding true proportion.
   (c) If we use $\dfrac{328}{400} = 0.82$ as an estimate of the corresponding true proportion, what can we say with a probability of 0.95 about the possible size of our error?

11. Solving the double inequality on page 303 for p, we obtain the confidence limits

$$\frac{x + \dfrac{1}{2} z_{\alpha/2}^2 \pm z_{\alpha/2} \sqrt{\dfrac{x(n - x)}{n} + \dfrac{1}{4} z_{\alpha/2}^2}}{n + z_{\alpha/2}^2}$$

   (a) Use this formula to rework part (b) of Exercise 5.
   (b) Use this formula to rework part (b) of Exercise 6.
   (c) Use this formula to rework part (b) of Exercise 10.

12. If a sample constitutes a substantial portion (say, 5 per cent or more) of a population, the methods of this section should not be used without making an appropriate modification. If the sample itself is large, we can use the same *correction factor* as in the estimation of means, and write approximate large-sample $1 - \alpha$ confidence limits for p as

$$\frac{x}{n} \pm z_{\alpha/2} \sqrt{\frac{\dfrac{x}{n}\left(1 - \dfrac{x}{n}\right)}{n}} \cdot \sqrt{\frac{N - n}{N - 1}}$$

where N is, as before, the size of the population.
   (a) Rework part (b) of Exercise 5 assuming that there are 1,000 persons with incomes of $10,000 or more in the area being surveyed.
   (b) Rework part (b) of Exercise 6 assuming that the shipment contains 600 cans of these nuts.

13. A private opinion poll is hired by a political leader to estimate what *percentage* of the voters in his state favor more lenient abortion laws. How large a sample will they have to take to be able to assert with a probability of at least 0.95 that their estimate, the sample *percentage*, will be within 5 per cent of the correct value?

14. In a study of advertising campaigns, a national manufacturer wants to determine what proportion of shirts purchased for use by men are actually purchased by women. How large a sample will they need to be able to assert with a probability of 0.98 that their estimate, the proportion they get in the sample, will not be "off" by more than 0.03?

15. A bank wants to estimate from a sample of its thousands of accounts what proportion of its customers will require financing for a new car within the coming year.
    (a) How large a sample will they need to be able to assert with a probability of at least 0.95 that the sample proportion they get will be within 0.06 of the correct value?
    (b) If they have reason to believe that the proportion will be close to 0.25, how would this affect the required size of the sample?

16. A life insurance company, considering a "non-smoker plan" (with reduced rates for persons who do not smoke), wants to estimate what proportion of its many policy holders would qualify for the plan.
    (a) How large a·sample will they have to take from their files to be able to assert with a probability of 0.99 that the sample proportion and the true proportion will differ by less than 0.02?
    (b) If they have reason to believe that the proportion will be between 0.25 and 0.35, how would this affect the required size of the sample?

17. In planning his faculty for a new school, a high school principal claims that 3 out of 5 newly hired teachers will stay for more than a year, while his assistant feels that it should be 4 out of 5.
    (a) If the principal is regarded to be "three times as good" as his assistant in making such estimates, what *prior probability* should we assign to the respective claims?
    (b) What *posterior probabilities* would we assign to their claims if it was found that among 12 newly hired teachers 10 stayed for more than a year? (*Hint:* use Table VI.) Also find the mean of this *posterior distribution.*

18. The landscaping plans for a new office building call for a row of palm trees along the street side. The landscape designer tells the owner that if they plant *Washingtonia filifera,* 20 per cent of the palms will fail to survive the first heavy frost; the manager of the nursery which supplies the palms tells the owner that only 10 per cent of the palms will fail to survive the first heavy frost; and the owner's wife tells him that 40 per cent of the trees will fail to survive the first heavy frost.
    (a) If the owner feels that in this matter the landscape designer is 14 times as reliable as his wife while the manager of the nursery is 10 times as reliable as his wife, what *prior probabilities* would he assign to the three percentages?
    (b) If 12 of these palms are planted and three fail to survive the

first heavy frost, what *posterior probabilities* would the owner assign to the three percentages? (*Hint:* use Table VI.)

19. The method of Section 12.3 can be used also to find the posterior distribution of the parameter λ of the Poisson distribution. Suppose, for instance, that in a discussion between the executives of several employment agencies, one of them expresses the opinion that an ad in a local newspaper should produce 3 serious inquiries about an administrative job paying $20,000 a year, another expresses the opinion that such an ad should produce 4 serious inquiries about the job, while a third expresses the opinion that such an ad should produce 6 serious inquiries about the job.

    (a) If the three executives are about equally reliable in things like this, what *prior probabilities* should we assign to their respective claims?

    (b) If it turns out that only one person inquires about the job in response to the ad, what *posterior probabilities* should we assign to the three executives' claims? Assume that the number of inquiries is a random variable having the Poisson distribution with $\lambda = 3$, $\lambda = 4$, and $\lambda = 6$ according to the three claims.

20. The method of Section 12.3 can be used also to find posterior probabilities relating to the parameter a (the actual number of "successes") of the hypergeometric distribution. Suppose, for instance, that a coin dealer receives a shipment of four ancient gold coins from abroad, and (on the basis of past experience) he feels that the *prior probabilities* that 0, 1, 2, 3, or all 4 of them are counterfeits are, respectively, 0.80, 0.05, 0.03, 0.02, and 0.10. Since the cost of authentication is high, the dealer decides to select one of the four coins at random and have it checked. If it turns out that this coin is a counterfeit, what *posterior probabilities* will he assign to the possibilities that 0, 1, 2, or all 3 of the remaining coins are counterfeits? Also find the mean of this *posterior distribution* as an estimate of how many of the remaining coins the dealer can *expect* to be counterfeits.

12.4 Tests Concerning Proportions

Once the reader understands the fundamental ideas underlying tests of hypotheses, the various tests we shall study in the remainder of this book should not present any conceptual difficulties. In this section we shall be concerned with tests which enable us to decide, on the basis of sample data, whether the true value of a proportion (percentage, or probability) equals, is greater than, or is less than a given constant. These tests will enable us to check, say, whether the true proportion of fifth graders who can name the governor of their state is 0.40, whether it is true that at least 30 per cent of all families who leave California move to Arizona, or whether the probability that a person will buy a new car from the same dealer from whom he bought his last car is less than 0.60.

Questions of this kind are usually decided on the basis of the observed number, or proportion, of "successes" in n trials, and it will be assumed throughout this section that these trials are independent and that the probability of a success is the same for each trial. In other words, we shall assume that we can use the binomial distribution. When n is small, tests concerning "true" proportions (namely, tests concerning the parameter p of binomial distributions) can be based directly on tables of binomial probabilities such as Table VI. To illustrate, let us refer to the third of the examples of the preceding paragraph, and let us suppose that we have a sample of size $n = 14$ to test the null hypothesis $p = 0.60$ against the alternative $p < 0.60$ at the level of significance $\alpha = 0.05$. In the criteria of Figure 10.7, where we dealt with *continuous* normal distributions, we were able to draw the dividing lines so that the probabilities associated with the "tails" were *exactly* α or $\alpha/2$. Since this cannot be done when we deal with binomial distributions, we modify the criteria as follows: *We draw the dividing lines so that the probability of getting a value in the "tail" is as close as possible to the level of significance α (or to $\alpha/2$ in a two-tail test) without exceeding it.* Thus, for our example we observe (from Table VI) that for $p = 0.60$ and $n = 14$ the probability of getting *at most 4 successes* is

$$0.001 + 0.003 + 0.014 = 0.018$$

and the probability of getting *at most 5 successes* is

$$0.001 + 0.003 + 0.014 + 0.041 = 0.059$$

Since this last value exceeds 0.05, we reject the null hypothesis $p = 0.60$ (and accept the alternative $p < 0.60$) when in a random sample of $n = 14$ persons interviewed there are *at most 4* who will buy a new car from the same dealer from whom they bought their last car (see also Figure 12.3).

To give an example where we use a two-tail test, suppose we want to decide, on the basis of a random sample of size $n = 12$, whether the true proportion of fifth graders who can name the governor of their state is 0.40, as claimed in the first paragraph of this section. The level of significance is to be $\alpha = 0.05$. Thus, we observe from Table VI that for $p = 0.40$ and $n = 12$ the probability of getting *at most one success* is

$$0.002 + 0.017 = 0.019$$

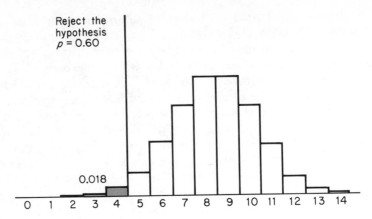

Figure 12.3. Binomial distribution with $p = 0.60$ and $n = 14$.

the probability of getting *at most 2 successes* is

$$0.002 + 0.017 + 0.064 = 0.083$$

the probability of getting *9 or more successes* is

$$0.002 + 0.012 = 0.014$$

the probability of getting *8 or more successes* is

$$0.002 + 0.012 + 0.042 = 0.056$$

and since the second and fourth of these values exceed 0.025, we find that the null hypothesis $p = 0.40$ will have to be rejected when fewer than 2 or more than 8, namely, when 0, 1, 9, 10, 11, or 12 of the 12 fifth graders can actually name the governor of their state (see also Figure 12.4).

When n is large, tests concerning "true" proportions (percentages, or probabilities) are usually based on the normal curve approximation of the binomial distribution, and as we indicated on page 213, this approximation is satisfactory so long as np and $n(1 - p)$ both exceed 5. Since the mean and the standard deviation of the binomial distribution are given by np and $\sqrt{np(1 - p)}$, respectively, we can base the test of the null hypothesis $p = p_0$ on the statistic*

$$z = \frac{x - np_0}{\sqrt{np_0(1 - p_0)}}$$

---

* Some authors write the numerator of this formula for z as $x - np_0 \pm \frac{1}{2}$, whichever is *numerically* smaller, but there is generally no need for this *continuity correction* so long as n is larger.

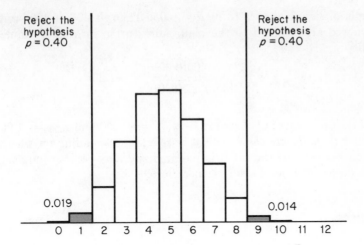

Figure 12.4. Binomial distribution with $p = 0.40$ and $n = 12$.

having (approximately) the standard normal distribution. The actual test criteria are again the ones shown in Figure 10.7 on page 277: for the one-sided alternative $p < p_0$ we reject the null hypothesis when $z < -z_\alpha$; for the one-sided alternative $p > p_0$ we reject the null hypothesis when $z > z_\alpha$; and for the two-sided alternative $p \neq p_0$ we reject the null hypothesis when $z < -z_{\alpha/2}$ or $z > z_{\alpha/2}$. As before, α is the level of significance, and z_α and $z_{\alpha/2}$ are as defined on page 250.

To illustrate this kind of test, suppose that in the second of the examples mentioned in the beginning of this section, a sample check of the records of several large van lines reveals that the belongings of 153 of 600 families leaving California are being shipped to Arizona. To test the null hypothesis $p = 0.30$ against the alternative $p < 0.30$ at the level of significance $\alpha = 0.01$, we substitute $x = 153$, $n = 600$, and $p_0 = 0.30$ into the above formula for z, getting

$$z = \frac{153 - 600(0.30)}{\sqrt{600(0.30)(0.70)}} = -2.41$$

Since this is less than $-z_{0.01} = -2.33$, we find that *the null hypothesis will have to be rejected;* that is, we can conclude that less than 30 per cent of all families who leave California move to Arizona.

To give an example where we use a two-tail test, suppose we want to investigate the claim that 60 per cent of the students attending a large university are opposed to the administration's plan to increase student fees in order to build new parking facilities. Suppose, furthermore, that the level of significance is to be $\alpha = 0.05$. Then, if 228 of 400 students interviewed are opposed to the plan, substitution into the formula for z yields

$$z = \frac{228 - 400(0.60)}{\sqrt{400(0.60)(0.40)}} = -1.22$$

and since this value falls between $-z_{0.025} = -1.96$ and $z_{0.025} = 1.96$, we find that *the null hypothesis cannot be rejected*. Depending on whether a decision one way or the other is required, we may reserve judgment or accept 60 per cent as the correct percentage.

EXERCISES

1. The null hypothesis $p = 0.10$ is to be tested against the alternative hypothesis $p > 0.10$ at the level of significance $\alpha = 0.05$, where p is the parameter of a binomial distribution with $n = 10$. Use Table VI to find the values of the random variable having this distribution for which the null hypothesis will have to be rejected. What is the *actual* probability of committing a Type I error?

2. Suppose we want to decide on the basis of 15 flips of a coin whether it may be regarded as being "honest." How many heads would we have to get to be able to reject the null hypothesis that the probability of heads is 0.50, if the level of significance is to be 0.05.

3. The null hypothesis $p = 0.80$ is to be tested against the alternative hypothesis $p < 0.80$ at the level of significance $\alpha = 0.01$, where p is the parameter of a binomial distribution with $n = 13$. Use Table VI to find the values of the random variable having this distribution for which the null hypothesis will have to be rejected. What is the *actual* level of significance of this test criterion?

4. Suppose that a physicist claims that at most 5 per cent of all persons exposed to a certain amount of radiation will feel any effects. If 9 persons are exposed to this much radiation, how many of them will have to feel some effects before the null hypothesis $p = 0.05$ can be rejected at the level of significance $\alpha = 0.01$.

5. The manufacturer of a spot remover claims that his product removes at least 85 per cent of all spots. What can we conclude about this claim at the level of significance $\alpha = 0.05$, if the spot remover removed only 164 of 200 spots chosen at random from spots on clothes brought to a cleaning establishment?

6. A nutritionist claims that 45 per cent of the pre-school children in a certain region of the country have protein-deficient diets. Test this

claim at the level of significance $\alpha = 0.01$, if a sample survey showed that among 500 such children 289 had protein-deficient diets.

7. A fund-raising organization claims that it gets an 8 per cent response to its solicitations. Test this claim against the alternative that their figure is too high, if only 137 responses are received to 2,000 letters sent out to raise funds for a certain charity. Use the level of significance $\alpha = 0.05$.

8. A social scientist claimed that among persons living in rural areas 51 per cent are in favor of capital punishment (while 40 per cent are against it and 9 per cent are undecided). If 137 of 250 persons (interviewed in rural areas as part of a sample survey) say that they favor capital punishment, test the null hypothesis $p = 0.51$ against the alternative $p \neq 0.51$ at the level of significance 0.05.

9. In a random sample of 300 industrial accidents, it was found that 173 were due at least partially to unsafe working conditions. Use the level of significance $\alpha = 0.01$ to check whether this is consistent with the claim that 65 per cent of such accidents are due at least partially to unsafe working conditions.

10. In the construction of tables of random numbers (see discussion on page 221), there are various ways of checking on possible departures from randomness. For instance, there should be about as many even digits (0, 2, 4, 6, or 8) as there are odd digits (1, 3, 5, 7, or 9). Count the number of even digits among the 350 digits constituting the first ten rows of the table on page 492, and test at the level of significance $\alpha = 0.05$ whether we should be concerned about the possibility that these random numbers are "not random?"

11. In a sample survey of retired persons, 627 of 800 stated that they preferred living in an apartment to living in a one-family home. Test the null hypothesis that the "true" proportion of retired persons who feel this way is 0.75 against the alternative hypothesis that this figure is incorrect one way or the other, using (a) the level of significance $\alpha = 0.05$, and (b) the level of significance $\alpha = 0.01$.

12. A college magazine maintains that 70 per cent of all college men expect their dates to furnish their own cigarets. Test this claim against the alternative hypothesis $p < 0.70$, if 500 college men interviewed as part of a sample survey included 318 who said that they expect their dates to furnish their own cigarets. Use the level of significance $\alpha = 0.01$.

12.5 Differences Between Proportions

There are many problems in which we must decide whether an observed difference between two sample proportions, or percentages, is significant or whether it may reasonably be attributed to chance. For instance, if 132 of 300 randomly selected housewives in one city say that they take empty aluminum cans to places where they are picked up for recycling, but only 148 of 400 randomly selected housewives in another city make this claim, we may want to decide whether the difference between the corresponding proportions, namely, $\dfrac{132}{300} = 0.44$ and $\dfrac{148}{400} = 0.37$, can be attributed to chance or whether it implies that housewives in the first city are more ecology minded than those in the other city. Similarly, if 16 of 200 tractors produced on one assembly line required extensive adjustments before they could be shipped, while the same was true for 20 of 400 tractors produced on another assembly line, it may be of interest to know whether the difference between $\dfrac{16}{200} = 0.08$ and $\dfrac{20}{400} = 0.05$ can reasonably be attributed to chance.

Questions of this kind are usually decided on the basis of the theory that if x_1 and x_2 are the number of "successes" observed, respectively, in large independent random samples of size n_1 and n_2, and if p_1 and p_2 are the corresponding probabilities for success on an individual trial, then *the sampling distribution of the difference between the two sample proportions, $x_1/n_1 - x_2/n_2$, can be approximated closely with a normal distribution having the mean $p_1 - p_2$ and the standard deviation*

$$\sqrt{\frac{p_1(1 - p_1)}{n_1} + \frac{p_2(1 - p_2)}{n_2}}$$

In accordance with the terminology introduced in Chapter 9, we refer to the standard deviation of this sampling distribution as the *standard error of the difference between two proportions*. Discussions of this theory may be found in most textbooks on mathematical statistics; for instance, in those listed on page 298.

To illustrate what is meant by this sampling distribution, suppose that p_1 is the actual proportion of male voters in Vancouver favoring the construction of a new bridge and that p_2 is the corresponding proportion of female voters. If we sent out a large number of interviewers, telling each to interview random samples of n_1 male voters and n_2 female voters in the given city, we would expect that the figures they get for x_1 and x_2 (the number of male and female voters favoring the construction of the new bridge) are not all the same, nor are the values they obtain for the difference between x_1/n_1 and x_2/n_2. If we then constructed a distribution of the differences $x_1/n_1 - x_2/n_2$ obtained by these interviewers, we would get an *experimental sampling distribution of the difference between two proportions*. The sampling distribution referred to above is the corresponding *theoretical sampling distribution*.

This example also serves to illustrate why we stated that the two samples must be *independent*, that they must, so to speak, be selected separately. It stands to reason that we might well get very misleading results if we interviewed married couples, whose views on most subjects are apt to be influenced by one another.

Using the above theory concerning the sampling distribution of the difference between two proportions, we can now base *large-sample tests* concerning the equality of two population proportions, namely, *large-sample tests* of the null hypothesis $p_1 = p_2$, on the statistic

$$\blacktriangle \qquad z = \frac{\dfrac{x_1}{n_1} - \dfrac{x_2}{n_2}}{\sqrt{p(1-p)\left(\dfrac{1}{n_1} + \dfrac{1}{n_2}\right)}} \qquad \text{with } p = \frac{x_1 + x_2}{n_1 + n_2} \qquad \blacktriangle$$

having approximately the standard normal distribution. We obtained this formula for z by subtracting from the observed difference between the two sample proportions the mean of its sampling distribution (which under the null hypothesis equals $p_1 - p_2 = 0$), and then dividing by the standard deviation with p substituted for p_1 and p_2. It is customary to refer to the expression given for p as a *pooled estimate* of the common value of p_1 and p_2.

Depending on the alternative hypothesis, appropriate tests of the null hypothesis $p_1 = p_2$ are the ones shown in Figure 10.7 on page 277: we reject the null hypothesis for values of z falling into the left-hand tail, the right-hand tail, or both tails of the sampling distribution, depending on whether the alternative hypothesis is, respectively, $p_1 < p_2$, $p_1 > p_2$, or $p_1 \neq p_2$.

To illustrate this kind of test, let us refer back to the tractor-assembly example on page 318, where 16 tractors required extensive adjustments in a random sample of 200 tractors from one assembly line, and 20 tractors required extensive adjustments in a random sample of 400 tractors from another assembly line. Let us suppose, furthermore, that the null hypothesis $p_1 = p_2$, namely, the null hypothesis that there is no difference in the quality of the work of the two assembly lines, is to be tested against the two-sided alternative $p_1 \neq p_2$ at the level of significance

$\alpha = 0.05$. Substituting $x_1 = 16$, $n_1 = 200$, $x_2 = 20$, and $n_2 = 400$ first into the expression for p and then into the formula for z, we obtain

$$p = \frac{16 + 20}{200 + 400} = 0.06$$

and

$$z = \frac{\dfrac{16}{200} - \dfrac{20}{400}}{\sqrt{(0.06)(0.94)\left(\dfrac{1}{200} + \dfrac{1}{400}\right)}} = 1.46$$

Since 1.46 falls between $-z_{0.025} = -1.96$ and $z_{0.025} = 1.96$, *the null hypothesis cannot be rejected;* that is, we cannot conclude from the given data that one of the assembly lines is actually better than the other.

The example we have given here is typical of tests concerning the equality of two population proportions. In Exercise 6 on page 324 we shall see how the test can be modified to apply also to tests of the null hypothesis that the difference between the population proportions equals a given constant, that is, the hypothesis $p_1 - p_2 = \delta$, where δ (*delta*) is not necessarily zero. A test which serves to compare more than two sample proportions will be discussed in the next section.

12.6 Differences Among k Proportions

In many applications we must decide whether observed differences among more than two sample proportions, or percentages, are significant or whether they can be attributed to chance. For instance, if in January, 1973, insurance salesman A contacted 110 customers and sold insurance policies to 13, insurance salesman B contacted 80 customers and sold insurance policies to 17, while insurance salesman C contacted 160 customers and sold insurance policies to 20, their office manager may want to know whether the differences among the corresponding propor- tions of sales can be attributed to chance. Similarly, if random samples of registered voters (100 without high school diplomas, 300 with high school diplomas but no college education, and 200 college graduates) are asked how they would vote on a certain piece of legislation, it may be of interest to know whether the *actual* proportions of favorable votes are the same for all three groups. Suppose, thus, that the results of this sam- ple survey are as shown in the following table:

	No High School Diploma	High School Diploma But No College	College Graduates
For the Legislation	33	147	114
Against the Legislation	67	153	86

The proportions of favorable votes for the three groups are, respectively, 0.33, 0.49, and 0.57, and it will be of interest to see whether the differences among them can be attributed to chance.

To illustrate how we handle this kind of problem, let us denote the true proportions of voters favoring the legislation in the three groups p_1, p_2, and p_3. *Thus, we shall want to test the null hypothesis $p_1 = p_2 = p_3$ against the alternative that p_1, p_2, and p_3 are not all alike.* If the null hypothesis is true, we can combine the three samples and estimate the common proportion of voters favoring the legislation as

$$\frac{33 + 147 + 114}{100 + 300 + 200} = \frac{294}{600} = 0.49$$

With this estimate, we could *expect* $100(0.49) = 49$ votes for the legislation in the no-high-school-diploma group, $300(0.49) = 147$ votes for the legislation in the high-school-diploma-but-no-college group, and $200(0.49) = 98$ votes for the legislation in the college-graduates group. Subtracting these figures from the sizes of the respective samples, we would correspondingly expect $100 - 49 = 51$, $300 - 147 = 153$, and $200 - 98 = 102$ votes against the legislation. These results are summarized in the following table, where the *expected frequencies* are shown in parentheses below the ones that were actually observed:

	No High School Diploma	High School Diploma But No College	College Graduates
For the Legislation	33 (49)	147 (147)	114 (98)
Against the Legislation	67 (51)	153 (153)	86 (102)

To test the null hypothesis $p_1 = p_2 = p_3$, we shall now compare the frequencies that were actually observed with those we could *expect* if the null hypothesis were true. It stands to reason that the null hypothesis should be accepted if the two sets of frequencies are very much alike; after all, we would then have obtained almost exactly what we could have expected if the null hypothesis were true. On the other hand, if the discrepancies between the two sets of frequencies are large, the observed frequencies do not agree with what we could have expected, and this is an indication that the null hypothesis must be false.

Writing the observed frequencies as f's and the expected frequencies as e's, we can put this comparison on a precise basis by using the statistic

▲
$$\chi^2 = \Sigma \frac{(f - e)^2}{e}$$
▲

called "*chi-square.*" Since the notation used in defining this statistic is somewhat abbreviated, let us state in words that χ^2 *is the sum of the quantities obtained by dividing* $(f - e)^2$ *by e separately for each "cell" of the table.*

The sampling distribution of this χ^2 statistic can be approximated closely with the *chi-square distribution* introduced first in Chapter 11. In our particular example, the number of degrees of freedom (the parameter of the chi-square distribution) is 2, but in general *when we compare k sample proportions the number of degrees of freedom is k − 1*. If there is a close agreement between the f's and the e's, the differences $f - e$ and, hence, χ^2 will be small; otherwise, some of the differences $f - e$ and, hence, χ^2 will be large. Consequently, we reject the null hypothesis at the level of significance α *if the value we obtain for the χ^2 statistic exceeds χ^2_α* (see also Figure 11.1).

Returning now to our numerical example, letting $\alpha = 0.05$, and substituting into the formula for χ^2, we obtain

$$\chi^2 = \frac{(33 - 49)^2}{49} + \frac{(147 - 147)^2}{147} + \frac{(114 - 98)^2}{98} + \frac{(67 - 51)^2}{51}$$
$$+ \frac{(153 - 153)^2}{153} + \frac{(86 - 102)^2}{102}$$
$$= 15.36$$

and since 15.36 exceeds 5.991, the value of $\chi^2_{0.05}$ for 2 degrees of freedom (see Table III at the end of the book), *the null hypothesis will have to be rejected*. This means that there *is* a difference in attitude concerning the given piece of legislation depending on the extent of one's education.

In general, if we want to compare k sample proportions, we first calculate the expected frequencies as we did on page 321. Combining the data, we estimate p as

▲
$$\frac{x_1 + x_2 + \ldots + x_k}{n_1 + n_2 + \ldots + n_k}$$
▲

where the n's are the sizes of the respective samples and the x's are the corresponding numbers of successes. Multiplying the n's by this estimate of p, we obtain the expected frequencies for the first row of the table, and subtracting these values from the totals of the corresponding samples, we obtain the expected frequencies for the second row of the table. Note that the expected frequency for any one of the cells may also be obtained by multiplying the total of the row to which it belongs by the total of the column to which it belongs, and then dividing by the grand total $n_1 + n_2 + \ldots + n_k$. Next, χ^2 is calculated according to the formula

▲
$$\chi^2 = \Sigma \frac{(f - e)^2}{e}$$
▲

with $(f - e)^2/e$ determined separately for each of the $2k$ cells of the table. Then, we reject the null hypothesis $p_1 = p_2 = \ldots = p_k$ at the level of significance α (against the alternative that these p's are *not all equal*) if the value obtained for χ^2 exceeds χ^2_α for $k - 1$ degrees of freedom.

   Since the sampling distribution of this χ^2 statistic is only approximately the chi-square distribution with $k - 1$ degrees of freedom, it is best not to use this test when one or more of the expected frequencies is less than 5. It is also of interest to note that for $k = 2$ the χ^2 statistic of this section actually equals the *square* of the z statistic of Section 12.5 (see Exercises 7 and 8 below); thus, for $k = 2$ the two tests are *equivalent*, although the one of Section 12.5 can be used also when the alternative hypothesis is $p_1 < p_2$ or $p_1 > p_2$.

EXERCISES

1. Referring to the example on page 318, which dealt with the recycling of aluminum cans, test at the level of significance $\alpha = 0.05$ whether the difference between the two sample proportions is significant.

2. Two groups of 80 patients each took part in an experiment in which one group received pills containing an anti-allergy drug, while the other group received a placebo (a pill containing no drug). If in the group given the drug 23 exhibited allergic symptoms while in the group given the placebo 41 exhibited such symptoms, is this sufficient evidence to conclude at the level of significance $\alpha = 0.01$ that the drug is effective in reducing these symptoms?

3. A study showed that 84 of 200 persons who saw a deodorant advertised
    during the telecast of a football game and 96 of 200 persons who saw
    it advertised on a variety show remembered two hours later the name
    of the deodorant. Use the level of significance $\alpha = 0.05$ to test the
    null hypothesis that there is no difference between the corresponding
    "true" proportions.

4. In a true-false test, a test item is considered to be *good* if it discrimi-
    nates between well-prepared students and poorly-prepared students.
    What can we conclude about the merit of a test item which was ans-
    wered correctly by 205 of 250 well-prepared students and by 137 of
    250 poorly-prepared students. Use the level of significance $\alpha = 0.05$.

5. The manager of a motel, in trying to decide which of two supposedly
    equally good cigaret-vending machines to install, tests each machine
    500 times. If the first machine fails to work (that is, neither delivered
    the cigarets nor returned the coins) 26 times and the second machine
    fails to work 12 times, can he conclude at the level of significance
    $\alpha = 0.05$ that the second machine is really better?

6. If we want to test the null hypothesis that the difference between two
    population proportions equals some constant δ (*delta*), not necessarily
    0, we can use the theory on page 318 and base our decision on the
    statistic

    $$z = \frac{\dfrac{x_1}{n_1} - \dfrac{x_2}{n_2} - \delta}{\sqrt{\dfrac{p_1(1 - p_1)}{n_1} + \dfrac{p_2(1 - p_2)}{n_2}}}$$

    with x_1/n_1 and x_2/n_2 substituted for p_1 and p_2. The sampling distribu-
    tion of this statistic can be approximated with the standard normal
    distribution when the samples are sufficiently large.

    (a) Referring to Exercise 2, use this theory and the level of signifi-
        cance $\alpha = 0.05$ to test whether the percentage of patients
        exhibiting allergic symptoms is at least 8 per cent less for those
        who actually receive the drug.
    (b) Referring to Exercise 4, use this theory and the level of signifi-
        cance $\alpha = 0.05$ to test whether for the given item the propor-
        tion of correct answers is at least 20 per cent higher among
        well-prepared students than among poorly-prepared students.

7. Rework Exercise 1 using the chi-square statistic, and verify that χ^2
    equals the *square* of the value previously obtained for z.

8. Rework Exercise 3 using the chi-square statistic, and verify that χ^2
    equals the *square* of the value previously obtained for z.

9. Referring to the example on page 320, test at the level of significance
    $\alpha = 0.05$ whether the differences among the proportions of sales made
    by the three insurance salesmen can be attributed to chance.

10. If 26 of 200 tires of Brand A failed to last 20,000 miles, while the corresponding figures for 200 tires each of Brands B, C, and D were 23, 15, and 32, use the level of significance $\alpha = 0.05$ to test the null hypothesis that there is no difference in the quality of the four kinds of tires.

11. A market research study shows that among 100 men, 100 women, and 200 children interviewed, there were, respectively, 42, 51, and 135 who did not like the flavor of a new toothpaste. Test at the level of significance $\alpha = 0.01$ whether the differences among the corresponding sample proportions (that is, 0.42, 0.51, and 0.675) can be attributed to chance.

12. In studying problems related to its handling of reservations, an airline takes a random sample of 80 of the complaints about reservations filed in each of four cities. If 52 of the complaints from City A, 48 of the complaints from City B, 63 of the complaints from City C, and 57 of the complaints from City D concern overselling of available space, test at the level of significance $\alpha = 0.05$ whether the differences among the corresponding sample proportions can be attributed to chance.

12.7 Contingency Tables

The χ^2 statistic plays an important role in many other tests dealing with *count data*, or *enumeration data*, namely, in problems where information is obtained by counting rather than measuring. The method we shall treat in this section is an extension of the one discussed in the preceding section, and we shall apply it to two distinct kinds of problems. In the first kind of problem we deal with trials permitting *more than two possible outcomes*. For example, in the illustration on page 321 we might have given each person the choice of being for the legislation, against it, or undecided, and we might thus have obtained the following table:

	No High School Diploma	High School Diploma But No College	College Graduates
For the Legislation	25	103	88
Undecided	26	71	35
Against the Legislation	49	126	77

We refer to this kind of table as a 3-by-3 table because it contains 3 rows and 3 columns; more generally, when there are k samples and each trial permits r alternatives, we refer to the resulting table as an *r-by-k table*. Note that as in the examples of Section 12.6, the *column totals* are fixed (they are the sizes of the respective samples) while the row totals are values of random variables (they depend on chance).

In the second kind of problem, the column totals as well as the row totals are left to chance. Suppose, for instance, that a psychologist working in the personnel department of a large company wants to know whether there really is a relationship between a person's performance in the company's training program and his (or her) ultimate success in the job. Suppose, furthermore, that a sample of 400 cases taken from the company's very extensive files yielded the results shown in the following table:

	Performance in Training Program		
	Below Average	*Average*	*Above Average*
Poor	23	60	29
Average	28	79	60
Very Good	9	49	63

Success in Job (Employer's Rating)

This is also a 3-by-3 table, and it is mainly in connection with problems of this kind that *r-by-k* tables are referred to as *contingency tables*.

Before we illustrate the method of analysis used in problems of either kind, let us first examine what hypotheses we shall want to test. In the first problem, the one dealing with voter reaction to the given piece of legislation, we want to test the null hypothesis that the probabilities of getting a vote for the legislation, a vote against the legislation, or an undecided vote are the same for each group. In other words, we want to test the null hypothesis that a person's attitude toward the legislation is *independent* of the extent of his education. In the second problem we are also concerned with a null hypothesis of *independence*, namely, the null hypothesis that a company employee's success in his job is independent of his performance in the training program.

To illustrate the analysis of an *r-by-k* table, let us refer to the second example and let us begin by calculating (as in the example of Section 12.6) the *expected cell frequencies*. If the null hypothesis of independence is true, the probability of randomly selecting an employee whose performance was below average in the training program *and* who is doing poorly in his job is given by the *product* of the probability that his performance was below average in the training program and the probability that he is doing poorly in his job. (This is simply an application of the Special Multiplication Rule on page 139.) Using the totals of the first column and the first row to *estimate* these two probabilities, we get $\dfrac{23 + 28 + 9}{400} = \dfrac{60}{400}$ for the probability that an employee (randomly selected from among the company's employees) performed below average in the training program, and $\dfrac{23 + 60 + 29}{400} = \dfrac{112}{400}$ for the probability that he is doing poorly in his job. Hence, we estimate the probability of choosing an employee who performed poorly in the training program *and* is doing poorly in his job as $\dfrac{60}{400} \cdot \dfrac{112}{400}$, and in a sample of size 400 we would expect to find

$$400 \cdot \frac{60}{400} \cdot \frac{112}{400} = \frac{60 \cdot 112}{400} = 16.8$$

employees who fit this description. Note that after cancelling 400's, we obtained the result by multiplying the total of the first column by the total of the first row and then dividing by the grand total of 400. In general, we can (as on page 323) find the expected frequency for any one of the cells of a contingency table by *multiplying the total of the row to which it belongs by the total of the column to which it belongs and then dividing by the grand total for the whole table.* Using this procedure, we obtain an expected frequency of $\dfrac{112 \cdot 188}{400} = 52.6$ for the second cell of the first row, and $\dfrac{167 \cdot 60}{400} = 25.0$ and $\dfrac{167 \cdot 188}{400} = 78.5$ for the first two cells of the second row.

The remaining expected cell frequencies could be found in the same way, but making use of the fact that *the sum of the expected frequencies for each row or column must equal the sum of the corresponding observed frequencies* (see Exercise 8 on page 334), we obtain by subtraction

$$112 - 16.8 - 52.6 = 42.6$$

for the third cell of the first row, and

$$167 - 25.0 - 78.5 = 63.5$$

for the third cell of the second row, and

$$60 - 16.8 - 25.0 = 18.2$$
$$188 - 52.6 - 78.5 = 56.9$$

and

$$152 - 42.6 - 63.5 = 45.9$$

for the three cells of the third row. These results are summarized in the following table, where the expected frequencies are shown in parentheses below the corresponding observed cell frequencies:

		Performance in Training Program		
		Below Average	Average	Above Average
	Poor	23 (16.8)	60 (52.6)	29 (42.6)
Success in Job (Employer's Rating)	Average	28 (25.0)	79 (78.5)	60 (63.5)
	Very Good	9 (18.2)	49 (56.9)	63 (45.9)

From here on, the work is like that of the preceding section; we let f denote an observed frequency, e an expected frequency, and we calculate the χ^2 statistic according to the formula

▲
$$\chi^2 = \Sigma \frac{(f - e)^2}{e}$$
▲

with $(f - e)^2/e$ calculated separately for each cell of the table. Then *we reject the null hypothesis of independence at the level of significance α if the value obtained for χ^2 exceeds the value of χ_α^2 for $(r - 1)(k - 1)$ degrees of freedom*, where r and k are, respectively, the number of rows and the number of columns. In our example the number of degrees of freedom is $(3 - 1)(3 - 1) = 4$, and it should be observed that after we had calculated *four* of the expected frequencies, all the others were obtained by subtraction from the totals of appropriate rows and columns.

Actually calculating χ^2 for our example, we get

$$\chi^2 = \frac{(23 - 16.8)^2}{16.8} + \frac{(60 - 52.6)^2}{52.6} + \frac{(29 - 42.6)^2}{42.6}$$

$$+ \frac{(28 - 25.0)^2}{25.0} + \frac{(79 - 78.5)^2}{78.5} + \frac{(60 - 63.5)^2}{63.5}$$

$$+ \frac{(9 - 18.2)^2}{18.2} + \frac{(49 - 56.9)^2}{56.9} + \frac{(63 - 45.9)^2}{45.9}$$

$$= 20.34$$

and since this exceeds 9.488 as well as 13.277, the values of $\chi^2_{0.05}$ and $\chi^2_{0.01}$ for 4 degrees of freedom, we find that *the null hypothesis will have to be rejected at either level of significance.* Thus, we have shown that there *is* a dependence (or relationship) between an employee's performance in the training program and his success in his job.

It is customary in problems of this kind to round the expected cell frequencies to one decimal or to the nearest whole number. Most of the entries of Table III are given to three decimals, but there is seldom any need to carry more than two decimals when calculating χ^2 for an *r*-by-*k* table. As we pointed out on page 323, the statistic we are using here has only approximately a chi-square distribution and, hence, *it should not be used unless each of the expected frequencies is at least 5.* If one or more of the expected frequencies is less than 5, it may nevertheless be possible to use the test by combining some of the cells (and subtracting one degree of freedom for each cell thus eliminated).

12.8 Tests of Goodness of Fit

In this section we shall study another application of the χ^2 criterion, in which we compare an observed frequency distribution with a distribution we might *expect* according to theory or assumptions. To illustrate this method, let us take another look at the random digits of Table IX, which was supposed to have been constructed in such a way that each digit is a value of a random variable which takes on the values 0, 1, 2, 3, 4, 5, 6, 7, 8, and 9 with equal probabilities of $\frac{1}{10}$. To check whether this is so, we might count how many times each digit appears in the table, but to simplify our work, let us merely take those in the first five columns of the table on page 490. This yields the values shown in the "observed frequency" column of the following table:

Digit	Observed Frequency	Expected Frequency
0	23	25
1	25	25
2	20	25
3	23	25
4	23	25
5	22	25
6	29	25
7	25	25
8	33	25
9	27	25

The expected frequencies in the right-hand column were obtained by multiplying each of the probabilities of $\frac{1}{10}$ by 250, the total number of digits which we counted.

As can be seen by inspection, there are some discrepancies between the observed frequencies and those which we expect. For instance, there are not too many 2's and quite a few 6's and 8's. Also, as can easily be checked, the *mean* of the observed distribution is 4.768 whereas that of the expected distribution is 4.5. To check whether these discrepancies may be attributed to chance, we first determine χ^2 by means of the formula

$$\chi^2 = \Sigma \frac{(f - e)^2}{e}$$

calculating $(f - e)^2/e$ separately for each class of the distribution. Then, if it turns out that the value we obtain is *too large*, we reject the null hypothesis that the 250 digits are values of a random variable which takes on the values 0, 1, 2, 3, 4, 5, 6, 7, 8, and 9 with equal probabilities of $\frac{1}{10}$.

Substituting the observed and expected frequencies from the above table into the formula for χ^2, we get

$$\chi^2 = \frac{(23 - 25)^2}{25} + \frac{(25 - 25)^2}{25} + \frac{(20 - 25)^2}{25} + \frac{(23 - 25)^2}{25}$$

$$+ \frac{(23 - 25)^2}{25} + \frac{(22 - 25)^2}{25} + \frac{(29 - 25)^2}{25} + \frac{(25 - 25)^2}{25}$$

$$+ \frac{(33 - 25)^2}{25} + \frac{(27 - 25)^2}{25}$$

$$= 5.20$$

To decide whether this value is large enough to reject the null hypothesis (at the level of significance α), we have only to check whether it exceeds χ_α^2 for an appropriate number of degrees of freedom. In general, *the number of degrees of freedom for this kind of test is given by the number of terms $(f - e)^2/e$ added in obtaining χ^2 minus the number of quantities, obtained from the observed data, that are used in calculating the expected frequencies.*

In our example we had 10 terms in the formula for χ^2, and the only quantity needed from the observed data to calculate the expected frequencies was the total frequency of 250. Hence, the number of degrees of freedom is $10 - 1 = 9$, and for the level of significance $\alpha = 0.05$, Table III yields $\chi_{0.05}^2 = 16.919$. Since we had $\chi^2 = 5.20$, we find that *the null hypothesis cannot be rejected;* in other words, the table of random numbers has "passed the test."

The method we have illustrated in this section is used quite generally to test how well distributions we *expect* (on the basis of theory or assumptions) fit, or describe, observed data. In some of the exercises which follow, we shall thus test whether it is reasonable to treat an observed distribution as if it had (at least approximately) the shape of a normal distribution, and we shall also test whether given sets of data fit the pattern of binomial and Poisson distributions.

EXERCISES

1. Analyze the 3-by-3 table on page 325 and decide at the level of significance $\alpha = 0.05$ whether there *is* a difference in attitude toward the legislation depending on the extent of one's education.

2. A sample survey, designed to show where persons living in different parts of the country buy non-prescribed medicines, yielded the following results:

	Northeast	North Central	South	West
Drugstores	219	200	181	180
Grocery Stores	39	52	89	60
Others	42	48	30	60

Test the null hypothesis that, so far as non-prescribed medicines are concerned, the buying habits of persons living in the given parts of the country are the same. Use the level of significance $\alpha = 0.05$.

3.   The following sample data pertain to the shipments received by a large
     firm from three different vendors:

	Number Rejected	Number Imperfect but Acceptable	Number Perfect
Vendor A	12	23	89
Vendor B	8	12	62
Vendor C	21	30	119

Test at the level of significance $\alpha = 0.01$ whether the three vendors
ship products of equal quality.

4.   The following sample data are based on a study of the majors of stu-
     dents from abroad enrolled in institutions of higher education in the
     United States in 1966:

	Europe	Latin America	Africa
Engineering	34	50	24
Humanities	62	66	19
Physical Sciences	30	31	26
Social Sciences	33	34	38
All Other Fields	41	69	43
Sample Sizes	200	250	150

Use the level of significance $\alpha = 0.01$ to test the null hypothesis that
the true proportions of students majoring in these areas are the same
for students from Europe, Latin America, and Africa.

5.   A large electronics firm which hires many handicapped workers wants
     to determine whether their handicaps affect such workers' perfor-
     mance. Use the level of significance $\alpha = 0.05$ to decide on the basis
     of the sample data shown in the following table whether it is reason-
     able to maintain that the handicaps have no effect on the workers'
     performance:

	Performance		
	Above Average	Average	Below Average
Blind	21	64	17
Deaf	16	49	14
No Handicap	29	93	28

6. Suppose that a social scientist who wants to determine whether there is a relationship between women's natural hair color and their self-control obtained the sample data shown in the following table:

	Very Self-Controlled	Average	Very Impulsive
Blonde	24	90	49
Brunette	33	106	45
Redhead	25	82	46

Use the level of significance $\alpha = 0.05$ to test the null hypothesis that there is no relationship.

7. If the analysis of a contingency table shows that there is a relationship between the two variables under consideration, the *strength* of this relationship may be measured by means of the *contingency coefficient*

$$C = \sqrt{\frac{\chi^2}{\chi^2 + n}}$$

where n is the total frequency for the entire table. This coefficient assumes values between 0 (corresponding to independence) and a maximum value less than 1, which depends on the number of rows and columns in the table. For example, for a 3-by-3 table the maximum value is 0.816. The larger C is for a given table, the stronger is the dependence between the two variables.

(a) Calculate C for the contingency table of Exercise 4.

(b) Calculate C for the contingency table of Exercise 2.

8. Letting f_{ij} and e_{ij} denote the observed and expected frequencies of the cell which is in the ith row and the jth column of an r-by-k contingency table, show that

   (a) $\displaystyle\sum_{j=1}^{k} f_{ij} = \sum_{j=1}^{k} e_{ij}$ (namely, that the sum of the observed frequencies for any given row equals the sum of the corresponding expected frequencies);

   (b) $\displaystyle\sum_{i=1}^{r} f_{ij} = \sum_{i=1}^{r} e_{ij}$ (namely, the sum of the observed frequencies for any given column equals the sum of the corresponding expected frequencies).

   (*Hint:* let R_i denote the sum of the observed frequencies in the ith row, C_j the sum of the observed frequencies in the jth column, and T the total of the observed frequencies for the entire table, and then make use of the fact that

   $$e_{ij} = \frac{R_i \cdot C_j}{T}$$

9. The following is the distribution of the number of females in 120 litters, each consisting of three guinea pigs:

Number of Females in Litter	Number of Litters
0	12
1	50
2	41
3	17

   (a) Assuming that there is a fifty-fifty chance for each guinea pig to be male or female and that the binomial distribution applies, find the corresponding probabilities for 0, 1, 2, or 3 females and the expected frequencies.

   (b) Use the χ^2 criterion and the level of significance $\alpha = 0.05$ to decide whether the binomial distribution with $n = 3$ and $p = \dfrac{1}{2}$ provides a good fit to the observed distribution.

10. In order to see whether a die is balanced, it was rolled 240 times and the following results were obtained: 1 occurred 37 times, 2 occurred 34 times, 3 occurred 47 times, 4 occurred 38 times, 5 occurred 35 times, and 6 occurred 49 times.

    (a) What are the corresponding expected frequencies for a balanced die?

(b) Calculate χ^2 and test at the level of significance $\alpha = 0.05$ whether the die is balanced.

11. The following is the distribution of calls which the switchboard of a government building received during 600 five-minute intervals:

Number of Calls	Frequency
0	34
1	131
2	160
3	136
4	72
5	37
6	22
7	8

(a) What would be the corresponding expected frequencies if the number of calls arriving at this switchboard in a five-minute interval is a Poisson random variable which takes on the values 0, 1, 2, 3, 4, 5, 6, and 7 with respective probabilities of 0.08, 0.21, 0.26, 0.21, 0.13, 0.07, 0.03, and 0.01?

(b) Calculate χ^2 and test at the level of significance $\alpha = 0.01$ whether the number of calls arriving at the switchboard in a five-minute interval is, in fact, this kind of Poisson random variable.

12. Referring to the results obtained in Exercise 3 on page 242, test the null hypothesis that the simulated coins *are* balanced. Use the level of significance $\alpha = 0.05$.

13. The following is the distribution of the scores first given on page 14, with the first and last classes relabeled as indicated:

Scores	Frequency
19 or less	1
20–29	6
30–39	9
40–49	31
50–59	42
60–69	32
70–79	17
80–89	10
90 or more	2
	150

As we showed in Chapters 3 and 4, the mean of the original distribution was 56.7 and its standard deviation was 15.4.

(a) Given a normal distribution with $\mu = 56.7$ and $\sigma = 15.4$, find the area under the curve to the left of 19.5, between 19.5 and 29.5, between 29.5 and 39.5, between 39.5 and 49.5, between 49.5 and 59.5, between 59.5 and 69.5, between 69.5 and 79.5, between 79.5 and 89.5, and to the right of 89.5.

(b) Multiplying by 150 (the total frequency) each of the normal curve areas obtained in part (a), determine the *expected normal curve frequencies* corresponding to the nine classes of the distribution.

(c) Calculate χ^2 for the observed frequencies and the expected frequencies obtained in part (b), and test the null hypothesis that the scores can be looked upon as a random sample from a normal population. Combine adjacent classes, if necessary, so that none of the expected frequencies will be less than 5, and use the level of significance $\alpha = 0.05$. Note that the number of degrees of freedom for testing the fit of a normal distribution by this method is $k - 3$, where k is the number of terms $(f - e)^2/e$ in the chi-square statistic. Does this agree with the rule for the number of degrees of freedom on page 331?

14. Use the method of Exercise 13 to test at the level of significance $\alpha = 0.01$ whether it is reasonable to look at the distribution of the 50 means of Figure 9.3 on page 228 as having roughly the shape of a normal distribution. Make use of the fact that the mean and the standard deviation of the distribution are 15.76 and 1.89, and relabel the first and last classes to read, respectively, "less than 10.5" and "20.5 or more." Also combine adjacent classes where necessary, so that none of the expected frequencies is less than 5.

BIBLIOGRAPHY

Derivations of the formula for the standard error of the difference between two proportions may be found in most textbooks in mathematical statistics; for instance, in the one by the author listed on page 298.

13

Analysis of Variance

13.1 Differences Among k Means

Let us now generalize the work of Section 10.8 and consider problems of deciding whether observed differences among *more than two* means can be attributed to chance, or whether they are indicative of actual differences among the means of the corresponding populations. For instance, we may want to decide on the basis of sample data whether there really is a difference in the effectiveness of three methods of teaching foreign languages, we may want to compare the average yield per acre of several varieties of wheat, we may want to see whether there really is a difference in the average mileage obtained with different kinds of gasoline, we may want to judge whether there really is a difference in the quality of several mattresses, and so on. To illustrate, suppose we want to compare the average yield of four varieties of wheat on the basis of the following data (in pounds per plot) which an agronomist obtained for three test plots of each variety:

Variety A: 51, 52, 47
Variety B: 50, 43, 42
Variety C: 46, 46, 43
Variety D: 49, 49, 46

The means of these four samples are, respectively, 50, 45, 45, and 48, and what we would like to know is whether the differences among these means are significant or whether they can be attributed to chance—after all, the size of the four samples is *very small*.

If we let μ_1, μ_2, μ_3, and μ_4 denote the true average yields (per test plot) of the four varieties of wheat, we can write the null hypothesis which we shall want to test as

$$\mu_1 = \mu_2 = \mu_3 = \mu_4$$

Correspondingly, the alternative hypothesis is that μ_1, μ_2, μ_3, and μ_4 are *not all equal.** In connection with these two hypotheses, let us point out that small differences among the $\bar{x}$'s may, as always, be attributed to chance, so that if the $\bar{x}$'s are very nearly the same size, this would support the null hypothesis that the population means are all equal. On the other hand, if the differences among the $\bar{x}$'s are too large to be reasonably attributed to chance, this would support the alternative hypothesis that the μ's are not all equal. Hence, we need a precise measure of the size of the discrepancies among the $\bar{x}$'s and, correspondingly, a well-defined criterion for testing the null hypothesis concerning the μ's.

An obvious measure of the discrepancies among the $\bar{x}$'s is their variance, and if we calculate it for our example according to the formula on page 67, we obtain

$$s_{\bar{x}}^2 = \frac{(50 - 47)^2 + (45 - 47)^2 + (45 - 47)^2 + (48 - 47)^2}{4 - 1}$$
$$= 6$$

Note that we used the subscript $\bar{x}$ to indicate that this quantity measures the variability of the $\bar{x}$'s. As we shall see later, this quantity plays an important role in our ultimate decision whether to accept or reject the null hypothesis concerning the four population means.

Let us now make an assumption which is critical to the method of analysis we shall employ: *it will be assumed that the populations from which the samples were obtained can be approximated closely with normal distributions having the same variance σ^2.* This means that we shall assume that if we planted many test plots with the four varieties of wheat, the distribution of the respective yields for each variety would follow the over-all pattern of a normal distribution with the same variance σ^2.

---

* In connection with more advanced work (for instance, that of Sections 13.4 and 13.5), it is desirable to write these means as $\mu_1 = \mu + \alpha_1$, $\mu_2 = \mu + \alpha_2$, $\mu_3 = \mu + \alpha_3$, and $\mu_4 = \mu + \alpha_4$. Here $\mu = \dfrac{\mu_1 + \mu_2 + \mu_3 + \mu_4}{4}$ is called the *grand mean*, and the α's (whose sum is zero, as the reader will be asked to show in Exercise 11 on page 348) are called the *treatment effects*. In this notation, the null hypothesis becomes

$$\alpha_1 = \alpha_2 = \alpha_3 = \alpha_4 = 0$$

and the alternative hypothesis is that the α's are *not all zero*.

If we combine this assumption with the assumption that the null hypothesis is true, we can now look upon the four samples as samples from *one and the same population.* Making use of the fact that for samples from infinite populations the standard error of the mean is given by $\sigma_{\bar{x}} = \sigma/\sqrt{n}$ (see page 230), we can look upon $s_{\bar{x}}^2$ as an estimate of $\sigma_{\bar{x}}^2 = \sigma^2/n$ and, hence, $n \cdot s_{\bar{x}}^2$ as an estimate of σ^2. Returning to our numerical example where we had $n = 3$, we can thus use $3 \cdot 6 = 18$ as an estimate of the common population variance σ^2.

If σ^2 were known, we could compare $n \cdot s_{\bar{x}}^2$ with σ^2 and reject the null hypothesis if $n \cdot s_{\bar{x}}^2$ were much larger than σ^2. However, in our example (as in most practical problems) σ^2 is unknown and we have no choice but to estimate it on the basis of the given data. Having assumed that the samples all come from populations with the same variance σ^2, we could use each of the sample variances (namely, s_1^2, s_2^2, s_3^2, and s_4^2) as an estimate of σ^2 and hence, we can also use their mean. Referring again to our numerical example, we shall thus estimate σ^2 by means of the quantity

$$\frac{s_1^2 + s_2^2 + s_3^2 + s_4^2}{4} = \frac{1}{4}\left[\frac{(51 - 50)^2 + (52 - 50)^2 + (47 - 50)^2}{3 - 1}\right.$$
$$+ \frac{(50 - 45)^2 + (43 - 45)^2 + (42 - 45)^2}{3 - 1}$$
$$+ \frac{(46 - 45)^2 + (46 - 45)^2 + (43 - 45)^2}{3 - 1}$$
$$+ \left.\frac{(49 - 48)^2 + (49 - 48)^2 + (46 - 48)^2}{3 - 1}\right]$$
$$= 8$$

We thus have two estimates of σ^2,

$$n \cdot s_{\bar{x}}^2 = 3 \cdot 6 = 18 \quad \text{and} \quad \frac{s_1^2 + s_2^2 + s_3^2 + s_4^2}{4} = 8$$

and comparing them, we can now assert that if the first (*which is based on the variation among the means of the samples*) is much larger than the second (*which is based on the variation within the samples*), it is reasonable to reject the null hypothesis. If the null hypothesis is not true, the first estimate is apt to be larger than the second, as it would reflect whatever differences there may exist among the population means as well as chance variation; the second estimate reflects only chance variation.

To put the comparison of these two estimates of σ^2 on a rigorous basis, we use the statistic

$$\blacktriangle \quad F = \frac{\text{Estimate of } \sigma^2 \text{ based on the variation among the } \bar{x}\text{'s}}{\text{Estimate of } \sigma^2 \text{ based on the variation within the samples}} \quad \blacktriangle$$

appropriately called a *variance ratio*. If the various assumptions made earlier (that the null hypothesis is true and that the samples come from normal populations with equal variances) are true, the sampling distribution of this statistic is the F distribution introduced in Chapter 11. Hence, if the value we get for F exceeds F_α (see page 295 and Figure 11.4), the null hypothesis can be rejected. The necessary values of F_α can be obtained from Table IV for $\alpha = 0.05$ or 0.01; if we compare the means of k samples of size n, the corresponding numerator and denominator degrees of freedom are $k - 1$ and $k(n - 1)$. Note that the numerator of the expression for F is an estimate of σ^2 based on $k - 1$ independent deviations from the mean, while the denominator is an estimate of σ^2 based on $k(n - 1)$ independent deviations from the mean, namely, $n - 1$ in each of the k samples.

Returning now to our numerical example, we find that $F_{0.05} = 4.07$ for $k - 1 = 4 - 1 = 3$ and $k(n - 1) = 4(3 - 1) = 8$ degrees of freedom. Since this value is *not* exceeded by

$$F = \frac{18}{8} = 2.25$$

the value of the F statistic for our example, we find that the null hypothesis *cannot be rejected*. Although there are sizeable differences among the means of the four samples, the variations within the samples are also large, and we conclude that the differences among the sample means may well be attributed to chance.

The technique we have just described is the simplest form of a powerful statistical tool called the *analysis of variance*, ANOVA for short. Although we could go ahead and perform F tests for differences among k means without further discussion, it will be instructive to look at the problem from an analysis-of-variance point of view, and we shall do so in the next section.

13.2 One-Way Analysis of Variance

The basic idea of analysis of variance is to express a measure of the total variability of a set of data as a sum of terms, each of which can be attributed to a specific source, or cause, of variation. With reference to the example of the preceding section, two such sources of variation might be (1) actual differences in the average yield of the four varieties of wheat, and (2) chance, which in problems of this kind is also referred to as the *experimental error*. The measure of the total variation which we shall use is the *total sum of squares**

$$SST = \sum_{i=1}^{k} \sum_{j=1}^{n} (x_{ij} - \bar{x}..)^2$$

where x_{ij} is the jth observation of the ith sample ($i = 1, 2, \ldots, k$ and $j = 1, 2, \ldots, n$), and $\bar{x}..$ is the *grand mean*, namely, the mean of all of the nk measurements or observations. Note that if we divided the total sum of squares SST by $nk - 1$, we would get the *variance* of the data; hence, the total sum of squares of a set of data is interpreted in much the same way as their sample variance.

Denoting by $\bar{x}_i.$ the mean of the ith sample, we can write as follows the formula, or identity, which forms the basis of a *one-way analysis of variance*

$$SST = n \cdot \sum_{i=1}^{k} (\bar{x}_i. - \bar{x}..)^2 + \sum_{i=1}^{k} \sum_{j=1}^{n} (x_{ij} - \bar{x}_i.)^2$$

If we look closely at the two terms into which the total sum of squares SST has been partitioned, we find that the first term is a measure of the variation among the sample means; in fact, if we divide it by $k - 1$ we get the quantity which, on page 339, was denoted $n \cdot s_{\bar{x}}^2$. Similarly, the second term is a measure of the variation within the individual samples, and if we divided this term by $k(n - 1)$ we would get the mean of the variances of the individual samples, namely, the quantity which we put into the denominator of F on page 340.

It is customary to refer to the first term, namely the quantity which measures the variation *among* the sample means, as the *treatment sum of squares* $SS(Tr)$; the second term, which measures the variation *within* the samples, is usually referred to as the *error sum of squares* SSE. This terminology is explained by the fact that most analysis of-variance techniques were originally developed in connection with agricultural experiments where different fertilizers, for example, were regarded as different *treatments* applied to the soil. (This also explains the terminology used in the footnote to page 338.) The word "error" in "error sum of squares" refers to the *experimental error*, namely what we referred to in Section 13.1 as *chance*. Although this may sound confusing at first, we shall refer to the four varieties of wheat in our example as four different treatments,

* The use of double subscripts and double summations was explained briefly in Section 3.6.

and in other experiments we may refer to three different teaching techniques as three different treatments, to five different dosages of a medicine as five different treatments, and so on. Before we go any further, let us verify the identity

$$SST = SS(Tr) + SSE$$

with reference to the numerical example of Section 13.1. Substituting into the formulas for the different sums of squares, we get

$$
\begin{aligned}
SST = {}& (51 - 47)^2 + (52 - 47)^2 + (47 - 47)^2 + (50 - 47)^2 \\
& + (43 - 47)^2 + (42 - 47)^2 + (46 - 47)^2 + (46 - 47)^2 \\
& + (43 - 47)^2 + (49 - 47)^2 + (49 - 47)^2 + (46 - 47)^2 \\
= {}& 118
\end{aligned}
$$

$$
\begin{aligned}
SS(Tr) = {}& 3[(50 - 47)^2 + (45 - 47)^2 + (45 - 47)^2 + (48 - 47)^2] \\
= {}& 54
\end{aligned}
$$

$$
\begin{aligned}
SSE = {}& (51 - 50)^2 + (52 - 50)^2 + (47 - 50)^2 + (50 - 45)^2 \\
& + (43 - 45)^2 + (42 - 45)^2 + (46 - 45)^2 + (46 - 45)^2 \\
& + (43 - 45)^2 + (49 - 48)^2 + (49 - 48)^2 + (46 - 48)^2 \\
= {}& 64
\end{aligned}
$$

and it follows that

$$SS(Tr) + SSE = 54 + 64 = 118 = SST$$

To test the null hypothesis $\mu_1 = \mu_2 = \ldots = \mu_k$ (or $\alpha_1 = \alpha_2 = \ldots = \alpha_k = 0$ in the notation of the footnote to page 338) against the alternative that *the treatment means are not all equal* (or that *the treatment effects are not all zero*), we now proceed as in Section 13.1 and compare $SS(Tr)$ with SSE by means of an appropriate F statistic. In practice, it has become the custom to exhibit the necessary work as follows in a so-called *analysis-of-variance table*:

Source of variation	Degrees of freedom	Sum of squares	Mean square	F
Treatments	$k - 1$	$SS(Tr)$	$MS(Tr) = \dfrac{SS(Tr)}{k - 1}$	$\dfrac{MS(Tr)}{MSE}$
Error	$k(n - 1)$	SSE	$MSE = \dfrac{SSE}{k(n - 1)}$	
Total	$kn - 1$	SST		

Here the second column lists the degrees of freedom (the number of *independent* deviations from the mean on which the respective sums of squares are based), the fourth column lists the *mean squares* $MS(Tr)$ and MSE which are obtained by dividing the corresponding sums of squares by their respective degrees of freedom, and the right-hand column gives the value of the F statistic as the ratio of the two mean squares. Note that the two mean squares are, in fact, the two estimates of σ^2 referred to in Section 13.1, and that the numerator and denominator degrees of freedom for the F test are $k - 1$ and $k(n - 1)$, namely, the figures corresponding to treatments and error in the "degrees of freedom" column.

Referring again to our numerical example dealing with the four varieties of wheat, we now get the following analysis-of-variance table:

Source of variation	Degrees of freedom	Sum of squares	Mean square	F
Treatments	3	54	18	2.25
Error	8	64	8	
Total	11	118		

Finally, the test of significance is performed by comparing the value obtained for F with F_α for $k - 1 = 3$ and $k(n - 1) = 8$ degrees of freedom. Since $F = 2.25$ is less than $F_{0.05} = 4.07$ for the given degrees of freedom, we find that (as before) the null hypothesis cannot be rejected.

The numbers used in our illustration were chosen so that the calculations would be very simple. In actual practice, the calculations of the sums of squares can be quite tedious unless we use the following short-cut formulas, where $T_{i\cdot}$ denotes the total of the observations corresponding to the ith treatment (that is, the sum of the values in the ith sample), and $T_{\cdot\cdot}$ denotes the *grand total* of all the data:

$$SST = \sum_{i=1}^{k} \sum_{j=1}^{n} x_{ij}^2 - \frac{1}{kn} \cdot T_{\cdot\cdot}^2$$

$$SS(Tr) = \frac{1}{n} \cdot \sum_{i=1}^{k} T_{i\cdot}^2 - \frac{1}{kn} \cdot T_{\cdot\cdot}^2$$

and by subtraction

$$SSE = SST - SS(Tr)$$

It will be left to the reader to verify in Exercise 1 below that for the yield-of-wheat example these formulas give

$$SST = 26{,}626 - 26{,}508 = 118$$
$$SS(Tr) = \frac{1}{3}(150^2 + 135^2 + 135^2 + 144^2) - 26{,}508$$
$$= 26{,}562 - 26{,}508$$
$$= 54$$

and

$$SSE = 118 - 54 = 64$$

The method which we have discussed here applies only when each sample has the same number of observations, but minor modifications make it applicable also when the sample sizes are not all equal. If there are n_i observations for the ith treatment, the computing formulas for the sums of squares become

▲ $$SST = \sum_{i=1}^{k} \sum_{j=1}^{n_i} x_{ij}^2 - \frac{1}{N} \cdot T_{..}^2$$ ▲

▲ $$SS(Tr) = \sum_{i=1}^{k} \frac{T_{i.}^2}{n_i} - \frac{1}{N} \cdot T_{..}^2$$ ▲

and

▲ $$SSE = SST - SS(Tr)$$ ▲

where $N = n_1 + n_2 + \ldots + n_k$. The total number of degrees of freedom is $N - 1$, and the degrees of freedom for treatments and error are, respectively, $k - 1$ and $N - k$.

To illustrate this technique, suppose that a bowler wants to know whether the weight of the ball he uses will affect his game, and he plans to roll six games each with the ball he has been using, a lighter ball, and a heavier ball. Not having the time, though, to finish, he only gets the following results:

Lighter ball	Ball he has been using	Heavier ball
182	190	161
165	179	178
196	208	165
157	186	
164	178	
180	211	

The means of these three samples are, respectively, 174, 192, and 168, which *suggests* that he should keep playing with the same ball. Since the samples are small, however, the differences among the means may not be significant, and to perform the above test we first calculate the sums of squares

$$SST = 489,746 - \frac{1}{15}\,(2,700)^2 = 489,746 - 486,000$$

$$= 3,746$$

$$SS(Tr) = \frac{1,044^2}{6} + \frac{1,152^2}{6} + \frac{504^2}{3} - 486,000$$

$$= 1,512$$

and

$$SSE = 3,746 - 1,512 = 2,234$$

This leads to the following analysis-of-variance table:

Source of variation	Degrees of freedom	Sum of squares	Mean square	F
Treatments	2	1,512	756	4.06
Error	12	2,234	186.2	
Total	14	3,746		

Finally, since $F = 4.06$ exceeds 3.89, the value of $F_{0.05}$ for 2 and 12 degrees of freedom, we find that *the null hypothesis will have to be rejected*, and we conclude that the weight of the ball *does* have an effect on the bowler's game.

EXERCISES

1. Verify the values of SST, $SS(Tr)$, and SSE obtained on page 344 by means of the special computing formulas.

2. To find the best arrangement of instruments on a control panel of an airplane, three different arrangements were tested by simulating an emergency condition and observing the reaction time required to correct the condition. The reaction times (in *tenths* of a second) of

twelve pilots (randomly assigned to the different arrangements) were as follows:

Arrangement 1: 8, 15, 10, 11
Arrangement 2: 16, 11, 14, 19
Arrangement 3: 12, 7, 13, 8

(a) Use the method of Section 13.1 and the level of significance $\alpha = 0.05$ to test whether the differences among the three sample means can be attributed to chance.
(b) Use the computing formulas of Section 13.2 to calculate the necessary sums of squares, construct an analysis of variance table, and check the value of the F statistic against that obtained in part (a).

3. The following are the miles per gallon which a test driver got for four tankfuls each of five brands of gasoline:

Brand A	Brand B	Brand C	Brand D	Brand E
17	13	16	15	16
11	18	16	12	11
16	16	19	15	13
12	17	21	14	12

(a) Use the method of Section 13.1 and the level of significance $\alpha = 0.01$ to test whether the differences among the five sample means can be attributed to chance.
(b) Use the computing formulas of Section 13.2 to calculate the necessary sums of squares, construct an analysis of variance table, and check the value of the F statistic against that obtained in part (a).

4. Mr. Cooper can drive to a friend's house along four different routes, and the following are the number of minutes in which he timed himself on five different occasions for each route:

Route 1: 15, 17, 16, 19, 19
Route 2: 15, 18, 22, 19, 22
Route 3: 14, 16, 17, 15, 18
Route 4: 11, 15, 14, 14, 20

Use the level of significance $\alpha = 0.05$ to test the null hypothesis that there is no difference in the true average time it takes Mr. Cooper to drive to his friend's house along the four different routes.

5. Referring to the data on page 54, perform an analysis of variance to test the null hypothesis that on the average the three kinds of paint cover the same area per gallon. Use the level of significance $\alpha = 0.05$.

6. The following are the number of mistakes made in three successive days by four technicians working for a photographic laboratory:

Technician A	Technician B	Technician C	Technician D
4	5	12	5
12	13	20	11
10	8	17	12

Test at the level of significance $\alpha = 0.01$ whether the differences among the four sample means can be attributed to chance.

7. To study the effectiveness of five different kinds of packaging, a processor of a breakfast food puts each kind into five different supermarkets. Use the following data, representing the number of sales of the breakfast food between 9 A.M. and noon on a given day, to test the null hypothesis that packaging has no effect on sales at the level of significance $\alpha = 0.05$:

> Packaging 1: 45, 32, 36, 32, 40
> Packaging 2: 37, 34, 46, 44, 34
> Packaging 3: 35, 37, 48, 46, 35
> Packaging 4: 36, 38, 50, 36, 45
> Packaging 5: 42, 39, 40, 45, 51

8. The following are the scores obtained by random samples of 12th graders from four different high schools in a test designed to measure their knowledge of current events:

> High School A: 87, 98, 95, 67, 84, 89, 98, 75
> High School B: 77, 61, 97, 82, 60, 87, 73, 58, 72
> High School C: 79, 60, 73, 91, 77, 53, 76, 82
> High School D: 70, 64, 89, 45, 52, 66

Perform an analysis of variance to test the null hypothesis that 12th graders are equally knowledgeable about current events in all four schools. Use the level of significance $\alpha = 0.05$.

9. To study the performance of a newly-designed motorboat it was timed over a marked course under various wind and water conditions. Use the following data (in minutes) to test the null hypothesis that the boat's performance is not affected by the differences in wind and water conditions:

> Calm conditions: 20, 17, 14, 24
> Moderate conditions: 21, 23, 16, 25, 18, 23
> Choppy conditions: 26, 24, 23, 29, 21

Use the level of significance $\alpha = 0.05$.

10. Three groups of guinea pigs were injected, respectively, with 0.5 mg. 1.0 mg, and 1.5 mg of a new tranquilizer, and the following are the number of seconds it took them to fall asleep:

   The 0.5-mg group: 11, 14, 11, 13, 9, 14, 15, 13
   The 1.0-mg group: 9, 10, 11, 12, 10, 8, 12, 10, 9, 13, 7, 9
   The 1.5-mg group: 7, 11, 8, 5, 8, 9, 10, 6

   Test at the level of significance $\alpha = 0.01$ whether we can reject the null hypothesis that differences in dosage have no effect.

11. With reference to the notation introduced in the footnote to page 338, show that $\alpha_1 + \alpha_2 + \alpha_3 + \alpha_4 = 0$, namely, that the sum of the *treatment effects* is (by assumption) zero.

13.3 Two-Way Analysis of Variance

When we analyzed the wheat-yield experiment on page 340, we observed that there were considerable, though not significant, differences among the four sample means. The results were not significant because there were also considerable differences among the values *within* each of the four samples and, hence, a large *experimental error*. Since this is the quantity which went into the denominator of the F statistic, the resulting F value was small and not significant. Let us now suppose that an interview with the person who conducted the experiment reveals that the first value for each variety of wheat was obtained from a test plot fertilized with Fertilizer X, the second value was obtained from a test plot fertilized with Fertilizer Y, and the third value was obtained from a test plot fertilized with Fertilizer Z. *This puts the whole experiment in a new light:* Rechecking the data we find that the four varieties of wheat averaged 49.0 pounds per plot with fertilizer X, 47.5 pounds per plot with Fertilizer Y, and 44.5 pounds per plot with Fertilizer Z. This suggests that *what we referred to as chance variation, or experimental error, in the analyses of Sections 13.1 and 13.2 may well have been caused (at least in part) by the use of different fertilizers.*

It also suggests that we should really have performed a *two-way analysis of variance*, in which the *total variability of the data* is partitioned into one component which we ascribe to possible differences due to one variable (the four varieties of wheat which we referred to as *treatments* in Section 13.2), a second component which we ascribe to possible differences due to a second variable (the three kinds of fertilizer which we refer to as *blocks*, again due to the origin of this method in agricultural research), while the remainder of the variability is ascribed to chance (referred to as the *experimental error* as in Section 13.2).

Before we go into any details, let us point out that there are essentially two different ways of analyzing such two-variable experiments—they depend on whether the two variables are *independent* or whether they *interact*. To illustrate what we mean here by "interact," suppose that a tire manufacturer is experimenting with different kinds of treads, and that he finds that one kind is especially good for use on dirt roads while another kind is especially good on icy roads. If this is the case, we say that there is an *interaction* between road conditions and the design of the treads. On the other hand, if all of the treads behaved equally well (or equally poorly) under all kinds of road conditions, we would say that there is *no interaction*, namely, that the two variables (tread design and road conditions) are independent. The *no interaction* case will be taken up first in Section 13.4, while a method which is suitable for testing for *interactions* will be described in Section 13.5.

13.4 Two-Way Analysis Without Interaction

To explain what hypotheses we shall want to test in this two-variable case, let us write μ_{ij} for the *true* mean which corresponds to the ith treatment and the jth block (in our example it is the true average yield of the ith variety of wheat used with the jth fertilizer) and express it as

$$\mu_{ij} = \mu + \alpha_i + \beta_j$$

As in the notation of the footnote to page 338, μ is the *grand mean* (the average of *all* the μ_{ij}) and the α_i are the *treatment effects* (whose sum is zero). Correspondingly, we refer to the β_j as the *block effects* (for which it can also be shown that their sum must be zero), and write the *two* null hypotheses which we shall want to test as

$$\alpha_1 = \alpha_2 = \ldots = \alpha_k = 0$$

and

$$\beta_1 = \beta_2 = \ldots = \beta_n = 0$$

The alternative to the *first null hypothesis* (which in our example amounts to the hypothesis that differences in variety have no effect on the yield) is that the treatment effects α_i are *not all zero*. Correspondingly, the alternative to the *second null hypothesis* (which in our example amounts to the hypothesis that the different fertilizers have no effect on the yield) is that the block effects β_j are *not all zero*.

To test the second of these null hypotheses we shall need a quantity which is analogous to the treatment sum of squares, but measures the variation of the means obtained for the different blocks (the means of 49.0, 47.5, and 44.5 which we obtained for the three fertilizers in our example) instead of the variation of the means obtained for the different treatments. Thus, if we let $T_{\cdot j}$ denote the total of the values of the jth block, substitute it for $T_{i\cdot}$ in the formula for $SS(Tr)$, sum on j instead of i, and interchange n and k, we obtain, analogous to $SS(Tr)$, the *block sum of squares*

$$\blacktriangle \qquad SSB = \frac{1}{k} \cdot \sum_{j=1}^{n} T_{\cdot j}^2 - \frac{1}{kn} \cdot T_{\cdot\cdot}^2. \qquad \blacktriangle$$

In a *two-way analysis of variance* (without interactions), we compute SST and $SS(Tr)$ according to the formulas on page 343, SSB according to the formula of the preceding paragraph, and we then obtain SSE by subtraction, namely, by making use of the fact that $SST = SS(Tr) + SSB + SSE$. This yields the formula

$$\blacktriangle \qquad SSE = SST - SS(Tr) - SSB \qquad \blacktriangle$$

Note that the error sum of squares for a two-way analysis of variance does *not* equal the error sum of squares for a corresponding one-way analysis, even though we denote it with the same symbol, SSE. In fact, what we have been doing in this section has been to partition the error sum of squares for the one-way analysis into *two terms:* the block sum of squares SSB and whatever is left, which we look upon as the *new* error sum of squares, and which we still write as SSE.

The discussion we have presented here leads to the following *analysis-of-variance* table for a *two-way analysis* in the no interaction case:

Source of variation	Degrees of freedom	Sum of squares	Mean square	F
Treatments	$k-1$	$SS(Tr)$	$MS(Tr) = \dfrac{SS(Tr)}{k-1}$	$\dfrac{MS(Tr)}{MSE}$
Blocks	$n-1$	SSB	$MSB = \dfrac{SSB}{n-1}$	$\dfrac{MSB}{MSE}$
Error	$(n-1)(k-1)$	SSE	$MSE = \dfrac{SSE}{(n-1)(k-1)}$	
Total	$nk-1$	SST		

The mean squares are again given by the corresponding sums of squares divided by their degrees of freedom, and the two F values are obtained by dividing, respectively, the mean squares for treatments and blocks by the mean square for error. Also, the degrees of freedom for blocks is $n - 1$ (like those for treatments with n substituted for k), and the degrees of freedom for error can be obtained by subtracting the degrees of freedom for treatments and blocks from the total number of degrees of freedom, namely, from $nk - 1$. Thus, in the *significance test for treatments* the numerator and denominator degrees of freedom for F are $k - 1$ and $(n - 1)(k - 1)$, and in the *significance test for blocks* the numerator and denominator degrees of freedom for F are $n - 1$ and $(n - 1)(k - 1)$.

Returning now to our numerical example, we find that the totals of the yields of wheat obtained with the three fertilizers are, respectively, 196, 190, and 178, so that

$$SSB = \frac{1}{4}(196^2 + 190^2 + 178^2) - \frac{1}{4 \cdot 3} \cdot 564^2$$
$$= 26,550 - 26,508$$
$$= 42$$

and, hence, the *new* error sum of squares is

$$SSE = 118 - 54 - 42 = 22$$

where the values of SST and $SS(Tr)$ were copied from page 344. We thus arrive at the following *analysis-of-variance table:*

Source of variation	Degrees of freedom	Sum of squares	Mean square	F
Treatments	3	54	18	4.90
Blocks	2	42	21	5.72
Error	6	22	3.67	
Total	11	118		

Using, as before, the level of significance $\alpha = 0.05$, we find that for 3 and 6 degrees of freedom $F_{0.05} = 4.76$, and that for 2 and 6 degrees of freedom $F_{0.05} = 5.14$. Since the first of these two values is exceeded by 4.90, we can *reject the null hypothesis concerning the treatments,* and since the second is exceeded by 5.72, we can also *reject the null hypothesis concerning the blocks.* In other words, we conclude that there *is* a difference in the true average yield of the four varieties of wheat and, furthermore, that there *is* a difference in the effects of the three fertilizers.

An interesting feature of our example is that the *proper* analysis led to significant results, whereas the analysis which failed to account for the use of different fertilizers did not. Of course, which kind of analysis is the proper one in any given situation will have to depend on the way in which the experiment was planned and conducted in the first place.

13.5 Two-Way Analysis with Interaction*

Looking again at the data on page 337, we cannot rule out the possibility that Variety A will give a high yield *only* with Fertilizer Y or that Variety B will have a high yield *only* with Fertilizer X. To detect such possible *interactions*, we need another way of measuring the experimental error, and this is usually done by *repeating* (statisticians call it "replicating") all or part of the experiment several times. To illustrate, suppose that the following are the number of defective pieces produced by four machine operators, working each of three different machines on three different occasions:

	Machine 1	*Machine 2*	*Machine 3*
Operator A	16, 13, 19	9, 15, 11	22, 25, 17
Operator B	18, 17, 21	15, 13, 12	14, 16, 12
Operator C	14, 16, 13	7, 12, 9	11, 14, 12
Operator D	13, 14, 16	3, 1, 9	13, 17, 14

In the notation which we shall use to analyze experiments like this x_{ijh} is the hth observation for the ith treatment and the jth block; $T_{...}$ is the grand total of all the observations; $T_{i..}$ is the total of all the observations for the ith treatment; $T_{.j.}$ is the total of all the observations for the jth block; and $T_{ij.}$ and $\bar{x}_{ij.}$ are, respectively, the total and the mean of all the observations for the ith treatment used in combination with the jth block. Also, there are k treatments and n blocks, and r observations for each of the $k \cdot n$ combinations of treatments and blocks. The *total, treatment,* and *block sums of squares* are calculated in the same way as before,

---

* The material in this section is somewhat more difficult and it may be omitted without loss of continuity.

although their formulas must now be written as

▲
$$SST = \sum_{i=1}^{k} \sum_{j=1}^{n} \sum_{h=1}^{r} x_{ijh}^2 - \frac{1}{knr} \cdot T_{...}^2$$
▲

▲
$$SS(Tr) = \frac{1}{nr} \cdot \sum_{i=1}^{k} T_{i..}^2 - \frac{1}{knr} \cdot T_{...}^2$$
▲

and

▲
$$SSB = \frac{1}{kr} \cdot \sum_{j=1}^{n} T_{.j.}^2 - \frac{1}{knr} \cdot T_{...}^2$$
▲

To obtain a formula for the variability among the means $\bar{x}_{ij.}$ that can be attributed to *interaction*, we make use of the fact that $\Sigma \Sigma \Sigma (\bar{x}_{ij.} - \bar{x})^2$, where $\bar{x}$ is the mean of all the $k \cdot n \cdot r$ observations, equals the *sum* of the treatment, block, and interaction sum of squares. By subtraction, we thus obtain the following computing formula for the *interaction sum of squares:*

▲
$$SSI = \frac{1}{r} \cdot \sum_{i=1}^{k} \sum_{j=1}^{n} T_{ij.}^2 - \frac{1}{knr} \cdot T_{...}^2 - SS(Tr) - SSB$$
▲

Finally, the *error sum of squares* is obtained, as before, by subtracting the various sums of squares from the *total sum of squares*, and we write

▲
$$SSE = SST - SS(Tr) - SSB - SSI$$
▲

In terms of these quantities, the analysis is then performed as indicated in the following *analysis-of-variance table:*

Source of variation	Degrees of freedom	Sum of squares	Mean square	F
Treatments	$k - 1$	$SS(Tr)$	$MS(Tr) = \dfrac{SS(Tr)}{k - 1}$	$\dfrac{MS(Tr)}{MSE}$
Blocks	$n - 1$	SSB	$MSB = \dfrac{SSB}{n - 1}$	$\dfrac{MSB}{MSE}$
Interaction	$(n - 1)(k - 1)$	SSI	$MSI = \dfrac{SSI}{(n - 1)(k - 1)}$	$\dfrac{MSI}{MSE}$
Error	$kn(r - 1)$	SSE	$MSE = \dfrac{SSE}{kn(r - 1)}$	
Total	$rkn - 1$	SST		

and we conclude that the variation among the "cell" means $\bar{x}_{ij}$ attributed to treatments, blocks, or interaction is *not significant* unless the corresponding value obtained for F exceeds $F_{0.05}$ or $F_{0.01}$ for $k - 1$ and $kn(r - 1)$ degrees of freedom, $n - 1$ and $kn(r - 1)$ degrees of freedom, or $(n - 1)(k - 1)$ and $kn(r - 1)$ degrees of freedom, respectively.*

Returning now to our numerical example, where we look upon the operators as treatments and the machines as blocks, we obtain

$$SST = 7{,}521 - \frac{1}{4 \cdot 3 \cdot 3} (493)^2 = 7{,}521 - 6{,}751.4 = 769.6$$

$$SS(Tr) = \frac{1}{9} (147^2 + 138^2 + 108^2 + 100^2) - 6{,}751.4$$

$$= 6{,}924.1 - 6{,}751.4 = 172.7$$

$$SSB = \frac{1}{12} (190^2 + 116^2 + 187^2) - 6{,}751.4$$

$$= 7{,}043.8 - 6{,}751.4 = 292.4$$

$$SSI = \frac{1}{3} (48^2 + 35^2 + 64^2 + 56^2 + 40^2 + 42^2 + 43^2 + 28^2 + 37^2$$

$$+ 43^2 + 13^2 + 44^2) - 6{,}751.4 - 172.7 - 292.4$$

$$= 7{,}360.3 - 6{,}751.4 - 172.7 - 292.4$$

$$= 143.8$$

$$SSE = 769.6 - 172.7 - 292.4 - 143.8 = 160.7$$

and the *analysis-of-variance table* with $k = 4$, $n = 3$, and $r = 3$ is shown on the next page. Since $F_{0.05}$ equals 3.01 for 3 and 24 degrees of freedom, 3.40 for 2 and 24 degrees of freedom, and 2.51 for 6 and 24 degrees of freedom, we find that *all three of the null hypotheses will have to be rejected* at this level of significance.

The fact that there are significant interactions is most important, as it requires that we be very careful before we judge that one machine operator is better than another, or that one machine performs better (or worse) than the rest. However, the significance of the interaction suggests that Mr. A should be kept away from Machine 3 and that Machine

---

* Actually, we now write the true mean corresponding to the ith treatment and the jth block as

$$\mu_{ij} = \mu + \alpha_i + \beta_j + \gamma_{ij}$$

where μ and the α's and β's are defined as before, and the γ's (*gammas*) are the *interaction effects*, whose sum is assumed to be zero when summed on either i or j. Thus, in addition to the two null hypotheses on page 349, we also test the null hypothesis that there is *no interaction*, namely, that *the γ's are all equal to zero*.

Source of variation	Degrees of freedom	Sum of squares	Mean square	F
Treatments	3	172.7	$\dfrac{172.7}{3} = 57.6$	$\dfrac{57.6}{6.7} = 8.6$
Blocks	2	292.4	$\dfrac{292.4}{2} = 146.2$	$\dfrac{146.2}{6.7} = 21.8$
Interaction	6	143.8	$\dfrac{143.8}{6} = 24.0$	$\dfrac{24.0}{6.7} = 3.6$
Error	24	160.7	$\dfrac{160.7}{24} = 6.7$	
Total	35	769.6		

2 should be reserved for Mr. D. (These things do not follow from the F tests, but they are observations which seem to be on fairly safe grounds.)

13.6 Some Further Considerations

The methods presented in this chapter have given us a brief introduction to the analysis of variance. The scope of this subject is vast, and new methods are constantly being developed as their need arises in experiments which do not quite conform to "textbook types."

All of the methods which we have discussed had the special feature that there were observations corresponding to *all possible combinations* of treatments and blocks, namely, observations for all possible combinations of the values (levels) of the variables under consideration. To show that this can be very impractical, we have only to consider an experiment in which we want to compare the yield of 35 varieties of corn and the effect of 12 different fertilizers. To use the method of Section 13.4 we would require $35 \cdot 12 = 420$ plots, and it does not take much imagination to see how difficult it would be to find that many plots for which soil composition, irrigation, slope, ..., are constant or otherwise controllable. The situation also gets worse when there are more than two variables. For instance, in the example of Section 13.5 many more observations would have been required if we also had to consider the possibility that the performance of the operators might be different if they make different

parts or work in different shifts (see Exercise 2 on page 466). Consequently, there is a need for methods of analysis which apply when we
cannot obtain observations for *all* possible combinations of the values
(levels) of *all* variables under consideration and are interested only in
hypotheses concerning *some*, and not all, of the parameters. This leads
to so-called incomplete block designs, which are mentioned briefly in
Chapter 17, and the reader may wish to study this material before continuing with Chapter 14.

EXERCISES

1. To study the performance of three different detergents at three different water temperatures, the following "whiteness" readings were
   obtained with specially designed equipment for nine loads of washing:

	Detergent A	Detergent B	Detergent C
Cold Water	45	43	55
Warm Water	37	40	56
Hot Water	42	44	46

   Perform a two-way analysis of variance, using the level of significance
   $\alpha = 0.05$.

2. Referring to Exercise 4 on page 346, suppose that for each route the
   first observation was obtained on a Monday, the second on a Tuesday,
   the third on a Wednesday, the fourth on a Thursday, and the fifth on
   a Friday. Use the level of significance $\alpha = 0.05$ to test
   (a) the null hypothesis that there is no difference in the true average
       time it takes Mr. Cooper to drive to his friend's house along
       the four different routes;
   (b) the null hypothesis that there is no difference in the true average
       time it takes Mr. Cooper to drive to his friend's house on the
       different days of the week.

3. Referring to Exercise 7 on page 347, suppose that the first value of each
   sample (that is, for each kind of packaging) represents sales at Supermarket A, the second value represents sales at Supermarket B, the
   third value represents sales at Supermarket C, the fourth value represents sales at Supermarket D, and the fifth value represents sales at
   Supermarket E. Use the level of significance $\alpha = 0.05$ to test null

hypotheses about the effects of the different kinds of packaging and the different markets on the sales of the breakfast food.

4. Four different, though supposedly equivalent, forms of a standardized reading achievement test were given to each of five students, and the following are the scores which they obtained:

	Student 1	Student 2	Student 3	Student 4	Student 5
Form A	75	73	59	69	84
Form B	83	72	56	70	92
Form C	86	61	53	72	88
Form D	73	67	62	79	95

Perform a two-way analysis of variance to test at the level of significance $\alpha = 0.01$ whether it is reasonable to treat the four forms as equivalent.

5. A laboratory technician measures the breaking strength of each of five kinds of linen threads by means of four different instruments, and obtains the following results (in ounces):

	Measuring Instruments			
	I_1	I_2	I_3	I_4
Thread 1	20.6	20.7	20.0	21.4
Thread 2	24.7	26.5	27.1	24.3
Thread 3	25.2	23.4	21.6	23.9
Thread 4	24.5	21.5	23.6	25.2
Thread 5	19.3	21.5	22.2	20.6

Perform an analysis of variance to test whether there are differences in breaking strength among the five kinds of threads and whether there are *systematic* differences in the settings of the measuring instruments. Use the level of significance $\alpha = 0.05$.

6. Verify symbolically that
    (a) in a two-way analysis of variance without interaction, the *sum* of the degrees of freedom given for treatments, blocks, and error equals $nk - 1$, namely, the total number of degrees of freedom;
    (b) in a two-way analysis of variance with interaction, the *sum* of the degrees of freedom given for treatments, blocks, interaction, and error equals $rkn - 1$, namely, the total number of degrees of freedom.

7. Suppose that in the experiment of Exercise 1 they tested three loads of washing at each combination of detergents and water temperature, and that they obtained the following results:

	Detergent A	Detergent B	Detergent C
Cold Water	45, 39, 46	43, 46, 41	55, 48, 53
Warm Water	37, 32, 43	40, 37, 46	56, 51, 53
Hot Water	42, 42, 46	44, 45, 38	46, 49, 42

Use the level of significance $\alpha = 0.01$ to test for differences among the detergents, differences due to water temperature, and interactions. Compare the results obtained here with those of Exercise 1.

8. An experiment was performed to judge the effect of four different fuels and two different types of launchers on the range of a certain rocket. Test, on the basis of the following pairs of observations (in nautical miles) for each combination of fuels and launchers, whether there is a significant effect due to differences in fuel, whether there is a significant effect due to differences in launchers, and whether there is an interaction:

	Fuel I	Fuel II	Fuel III	Fuel IV
Launcher X	62.5, 66.8	49.3, 51.1	33.8, 40.1	43.6, 49.2
Launcher Y	32.4, 39.7	31.7, 30.2	39.4, 42.6	51.8, 56.3

Use the level of significance $\alpha = 0.05$.

9. A consumer-products-testing service wants to compare the quality of 24 cakes baked in its kitchen with each of four different mixes pre-

pared according to three different recipes (varying in the amounts of fresh ingredients added), once by Chef X and once by Chef Y. They ask a taste tester to rate the cakes subjectively on a scale from 0 to 100 and obtain the following results, where in each case the first figure pertains to the cake baked by Chef X and the second figure pertains to the cake baked by Chef Y:

	Mix A	Mix B	Mix C	Mix D
Recipe 1	66, 62	70, 68	74, 68	73, 67
Recipe 2	68, 61	71, 73	74, 70	66, 61
Recipe 3	75, 68	69, 71	67, 63	70, 66

Using the level of significance $\alpha = 0.05$, analyze this experiment by the method of Section 13.5.

10. (*Continuation of Exercise 9*) Looking at the data of the preceding exercise, the reader may have noted that *ten times out of twelve* the cakes of Chef X got a higher rating than the corresponding cakes of Chef Y. This suggests that the variations which we looked upon as *experimental error* in the analysis of Exercise 9, may be due, partially at least, to differences in the conditions (different chefs) under which the experiment is repeated. In our example, this variation is reflected by the difference between the mean ratings of $\dfrac{843}{12} = 70.25$ and $\dfrac{798}{12} = 66.5$ received by the cakes of Chefs X and Y, and in general the variation that is thus attributed to *replication* (or repetition) is measured by the *replication sum of squares*

$$\blacktriangle \qquad SSR = \frac{1}{kn} \cdot \sum_{h=1}^{r} T_{\cdot\cdot h}^2 - \frac{1}{knr} \cdot T_{\cdot\cdot\cdot}^2 \qquad \blacktriangle$$

where $T_{\cdot\cdot h}$ is the total of the observations in the hth replication. So far as the analysis is concerned, SST, $SS(Tr)$, SSB, and SSI are computed in the same way as before, the error sum of squares is now given by

$$\blacktriangle \qquad SSE = SST - SS(Tr) - SSB - SSI - SSR \qquad \blacktriangle$$

and the remainder of the work is as shown in the following table:

Source of variation	Degrees of freedom	Sum of squares	Mean square	F
Treatments	$k-1$	$SS(Tr)$	$MS(Tr) = \dfrac{SS(Tr)}{k-1}$	$\dfrac{MS(Tr)}{MSE}$
Blocks	$n-1$	SSB	$MSB = \dfrac{SSB}{n-1}$	$\dfrac{MSB}{MSE}$
Interaction	$(n-1)(k-1)$	SSI	$MSI = \dfrac{SSI}{(n-1)(k-1)}$	$\dfrac{MSI}{MSE}$
Replication	$r-1$	SSR	$MSR = \dfrac{SSR}{r-1}$	$\dfrac{MSR}{MSE}$
Error	$(r-1)(kn-1)$	SSE	$MSE = \dfrac{SSE}{(r-1)(kn-1)}$	
Total	$rkn-1$	SST		

Note that the degrees of freedom for error are now $(r-1)(kn-1)$ instead of $kn(r-1)$, and that we can perform a fourth test of significance. The differences among the repetitions (replications) of the experiment are looked upon as significant if the value obtained for $\dfrac{MSR}{MSE}$ exceeds $F_{0.05}$ or $F_{0.01}$ with $r-1$ and $(r-1)(kn-1)$ degrees of freedom.

(a) Use this kind of analysis to rework Exercise 9 and discuss the results.
(b) Use this kind of analysis to rework the example in the text, which dealt with the number of defective pieces produced by four machine operators, working each of three different machines on three different occasions.
(c) Use this kind of analysis to rework Exercise 8 and discuss the results.

BIBLIOGRAPHY

The following are some of the many books that have been written on the subject of analysis of variance:

GUENTHER, W. C., *Analysis of Variance*. Englewood Cliffs, N.J.: Prentice-Hall, Inc., 1964.

HICKS, C. R., *Fundamental Concepts in the Design of Experiments*. New York: Holt, Rinehart & Winston, Inc., 1964.

LI, C. C., *Introduction to Experimental Statistics*. New York: McGraw-Hill, Inc., 1964.

SNEDECOR, G. W., and COCHRAN, W. G., *Statistical Methods*, 6th ed. Ames, Iowa: Iowa State University Press, 1967.

14

Nonparametric Methods

14.1 Introduction

Most of the methods we have discussed in the last four chapters required specific assumptions about the population (or populations) from which our data were obtained. Among other things, we often had to assume that these populations can be approximated closely with normal distributions, that a population has a given standard deviation σ, or that several populations have the same standard deviation. Since there are many situations where one or more of these assumptions cannot be met, statisticians have developed alternate techniques based on less stringent assumptions. This includes *nonparametric methods* as well as *distribution-free* methods, where the first term applies when we are not concerned with the parameters of populations *of a given kind*, and the second term applies when we make no assumptions about the populations from which we are sampling, except perhaps that they must be continuous. Actually, since this distinction is rather fine, it has become the custom to refer to either kind of method simply as nonparametric.

Many nonparametric methods have great intuitive appeal since they also fall under the heading of "quick and easy" or "short-cut" techniques. Not only are most of these methods simpler so far as arithmetical details are concerned, but many of them are easier to understand or explain than the standard techniques which they replace. On the other hand, it should be recognized that methods which require no (or virtually no) assumptions about the populations from which we are sampling are apt to be *less efficient* than the corresponding standard methods—in tests of hypotheses

362

they may expose us to greater risks of committing Type II errors, and in estimation problems they may provide us with intervals which are much wider. Generally, it is true that *the more we are willing to assume the more we can infer from a sample*, and to illustrate this point we have only to refer to the example on page 232, where Chebyshev's Theorem led to the result that "the probability is at most 0.25 that the mean of a random sample of size 64 from an infinite population with $\sigma = 20$ will differ from the mean of the population by 5 or more." Now, if we assume that the population has the shape of a normal distribution, we can use the method of Section 9.3 and make the same assertion with the probability 0.9544 (according to Table I). As we said, the more we are willing to assume the more we can infer from a sample, but we should add that *the more we assume the further we may stick out our neck*.

14.2 The One-Sample Sign Test

Except for the large-sample tests, all of the tests concerning means which we studied in Chapter 10 were based on the assumption that the populations from which we are sampling have roughly the shape of normal distributions. When this assumption is untenable, these tests can be replaced by a variety of nonparametric alternatives, among them the *sign test*, which we shall study in this section and in Section 14.3. To illustrate the *one-sample case* (that is, the nonparametric substitute for the various tests of Section 10.7), suppose we are dealing with a population having a continuous *symmetrical* distribution, so that the probability of getting a value less than the mean equals the probability of getting a value greater than the mean. To test the null hypothesis $\mu = \mu_0$ against an appropriate alternative, we replace each sample value exceeding μ_0 with a *plus sign* and each sample value less than μ_0 with a *minus sign*, and we can then test instead the null hypothesis that the probabilities of getting plus and minus signs are both $\dfrac{1}{2}$. (If a sample value happens to equal μ_0, which is possible since we always round the values of continuous random variables, it is simply discarded.)

To perform the actual tests, we use either of the two methods described in Section 12.4: When the sample is *small* we refer directly to Table VI (or some more extensive table of binomial probabilities), and when the sample is *large* we use the normal approximation of the binomial distribution. To illustrate the first of these techniques, let us refer again to the oceanographer of the example on page 278, and let us suppose that in 15 soundings he obtained the following ocean depths in fathoms: 45.2, 47.7, 42.0, 51.3, 43.9, 53.2, 40.0, 51.1, 44.6, 45.5, 37.8, 50.9, 47.3, 45.4, and

43.9. To test the null hypothesis that $\mu = 42.0$ fathoms against the alternative $\mu \neq 42.0$ fathoms at the level of significance $\alpha = 0.05$, we first replace each value less than 42.0 with a minus sign, each value greater than 42.0 with a plus sign, and discard the one value which actually equalled 42.0. Thus, we get

$$+ \quad + \quad + \quad + \quad + \quad - \quad + \quad + \quad + \quad - \quad + \quad + \quad + \quad +$$

and it remains to be seen whether "12 successes in 14 trials" supports the null hypothesis $p = \dfrac{1}{2}$ or the alternative hypothesis $p \neq \dfrac{1}{2}$. Using Table VI, we find that the probability of "12 or more successes" is $0.006 + 0.001 = 0.007$ for $n = 14$ and $p = \dfrac{1}{2}$, and since this value is less than $\dfrac{\alpha}{2} = 0.025$, we find that *the null hypothesis will have to be rejected.*

We could have used the normal approximation in this last problem, since np and $n(1 - p)$ both exceed 5 in accordance with the rule on page 213, but let us use a different example. Suppose, for instance, that in the example on page 249 we are interested in determining whether the pulse rate of the astronauts performing the given task is on the average increased by at least 22 beats per minute; that is, we shall want to test the null hypothesis $\mu = 22$ against the alternative hypothesis $\mu > 22$. Getting 21 plus signs, 9 minus signs, and discarding the two observations which actually equalled 22, it remains to be seen whether we can reject the null hypothesis $p = \dfrac{1}{2}$ against the alternative hypothesis $p > \dfrac{1}{2}$, say, at the level of significance $\alpha = 0.05$. Getting $\mu = np = 30 \cdot \dfrac{1}{2} = 15$ and

$$\sigma = \sqrt{np(1 - p)} = \sqrt{30 \cdot \dfrac{1}{2} \cdot \dfrac{1}{2}} = 2.74$$ for the mean and the standard deviation of the corresponding binomial distribution, we find that for $x = 21$ (plus signs),

$$z = \frac{21 - 15}{2.74} = 2.19$$

Since this exceeds $z_{0.05} = 1.64$, it follows that the null hypothesis will have to be rejected, and we conclude that the average pulse rate is increased by more than 22 beats per minute. When n is small in problems like this, it may be desirable to use the *continuity correction* given in the footnote to page 314, but if we had substituted $21 - 15 - \dfrac{1}{2}$ instead of $21 - 15$ into the formula for z, we would have obtained $z = 2.01$, and the conclusion would have been the same. Before we go on to the two-sample sign test, let us point out that if the population distribution is *not symmetrical*, the method of this section can be used nevertheless to test the null hypothesis $\tilde{\mu} = \tilde{\mu}_0$, where $\tilde{\mu}$ is the *population median*.

14.3 The Two-Sample Sign Test

The sign test has important applications in problems where we are dealing with *paired data*, so that each pair can be replaced with a *plus sign* if the first value is greater than the second, a *minus sign* if the first value is smaller than the second, or be discarded if the two values are equal. In connection with this, there are essentially two kinds of situations, depending on whether the data are actually *given as pairs* as in Exercises 22, 23, and 24 on page 286, or whether they consist of two independent samples which are *randomly paired*.

To illustrate the *first case*, let us refer to the analysis of variance problem of Exercises 9 and 10 on page 359, where we observed that *ten times out of twelve* the cakes of Chef X got a higher rating than the corresponding cakes of Chef Y. Copying the data from page 359, we obtain the values shown in the following table, where we also indicated by means of a plus sign that Chef X got the higher rating and by means of a minus sign that Chef X got the lower rating:

Chef X	Chef Y	
66	62	+
70	68	+
74	68	+
73	67	+
68	61	+
71	73	−
74	70	+
66	61	+
75	68	+
69	71	−
67	63	+
70	66	+

There are 10 plus signs and 2 minus signs, and it remains to be seen whether "10 successes in 12 trials" supports the null hypothesis $p = \dfrac{1}{2}$ or the alternative hypothesis $p > \dfrac{1}{2}$ (namely, the alternative hypothesis that Chef X is a better baker than Chef Y). Referring to Table VI, we find that for $n = 12$ and $p = \dfrac{1}{2}$ the probability of "10 or more successes" is $0.016 + 0.003 = 0.019$, and it follows that *the null hypothesis can be rejected* at the level of significance $\alpha = 0.05$, but not at the level of significance $\alpha = 0.01$. If we use the first, we can conclude that Chef X is a better baker than Chef Y.

To illustrate the *second case*, where we deal with two independent random samples, let us consider the following data pertaining to the *down-times* (periods in which they were inoperative due to failures, in minutes) of two different computers:

Computer A: 62, 71, 38, 61, 57, 61, 58, 50, 52, 80, 56, 47, 55, 32, 49,
        70, 41, 86, 52, 48, 59, 54
Computer B: 53, 48, 46, 49, 82, 41, 50, 36, 55, 49, 50, 22, 44, 62, 74,
        32, 49, 59, 47, 49, 59, 46, 71, 47

Note that there are two more values in the second sample, so that if we randomly match one value of the second sample with each value of the first sample, two values of the second sample will be left over. Actually doing this with the use of random numbers, we obtain the pairs shown in the following table, where we also indicated by means of a plus sign that the value obtained for Computer A is larger, and by means of a minus sign that the value obtained for Computer B is larger:*

Computer A	Computer B		Computer A	Computer B	
62	22	+	47	46	+
71	46	+	55	32	+
38	49	−	32	53	−
61	59	+	49	41	+
57	44	+	70	74	−
61	49	+	41	49	−
58	48	+	86	47	+
50	47	+	52	50	+
52	71	−	48	36	+
80	62	+	59	50	+
56	82	−	54	55	−

There are 15 plus signs and 7 minus signs, and we shall have to see whether "15 successes in 22 trials" supports the null hypothesis $p = \dfrac{1}{2}$

* We did this by assigning three-digit random numbers to the values of the second sample and then pairing the one with the smallest random number with 62, the first value obtained for Computer A, the one with the second smallest random number with 71, the second value obtained for Computer A, and so forth.

(namely, the null hypothesis that the true average down-time is the same for Computers A and B) or the alternative hypothesis $p \neq \frac{1}{2}$.

As in the first example of this section, we could base this test on a table of binomial probabilities, but since the table in this book goes only to $n = 15$, let us use the normal approximation of the binomial distribution. This can be done since np and $n(1 - p)$ are both equal to 11 in this example. Substituting $n = 22$ and $p = \frac{1}{2}$ into the formulas for the mean and the standard deviation of the binomial distribution, we get $\mu = np = 22 \cdot \frac{1}{2} = 11$ and $\sigma = \sqrt{np(1 - p)} = \sqrt{22 \cdot \frac{1}{2} \frac{1}{2}} = 2.345$ and, hence,

$$z = \frac{15 - 11}{2.345} = 1.71$$

Since this value falls between $-z_{0.025} = -1.96$ and $z_{0.025} = 1.96$, we find that *the null hypothesis cannot be rejected*. This may surprise the reader since there seem to be substantial differences (in fact, the two sample means are, respectively, 56.3 and 50.8), but it illustrates the point that *the sign test can be quite wasteful of information*. Note also that if we had used the *continuity correction*, we would have obtained

$$z = \frac{3.5}{2.345} = 1.49$$

and the conclusion would have been the same.

14.4 The Two-Sample Median Test

The *two-sample median test* serves the same purpose as the two-sample sign test for independent samples, but it does not require that we discard some of the data when the sample sizes are unequal. To perform this test we first find the median of the *combined data*, determine how many of the values in each sample fall above and below the median, and then analyze the resulting two-by-two table by the method of Section 12.7.

To illustrate this technique, let us refer back to the second example of the preceding section, the one dealing with the down-times of the two computers. The median of the combined data is 51, as can easily be checked, and we find that for Computer A there are 7 values below 51 and 15 values above 51, while for Computer B the corresponding figures are 16 and 8.

All this information is summarized in the following table, in which we have indicated also the totals of the rows and columns:

	Below Median	*Above Median*	
Computer A	7	15	22
Computer B	16	8	24
	23	23	46

Proceeding as in Section 12.7, we determine the *expected cell frequencies* for the first row as $\dfrac{22 \cdot 23}{46} = 11$ and $22 - 11 = 11$, and (also by subtraction) those of the second row as $23 - 11 = 12$ and $23 - 11 = 12$. Of course, we could have obtained these results also by arguing that *half the values in each sample can be expected to fall above the median and the other half below the median.* Then, substituting into the formula for χ^2 on page 322, we get

$$\chi^2 = \frac{(7 - 11)^2}{11} + \frac{(15 - 11)^2}{11} + \frac{(16 - 12)^2}{12} + \frac{(8 - 12)^2}{12}$$
$$= 5.58$$

and since this exceeds 3.841, the value of $\chi^2_{0.05}$ for 1 degree of freedom, we find that *the null hypothesis will have to be rejected* at this level of significance—the true average (median) down-time of the two computers is *not* the same. (As we indicated in connection with the sign test, we test whether the two populations have the same median, or assuming that they are symmetrical, whether they have the same mean.) The fact that the two-sample median test led to the rejection of the null hypothesis, whereas the two-sample sign test (applied to the same data in Section 14.3) did not, may be construed as evidence that the median test is not quite so wasteful of information as the sign test. This may be the case, but in general it is quite difficult to make meaningful comparisons of the merits of two or more nonparametric tests that can be used for the same purpose.

14.5 The k-Sample Median Test

The median test can easily be generalized so that it applies to k samples. As before, we find the median of the *combined data*, determine how many of the values in each sample fall above and below the median, and then analyze the resulting k-by-two table by the method of Section 12.7.

To illustrate this technique, let us refer to the following grades which random samples of eighth graders from three different schools obtained in an achievement test in mathematics:

School 1: 60, 55, 82, 70, 46, 63, 88, 69, 61, 43,
        76, 54, 58, 65, 73, 52
School 2: 74, 67, 37, 80, 72, 92, 19, 52, 77, 40,
        83, 76, 68, 21, 90, 74, 49, 70, 65, 58
School 3: 46, 60, 58, 80, 66, 39, 56, 61, 81, 70,
        75, 48, 43, 64, 57, 59, 87, 50, 73, 62

To test whether the differences among the three sample means are significant, we would ordinarily perform a one-way analysis of variance, but we shall not do so here because there seems to be much more variability in the second sample than there is in the other two. This suggests that it would be unreasonable to assume that the three population standard deviations are the same and, hence, that the method of Section 13.2 cannot be used.

To perform a median test, we first determine the median of the *combined data,* which is 63.5, as can easily be checked. Then we count how many of the grades in each sample fall below and above the median, and obtain the results shown in the following table:

	Below Median	*Above Median*
School 1	9	7
School 2	7	13
School 3	12	8

Since the corresponding *expected frequencies* for School 1 are 8 and 8, those for School 2 are 10 and 10, and those for School 3 are 10 and 10, substitution into the formula for χ^2 yields

$$\chi^2 = \frac{(9-8)^2}{8} + \frac{(7-8)^2}{8} + \frac{(7-10)^2}{10} + \frac{(13-10)^2}{10}$$
$$+ \frac{(12-10)^2}{10} + \frac{(8-10)^2}{10}$$
$$= 2.85$$

Since this value does not exceed 5.991, the value of $\chi^2_{0.05}$ for $k - 1 = 3 - 1$ $= 2$ degrees of freedom, we find that *the null hypothesis cannot be rejected;* that is, we *cannot* conclude that there is a difference in the true average (median) grade which students from the three schools would get in the test. (Although there was no need for it in this example or the one of the preceding section, let us point out that values which are actually *equal* to the median are not counted in this kind of analysis.)

EXERCISES

1. Use the normal approximation of the binomial distribution to rework the illustration on page 363, which dealt with the 15 soundings of the ocean depth.

2. The following are sample data of the amount of money (in dollars) which a person spends visiting a certain amusement park: 7.50, 10.00, 13.75, 9.50, 10.35, 11.45, 8.85, 9.25, 6.65, 15.60, 11.10, 8.50, 13.85, 9.85, and 10.15. Use the sign test (based on Table VI) and the level of significance $\alpha = 0.05$ to test the null hypothesis that on the average a person spends $9.00 at the park against the alternative that this figure is too low.

3. The yield of alfalfa from nine test plots is 1.1, 2.1, 1.8, 1.0, 1.7, 1.3, 1.9, 1.2, and 1.5 tons per acre. Use the sign test (based on Table VI) and the level of significance $\alpha = 0.05$ to test the null hypothesis that the true average yield is 2.0 tons per acre against the two-sided alternative that this figure is incorrect.

4. Use the sign test (based on Table VI) and the level of significance $\alpha = 0.05$ to rework Exercise 9 on page 284, which dealt with the weight losses of eight overweight women.

5. Referring to the data of Exercise 6 on page 42, test the null hypothesis that on the average the insects will live 119 seconds after having been sprayed with the insecticide. Use the alternative hypothesis $\mu \neq 119$, the level of significance $\alpha = 0.05$, and base the decision
   (a) on Table VI;
   (b) on the normal approximation of the binomial distribution.

6. Referring to Exercise 7 on page 42, use the sign test and the level of significance $\alpha = 0.01$ to test the null hypothesis that the true average weight for the entire production lot of floor wax is 12.05 ounces per can against the one-sided alternative $\mu < 12.05$. (Use *continuity correction*).

7. Referring to Exercise 8 on page 42, use the sign test and the level of significance $\alpha = 0.01$ to test the null hypothesis that on the average

30 twists are required to break one of the bars. Use the alternative hypothesis that this figure is too low.

8. Use the two-sample sign test to rework Exercise 22 on page 286.

9. Use the two-sample sign test to rework Exercise 23 on page 286.

10. Use the two-sample sign test to rework Exercise 24 on page 287.

11. The following are the number of speeding tickets issued by two policemen on 30 days: 7 and 10, 11 and 13, 10 and 11, 14 and 14, 11 and 15, 12 and 9, 6 and 10, 9 and 13, 8 and 11, 10 and 11, 11 and 15, 13 and 11, 7 and 10, 6 and 12, 10 and 14, 8 and 8, 11 and 12, 9 and 14, 9 and 7, 10 and 12, 6 and 7, 12 and 14, 9 and 11, 12 and 10, 11 and 13, 12 and 15, 7 and 9, 10 and 9, 11 and 13, and 8 and 10. Use the sign test at the level of significance $\alpha = 0.01$ to test the null hypothesis that on the average the two policemen issue equally many speeding tickets against the alternative hypothesis that on the average the second policeman issues more speeding tickets than the first.

12. The following are the number of employees absent from two departments of a large firm on 25 days: 2 and 4, 6 and 3, 5 and 5, 7 and 2, 3 and 1, 4 and 3, 2 and 5, 3 and 1, 4 and 3, 5 and 6, 5 and 4, 3 and 8, 6 and 4, 5 and 2, 4 and 3, 3 and 0, 2 and 5, 6 and 4, 3 and 1, 2 and 4, 5 and 2, 3 and 2, 4 and 6, 6 and 3, and 4 and 3. Use the sign test at the level of significance $\alpha = 0.05$ to test the null hypothesis that on the average there are equally as many absences in the two departments against the alternative hypothesis that on the average there are more absences in the first department.

13. Random samples of freshmen from two colleges obtained the following scores in a test designed to measure their knowledge of American history:

College A: 77, 93, 70, 81, 98, 93, 75, 89, 87, 66, 59, 87, 94, 73, 95, 67, 88, 97, 94, 90, 84, 89, 55, 69, 74
College B: 73, 65, 60, 72, 31, 75, 72, 60, 34, 85, 60, 60, 85, 48, 59, 63, 85, 91, 97, 71, 68, 53, 42, 71, 86

Randomly pair each of the grades from College A with one of the grades from College B, and then perform a sign test of the null hypothesis that there is no difference in the average knowledge of American history between freshmen attending the two colleges. Use the two-sided alternative that there *is* a difference and perform the test at the level of significance $\alpha = 0.05$.

14. Rework Exercise 13 using the two-sample median test.

15. The following are data on the breaking strength (in pounds) of samples of two kinds of 2-inch cotton ribbon:

Type I Ribbon: 143, 180, 199, 186, 168, 170, 185, 193, 198, 196, 175, 181, 132, 192, 196, 164, 179, 197, 176, 180

Type II Ribbon: 176, 165, 173, 195, 177, 199, 155, 135, 172, 183, 170, 165, 186, 160, 162, 190, 171, 165, 189, 171, 180, 176, 193, 194

Randomly match each value obtained for the Type I ribbon with a value obtained for the Type II ribbon, and then perform a sign test of the null hypothesis that there is no difference in the true average strength of the two types of ribbon. Use the alternative hypothesis that there *is* a difference and the level of significance $\alpha = 0.05$.

16. Rework Exercise 15 using the two-sample median test.

17. For comparison, two kinds of feed are fed to samples of pigs, and the following are their gains in weight (in ounces) after a fixed period of time:

Feed C: 13.8, 13.4, 12.0, 12.9, 15.2, 14.2, 13.1, 14.6, 13.3, 12.5, 10.4, 14.1, 15.0, 12.4, 13.5, 14.8, 15.5, 13.2, 12.3, 12.0, 15.6, 14.1

Feed D: 11.4, 13.8, 11.7, 14.3, 12.6, 11.1, 13.6, 12.2, 13.8, 10.5, 12.8, 14.0, 10.2, 12.8, 12.7, 15.4, 11.3, 12.6, 11.8, 13.1, 10.8, 11.9, 14.0, 12.4, 12.7, 11.8, 13.1

(a) Randomly match each value obtained for Feed C with a value obtained for Feed D, and perform a sign test of the null hypothesis that the true average gain in weight is the same for both feeds. Use a two-sided alternative and the level of significance $\alpha = 0.01$.

(b) Use the two-sample median test to test the null hypothesis of part (a) at the level of significance $\alpha = 0.01$. Again, use the alternative that the true average gain in weight is not the same for both feeds.

18. The following are the lifetimes (in hours) of samples of four kinds of light bulbs in continuous use:

Brand A: 603, 625, 641, 622, 585, 593, 660, 600, 633, 580, 615, 648
Brand B: 620, 640, 646, 620, 652, 639, 590, 646, 631, 669, 610, 619
Brand C: 587, 602, 617, 650, 588, 612, 574, 628, 617, 598, 602, 657
Brand D: 626, 608, 596, 601, 637, 654, 601, 597, 644, 614, 582, 626

Use the median test at the level of significance $\alpha = 0.05$ to test the null hypothesis that the true average lifetimes of the four kinds of light bulbs are the same.

19. The following are the scores a golf pro made on the same eighteen-hole course with three different sets of clubs:

    Set 1: 73, 71, 69, 70, 72, 74, 74, 68, 70, 73, 72, 71, 75, 76, 67, 77
    Set 2: 69, 73, 70, 69, 70, 72, 72, 66, 71, 73, 78, 72, 70, 70, 75, 68, 73
    Set 3: 73, 71, 74, 74, 76, 75, 69, 78, 70, 72, 73, 71, 75, 73, 77

    Use the median test at the level of significance $\alpha = 0.05$ to test the null hypothesis that the pro plays equally well with the three sets of clubs.

14.6 Rank-Sum Tests: The Mann-Whitney Test

So far we have studied two nonparametric alternatives to the *two-sample t test* (that is, the small-sample test for the difference between two means which we discussed in Section 10.8). Another nonparametric alternative to this kind of test is the *Mann-Whitney test* (also called the *U test*), which is generally not quite so wasteful of information, but requires more work. This test is based on *rank sums;* that is, the combined data are ranked according to size (usually "low to high"), and the test is based on the sum of the ranks assigned to the observations in either sample.

To illustrate how the Mann-Whitney test is used to decide *whether two samples come from identical populations or whether these populations have unequal means,* suppose we want to compare the following gains in weight (in pounds) of two samples of turkeys, which are fed different diets but are otherwise kept under identical conditions:

    Diet I: 14.0, 13.2, 18.4, 14.5, 15.1, 23.6, 14.7, 12.0, 10.2, 14.3,
             11.8, 14.9, 13.5, 10.7, 16.3
    Diet II: 16.2, 15.8, 21.1, 20.7, 18.9, 14.8, 20.1, 15.3, 19.2, 18.8,
             13.9, 12.6, 19.6, 15.4, 21.3

The means of these two samples are 14.5 and 17.6, respectively, and the problem is to decide whether their difference is significant. Note that we may well be reluctant to use the standard test of Section 10.8 since the first sample shows considerably more variability than the second; its values range from 10.2 to 23.6, whereas those of the second sample range from 12.6 to 21.3. (Of course, we could put this to a rigorous test using the F test of Section 11.2.)

To perform the Mann-Whitney test, we first rank the data *jointly* (as if they were one sample) in an increasing (or decreasing) order of magnitude. For our data, we thus obtain the following array, where we used the numbers I and II to indicate whether the corresponding turkey was fed Diet I or Diet II:

10.2	10.7	11.8	12.0	12.6	13.2	13.5	13.9	14.0	14.3
I	I	I	I	II	I	I	II	I	I

14.5	14.7	14.8	14.9	15.1	15.3	15.4	15.8	16.2	16.3
I	I	II	I	I	II	II	II	II	I

18.4	18.8	18.9	19.2	19.6	20.1	20.7	21.1	21.3	23.6
I	II	II	II	II	II	II	II	II	I

Assigning the data *in this order* the ranks 1, 2, 3, ..., and 30, we find that the values of the first sample (Diet I) occupy ranks 1, 2, 3, 4, 6, 7, 9, 10, 11, 12, 14, 15, 20, 21, and 30, while those of the second sample (Diet II) occupy ranks 5, 8, 13, 16, 17, 18, 19, 22, 23, 24, 25, 26, 27, 28, and 29. There are no *ties* in this example among values belonging to different samples, but if there were, we would assign each of the tied observations the *mean* of the ranks which they jointly occupy. (Thus, when the fourth and fifth values are identical we assign each the rank $\dfrac{4+5}{2} = 4.5$, when the ninth, tenth, and eleventh values are identical we assign each the rank $\dfrac{9+10+11}{3} = 10$, and so on.)

The null hypothesis we shall want to test is that both samples come from identical populations, and it stands to reason that in that case the means of the ranks assigned to the values of the two samples should be more or less the same. The alternative hypothesis is that the populations have unequal means, and if this difference is pronounced, most of the smaller ranks will go to the values of one sample while most of the higher ranks will go to those of the other sample.

Using *rank sums* (rather than the means of the ranks), we shall base the test of this null hypothesis on the statistic

$$U = n_1 n_2 + \frac{n_1(n_1+1)}{2} - R_1$$

where n_1 and n_2 are the sizes of the two samples and R_1 is the sum of the ranks assigned to the values of the first sample. (In practice, we find whichever rank sum is most easily obtained, as it is immaterial which sample is referred to as the "first.") In our example we have $n_1 = 15$, $n_2 = 15$, $R_1 = 1 + 2 + 3 + 4 + 6 + 7 + 9 + 10 + 11 + 12 + 14 + 15 + 20 + 21 + 30 = 165$, and hence,

$$U = 15 \cdot 15 + \frac{15 \cdot 16}{2} - 165 = 180$$

Under the null hypothesis that the $n_1 + n_2$ observations may be regarded as coming from identical populations, or one population, it can be shown that the sampling distribution of U has the mean

$$\mu_U = \frac{n_1 n_2}{2}$$

and the standard deviation

$$\sigma_U = \sqrt{\frac{n_1 n_2 (n_1 + n_2 + 1)}{12}}$$

Furthermore, if n_1 and n_2 are both greater than 8 (some statisticians prefer that they be both greater than 10), the sampling distribution of U can be approximated closely with a normal distribution. Hence, we reject the null hypothesis (against the alternative that the two population means are *unequal*) if

$$z = \frac{U - \mu_U}{\sigma_U}$$

is less than $-z_{\alpha/2}$ or exceeds $z_{\alpha/2}$, where α is as always the level of significance. If either n_1 or n_2 is so small that the normal curve approximation cannot be used, the test will have to be based on special tables such as those referred to in the Bibliography at the end of this chapter.

Returning now to our numerical example and substituting $n_1 = 15$ and $n_2 = 15$ into the formulas for μ_U and σ_U, we find that

$$\mu_U = \frac{15 \cdot 15}{2} = 112.5$$

$$\sigma_U = \sqrt{\frac{15 \cdot 15 \cdot 31}{12}} = 24.1$$

and, hence, that

$$z = \frac{180 - 112.5}{24.1} = 2.80$$

Since this value exceeds $z_{0.005} = 2.58$, we can *reject the null hypothesis* at the level of significance $\alpha = 0.01$; in other words, we conclude that there *is* a difference between the two feeds so far as weight gain is concerned.

Like the sign and median tests, the Mann-Whitney test has the important advantage that it requires fewer assumptions than the corresponding "standard" test. In fact, the only assumption needed is that the populations from which we are sampling are continuous, and in actual practice even the violation of this assumption is not serious. (When there are many ties, it may be well, though, to make the corrections in the formulas for μ_U and σ_U indicated in the book by S. Siegel listed on page 385.)

An interesting feature of the Mann-Whitney test is that, with a slight modification, it can also be used to test the null hypothesis that two samples come from identical populations against the alternative that the two populations have *unequal dispersions*, namely, that they differ in variability or spread. As before, the values of the two samples are arranged jointly in an increasing (or decreasing) order of magnitude, but now they are ranked *from both ends toward the middle*. We assign Rank 1 to the smallest value, Ranks 2 and 3 to the largest and second largest values, Ranks 4 and 5 to the second and third smallest, Ranks 6 and 7 to the third and fourth largest, and so on. Subsequently, the calculation of U and the performance of the test are the same as before. The only difference is that with this kind of ranking a *small rank sum* tends to indicate that the population from which the sample was obtained has a *greater variation* than the other, because its values occupy the more extreme positions. We shall not attempt to illustrate this technique with reference to our turkey-feed example, since this test loses its *sensitivity* for detecting differences in variability when the means of the populations are not the same. The reader will be asked to use it, though, in Exercises 7 and 8 on page 379.

14.7 Rank-Sum Tests: The Kruskal-Wallis Test

The Kruskal-Wallis test (also called the *H test*) is a rank-sum test which is used to test the null hypothesis that k independent random samples come from identical populations against the alternative that the means of the populations are not all equal. Like the k-sample median test, it thus provides a nonparametric alternative for a one-way analysis of variance. As in the Mann-Whitney test, the combined data are ranked according to size; then if R_i is the sum of the ranks assigned to the n_i observations in the ith sample and $n = n_1 + n_2 + \ldots + n_k$, the test is based on the statistic

$$\blacktriangle \qquad H = \frac{12}{n(n+1)} \sum_{i=1}^{k} \frac{R_i^2}{n_i} - 3(n+1) \qquad \blacktriangle$$

If the null hypothesis is true and each sample is at least of size 5, the sampling distribution of this statistic can be approximated closely with a chi-square distribution with $k - 1$ degrees of freedom. Consequently, we can reject the null hypothesis at the level of significance α if H exceeds χ_α^2 for $k - 1$ degrees of freedom. If the size of one or more samples is too small to use this approximation, the test will have to be based on special tables such as those referred to in the Bibliography at the end of this chapter.

To illustrate the Kruskal-Wallis test, suppose that three groups of foreign-language students are taught German by three different methods (using records and a text, using records only, and using only a text), and that in a test given at the end of the term they obtained the following scores:

> *Method Q:* 94, 87, 91, 74, 86, 97
> *Method S:* 85, 82, 79, 84, 61, 72, 80
> *Method T:* 89, 67, 72, 76, 69

As can easily be verified, the observations in the first sample are assigned the ranks 6, 13, 14, 16, 17, and 18, so that $R_1 = 84$; those in the second sample are assigned the ranks 1, 4.5, 8, 9, 10, 11, and 12, so that $R_2 = 55.5$; and those in the third sample are assigned the ranks 2, 3, 4.5, 7, and 15, so that $R_3 = 31.5$. (Note that tied observations belonging to different samples are again assigned the *mean* of the ranks which they jointly occupy; for tied observations belonging to the same sample it does not matter how they are assigned the ranks which they jointly occupy.) Substituting the values obtained for R_1, R_2, and R_3 together with $n_1 = 6$, $n_2 = 7$, and $n_3 = 5$ into the formula for H, we get

$$H = \frac{12}{18\cdot19}\left(\frac{84^2}{6} + \frac{55.5^2}{7} + \frac{31.5^2}{5}\right) - 3\cdot19$$
$$= 6.6$$

Since this exceeds 5.991, the value of $\chi_{0.05}^2$ for $k - 1 = 3 - 1 = 2$ degrees of freedom, *the null hypothesis will have to be rejected.* In other words, we can conclude that there *is* a difference in the effectiveness of the three methods of teaching foreign languages.

EXERCISES

1. The following are the number of minutes it took a sample of 15 men and 12 women to complete the written part of the test to obtain their drivers' licenses:

    Men: 6.5, 10.0, 7.0, 9.8, 8.5, 9.2, 9.0, 8.2, 10.8, 8.7, 6.7, 8.1, 7.9, 6.4, 8.9
    Women: 8.6, 7.8, 8.3, 6.6, 10.5, 6.3, 9.3, 8.4, 9.7, 8.8, 9.9, 7.6

    Use the Mann-Whitney test at the level of significance $\alpha = 0.05$ to test the null hypothesis that the two samples come from identical populations against the alternative that the populations have unequal means.

2. Apply the Mann-Whitney test to the example on page 366, the one dealing with the down-times of the two computers, and perform the significance test at the level of significance $\alpha = 0.01$.

3. The following are the weekly food expenditures (in dollars) of 10 families with two children chosen at random from each of two suburbs of a large city:

    Suburb A: 34.78, 42.60, 31.89, 34.50, 36.00, 39.38, 28.19, 50.45, 35.15, 31.95
    Suburb B: 30.12, 24.63, 45.91, 52.16, 54.59, 19.35, 32.76, 58.19, 25.75, 48.72

    Use the Mann-Whitney test at the level of significance $\alpha = 0.05$ to test the null hypothesis that the two samples come from identical populations against the alternative that the populations have unequal means.

4. Apply the Mann-Whitney test to the data of Exercise 13 on page 371 to test the null hypothesis that there is no difference in the average knowledge of American history between freshmen attending the two colleges. Use the level of significance $\alpha = 0.05$.

5. Use the Mann-Whitney test instead of the sign test to rework Exercise 15 on page 372.

6. Use the Mann-Whitney test to compare the lifetimes of the Brand A and B light bulbs of Exercise 18 on page 372.

7. With reference to Exercise 1, use the Mann-Whitney test (modified as suggested on page 376) to test the null hypothesis that the two samples come from identical populations against the alternative that the populations have unequal dispersions. Use the level of significance $\alpha = 0.05$.

8. With reference to Exercise 3, use the Mann-Whitney test (modified as suggested on page 376) to test the null hypothesis that the two samples come from identical populations against the alternative that the populations have unequal dispersions. Use the level of significance $\alpha = 0.05$.

9. The following are the number of misprints counted on pages selected at random from three Sunday editions of a newspaper:

April 11: 4, 10, 2, 6, 4, 12
April 18: 8, 5, 13, 8, 8, 10
April 25: 7, 9, 11, 2, 14, 7

Use the Kruskal-Wallis test at the level of significance $\alpha = 0.05$ to test the null hypothesis that the three samples come from identical populations against the alternative that the compositors and/or proof readers who worked on the three editions are not equally good.

10. Apply the Kruskal-Wallis test to the data of Exercise 8 on page 347 to test the null hypothesis that the 12th graders in the four schools are equally knowledgeable about current events. Use the level of significance $\alpha = 0.05$.

11. Apply the Kruskal-Wallis test to the data of Exercise 10 on page 348 to test the null hypothesis that the differences in dosage have no effect on the average time it takes the tranquilizer to help the guinea pigs fall asleep. Use the level of significance $\alpha = 0.05$.

12. Use the Kruskal-Wallis test instead of the median test to rework Exercise 18 on page 372.

14.8 Tests of Randomness: Runs

Since all the methods of inference that we have discussed were based on the assumption that we are dealing with random samples, it is worth noting that there are many applications in which it is difficult to decide whether this assumption is justifiable. This is true, particularly, when we have little or no control over the selection of the data. For instance, if we wanted to predict a department store's volume of sales for a given month, we would have no choice but to use sales data from previous years and, perhaps, collateral information about economic conditions in general. None of this information constitutes a random sample in the sense that it was obtained with the use of random numbers or similar schemes. Also, we would have no choice but to rely on whatever records happen to be available if we wanted to make long-range predictions of the weather, if we wanted to estimate the mortality rate of a disease, or if we wanted to study traffic accidents at a dangerous intersection.

Several methods have been developed in recent years which make it possible to judge the randomness of a sample on the basis of the order in which the observations were obtained. We can thus test whether patterns that look suspiciously nonrandom may be attributed to chance and, what is most important, this is done *after the data have been collected.* The technique we shall describe in this section and in Section 14.9 is based on the *theory of runs;* alternate methods may be found in the books referred to in the Bibliography at the end of this chapter.

A run is a succession of identical letters (or other kinds of symbols) which is followed and preceded by different letters or no letters at all. To illustrate, let us consider the following arrangement of healthy, *H*, and diseased, *D*, pine trees observed next to each other in the given order by a Forestry Service crew:

$$H\,H\,H\,H\,H\,D\,D\,D\,H\,H\,H\,H\,H\,H\,H\,H\,H\,D\,D\,H\,H\,D\,D\,D\,D\,D\,H\,H\,H\,H$$

Using braces to combine the letters which constitute the runs, we find that there is first a run of five *H*'s, then a run of three *D*'s, then a run of nine *H*'s, then a run of two *D*'s, then a run of two *H*'s, then a run of five *D*'s, and finally a run of four *H*'s. In all, there are seven runs of varying lengths.

The total number of runs appearing in an arrangement of this kind is often a good indication of a possible lack of randomness. If there are too few runs, we might suspect a definite grouping or clustering, or perhaps a trend; if there are too many runs, we might suspect some sort of repeated alternating pattern. In our example there seems to be a definite clustering; that is, the diseased trees seem to come in groups. It remains to be seen, however, whether this is significant or whether it can be attributed to chance.

The test we shall use to put this decision on a precise basis utilizes the fact that for arrangements of n_1 letters of one kind and n_2 letters of another kind the sampling distribution of u, the total number of runs, has the mean

$$\mu_u = \frac{2n_1n_2}{n_1 + n_2} + 1$$

and the standard deviation

$$\sigma_u = \sqrt{\frac{2n_1 n_2 (2n_1 n_2 - n_1 - n_2)}{(n_1 + n_2)^2 (n_1 + n_2 - 1)}}$$

▲   ▲

Furthermore, the sampling distribution of this statistic can be approximated closely with a normal distribution provided that neither n_1 nor n_2 is less than 10; for small values of n_1 and/or n_2 the test will have to be based on special tables such as those referred to in the Bibliography at the end of this chapter. Thus, for sufficiently large values of n_1 and n_2 we base our decision on the statistic

$$z = \frac{u - \mu_u}{\sigma_u}$$

having approximately the standard normal distribution, and we reject the null hypothesis and accept the alternative that the arrangement is *nonrandom* if the value obtained for z is less than $-z_{\alpha/2}$ or exceeds $z_{\alpha/2}$. We would use an appropriate one-tail test if the alternative hypothesis asserted that there is a trend, a definite clustering, or perhaps a repeated cyclic pattern.

    Returning now to the numerical example on page 380, the one dealing with the healthy and diseased pine trees, we find that $n_1 = 20$, $n_2 = 10$, $u = 7$, and hence that

$$\mu_u = \frac{2 \cdot 20 \cdot 10}{20 + 10} + 1 = 14.33$$

$$\sigma_u = \sqrt{\frac{2 \cdot 20 \cdot 10 (2 \cdot 20 \cdot 10 - 20 - 10)}{(20 + 10)^2 (20 + 10 - 1)}} = 2.38$$

and

$$z = \frac{7 - 14.33}{2.38} = -3.08$$

Since this value is less than $-z_{0.005} = -2.58$, we can *reject the null hypothesis* at the level of significance $\alpha = 0.01$. The total number of runs is much smaller than expected, and there is a strong indication that the diseased trees come in a nonrandom arrangement consisting of clusters or groups.

14.9 Tests of Randomness: Runs Above and Below the Median

The method of the preceding section is not limited to tests of the randomness of series of attributes (such as the H's and D's of our example). Any sample consisting of numerical measurements or observations can be treated similarly by using the letters a and b to denote, respectively, values falling above and below the median of the sample. Numbers equaling the median are omitted. The resulting series of a's and b's (representing the data in their *original order*) can be tested for randomness on the basis of the total number of runs of a's and b's, namely, the total number of *runs above and below the median*.

To illustrate this technique, let us refer again to the example on page 249 where the increases of the pulse rate of 32 astronauts performing a given task were given as 26, 20, 17, 22, 23, 21, 25, 33, 32, 25, 30, 27, 28, 24, 12, 21, 20, 14, 29, 23, 22, 36, 25, 21, 23, 26, 24, 19, 27, 24, 30, and 26. The median of this sample is 24, as can easily be checked, and not counting the three values which actually equal 24, we thus get the following arrangement of a's and b's:

$$a\ b\ b\ b\ b\ b\ a\ a\ a\ a\ a\ a\ a\ b\ b\ b\ b\ b\ a\ b\ b\ a\ a\ b\ b\ a\ b\ a\ b\ a\ a\ a$$

Since $n_1 = 15$, $n_2 = 14$, and $u = 11$, we get

$$\mu_u = \frac{2 \cdot 15 \cdot 14}{15 + 14} + 1 = 15.48$$

$$\sigma_u = \sqrt{\frac{2 \cdot 15 \cdot 14 (2 \cdot 15 \cdot 14 - 15 - 14)}{(15 + 14)^2 (15 + 14 - 1)}} = 2.64$$

and

$$z = \frac{11 - 15.48}{2.64} = -1.70$$

which falls between $-z_{0.025} = -1.96$ and $z_{0.025} = 1.96$. Hence, *the null hypothesis cannot be rejected* at the level of significance $\alpha = 0.05$, and there is no reason to doubt the randomness of the original sample.

The method of runs above and below the median is especially useful in detecting trends and cyclic patterns in economic data. If there is a trend, there will be first mostly a's and later mostly b's (or vice versa) and if there is a repeated cyclic pattern there will be a systematic alternation of a's and b's and, probably, too many runs.

14.10 Some Further Considerations

The methods presented in this chapter have given us a brief introduction to the ever-growing subject of nonparametric methods. Since the methods we gave were all tests of hypotheses, let us add that there are such things as *nonparametric confidence intervals* and nonparametric alternatives to other kinds of statistical techniques.

EXERCISES

1. Simulate *mentally* 100 flips of a coin by writing down a series of 100 H's and T's (representing *heads* and *tails*). Test for randomness on the basis of the total number of runs using a level of significance of 0.05.

2. Choose any four complete columns of random digits from Table IX (200 digits in all), represent each even digit by the letter E, each odd digit by the letter O, and test for randomness on the basis of the total number of runs at $\alpha = 0.05$.

3. Use random numbers to simulate an experiment in which 100 voters are asked to choose between Candidate A (represented by the digits 0, 1, 2, 3, 4, and 5) and Candidate B (represented by the digits 6, 7, 8, and 9). Test for randomness at the level of significance $\alpha = 0.05$.

4. The following arrangement indicates whether the price of a stock went up, U, or down, D, on 50 consecutive trading days on which its price did not remain the same:

$$U\,U\,U\,D\,U\,U\,D\,U\,U\,U\,U\,D\,D\,U\,D\,U\,U\,U\,U\,U\,D\,D\,D\,D\,U$$

(cont.) $$D\,D\,D\,D\,U\,U\,D\,U\,D\,D\,D\,D\,D\,U\,U\,U\,D\,U\,U\,U\,U\,U\,D\,D\,U$$

Use the level of significance $\alpha = 0.05$ to test whether this arrangement of U's and D's may be regarded as random.

5. The following arrangement indicates whether sixty consecutive cars which went by the toll booth of a bridge had local plates, L, or out-of-state plates, O:

$$L\,L\,O\,L\,L\,L\,L\,O\,O\,L\,L\,L\,L\,O\,L\,O\,L\,O\,O\,L\,L\,L\,L\,O\,L\,O\,L\,O\,O\,L\,L\,L\,L\,L\,L$$

(cont.) $$O\,L\,L\,L\,O\,L\,O\,L\,O\,L\,L\,L\,L\,O\,O\,L\,O\,O\,O\,O\,L\,L\,L\,L\,O\,L\,O\,L\,O\,O\,L\,L\,L\,O$$

Use the level of significance $\alpha = 0.05$ to test whether this arrangement of L's and O's may be regarded as random.

6. The run test of Section 14.8 can also be used as a nonparametric alternative for testing the significance between two means. As in the Mann-Whitney test, we rank the data belonging to the two samples jointly, write a 1 below each value belonging to the first sample, a 2 below each value belonging to the second sample, and then test for the randomness of this arrangement of 1's and 2's. If there are *too few runs*, this may well be accounted for by the fact that the two samples came from populations with unequal means.

    (a) Apply this method to the illustration on page 373, which dealt with the weight gains of samples of turkeys fed two different diets.

    (b) Use this method to rework Exercise 1 on page 378.

    (c) Use this method to rework Exercise 3 on page 378.

    (d) Use this method to rework Exercise 15 on page 372. Resolve ties between observations belonging to different samples by flipping a coin or using random numbers.

7. Use the method of runs above and below the median to test at the level of significance $\alpha = 0.05$ whether the data on the productivity of the 100 workers in Exercise 12 on page 26 may be regarded as random. Read off successive rows.

8. Use the method of runs above and below the median to test at the level of significance $\alpha = 0.05$ whether the breaking strengths of Exercise 13 on page 27 may be regarded as a random sample. Read off successive rows.

9. Test whether there is a significant trend in the following data on the wholesale trade of durable goods in the United States (in billions of dollars) for the years 1952 through 1968: 41.9, 44.1, 42.6, 51.4, 56.3, 53.8, 50.4, 59.3, 58.6, 59.8, 64.5, 68.7, 75.7, 82.7, 91.0, 90.4, and 100.0. Use the level of significance $\alpha = 0.05$.

10. The following are the number of defective pieces turned out by a machine during 24 consecutive shifts: 15, 11, 17, 14, 16, 12, 19, 17, 21, 15, 17, 19, 21, 14, 22, 16, 19, 12, 16, 14, 18, 17, 24, and 13. Test for randomness at the level of significance $\alpha = 0.01$.

BIBLIOGRAPHY

The nonparametric tests discussed in this chapter and many others are treated in detail in

BRADLEY, J. V., *Distribution-Free Statistical Tests*. Englewood Cliffs, N.J.: Prentice-Hall, Inc., 1968.

KRAFT, C. H., and VAN EEDEN, C., *A Nonparametric Introduction to Statistics*. New York: The Macmillan Company, 1968.

NOETHER, G. E., *Elements of Nonparametric Statistics*. New York: John Wiley & Sons, Inc., 1967.

SIEGEL, S., *Nonparametric Statistics for the Behavioral Sciences*. New York: McGraw-Hill, Inc., 1956.

Tables needed to perform various nonparametric tests for very small samples are given in the aforementioned book by S. Siegel and, among others, in

OWEN, D. B., *Handbook of Statistical Tables*. Reading, Mass.: Addison-Wesley Publishing Company, Inc., 1962.

15

Regression

15.1 Introduction

A major objective of many statistical investigations is to establish relationships which make it possible to predict one or more variables in terms of others. Thus, studies are made to predict the potential sales of a new product in terms of its price, a patient's weight in terms of the number of weeks he or she has been on a diet, family expenditures on entertainment in terms of family income, the per capita consumption of certain foods in terms of their nutritional values and the amount of money spent advertising them on television, and so forth.

Although it is, of course, desirable to be able to predict one quantity *exactly* in terms of others, this is seldom possible, and in most instances we have to be satisfied with predicting averages or expectations. Thus, we may not be able to predict *exactly* how much money Mr. Brown will make ten years after graduating from college, but (given suitable data) we can predict the *average* income of a college graduate in terms of the number of years he has been out of college. Similarly, we can at best predict the *average* yield of a given variety of wheat in terms of data on the rainfall in July, and we can at best predict the *average* performance of students starting college in terms of their I.Q.'s. This problem of predicting the *average* value of one variable in terms of the known value of another variable (or the known values of other variables) *is* the problem of *regression*, a term which dates back to Francis Galton, who used it first in connection with a study of the relationship between the heights of fathers and sons.

Whenever possible, we try to express relationships between variables that are measured or observed and those that are to be predicted in terms of mathematical equations. This approach has been extremely successful in the physical sciences, where it is known, for instance, that at a constant temperature the relationship between the volume (v) and the pressure (p) of a gas is given by the formula

$$v \cdot p = k$$

where k is a numerical constant. Similarly, in biological science it has been discovered that the relationship between the size of a culture of bacteria (y) and the time it has been exposed to certain favorable conditions (x) may be written as

$$y = a \cdot b^x$$

where a and b are numerical constants.

Although the equations of the preceding paragraph referred to examples in physics and biology, they apply equally well to describe relationships in other fields. In economics, for instance, we could let x stand for price, y for demand, and write the equation of a so-called *demand curve* as $x \cdot y = k$; similarly, the second equation might be used to describe the growth of an industry or phenomena related to the process of learning.

Of the many equations that can be used for purposes of prediction, the simplest and the most widely used is the *linear equation* (in two unknowns) which is of the form

$$y = a + bx$$

where a and b are numerical constants. Once these constants are known (usually, they are determined on the basis of sample data), we can calculate a predicted value of y for any value of x simply by substitution. *Linear equations are important not only because there exist many relationships that are actually of this form, but also because they often provide close approximations (at least within a given range of interest) to relationships which would otherwise be difficult to describe in mathematical terms.*

The term "linear equation " arises from the fact that, when plotted on ordinary graph paper, all pairs of values of x and y which satisfy an equation of the form $y = a + bx$ will fall on a straight line. To illustrate, let us consider the equation

$$y = 1.36 - 0.04x$$

whose graph is shown in Figure 15.1. In this equation, x stands for road width (in yards) and y stands for the corresponding number of accidents

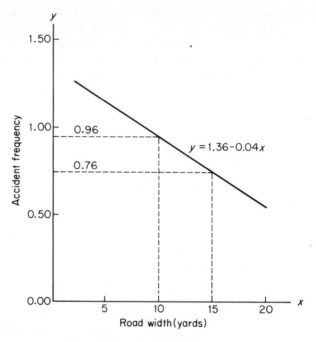

Figure 15.1. Graph of linear equation.

occurring per million vehicle miles; the constants 1.36 and -0.04 were obtained from a study by a state highway department. If the width of a road is 10 yards, the predicted number of accidents per million vehicle miles is $1.36 - 0.04(10) = 0.96$, and by similar substitution, if the width of a road is 15 yards, the predicted number of accidents per million vehicle miles is 0.76. Of course, these predictions will have to be interpreted as *averages*, or *expectations*, since two roads having the same width may well have different accident rates; in fact, for one and the same road the accident rate (number of accidents per million vehicle miles) will vary from month to month, or from year to year.

In any problem in which predictions are to be based on a mathematical equation, we must first decide what kind of curve to fit to the given data; that is, we must first decide whether it is to be a straight line, a *parabola* of the form

$$y = a + b_1x + b_2x^2$$

an exponential curve like the one on page 387 (the one concerning the size of the culture of bacteria), or one of many other kinds of curves. This question is sometimes decided for us by the *nature* of the data, namely, by theory pertaining to the underlying variables, but usually it is decided by direct inspection of the data. We plot the data on ordinary graph paper, sometimes on special logarithmic or log-log graph paper as will be illustrated in Section 15.3, and we thus decide upon the kind of curve which will best describe the over-all pattern of the data. If we are not interested in obtaining a mathematical equation which describes the relationship between the variables, we can indicate the relationship *graphically* by various means; for instance, *moving averages* (see Exercise 10 on page 399) are widely used for this purpose in connection with economic data.

15.2 Linear Regression

To illustrate the general procedure used in *fitting* a straight line to data consisting of paired observations of two variables x and y, let us consider the following data on the number of hours which some students studied for an examination and the grades which they received:

Number of Hours Studied x	Grade in Examination y
8	56
5	44
11	79
13	72
10	70
5	54
18	94
15	85
2	33
8	65

If we plot the ten points corresponding to these paired values of x and y as in Figure 15.2, we find that although the points do not actually fall on a straight line, the over-all pattern of the relationship is fairly well described by the dashed line. Thus, it would seem reasonable to express the relationship between the number of hours the students studied and their grades by means of a linear equation, that is, by means of an equation of the form $y = a + bx$.

We now face the problem of finding the equation of the line which in some sense provides the *best fit* to the data and which, it is hoped, will later

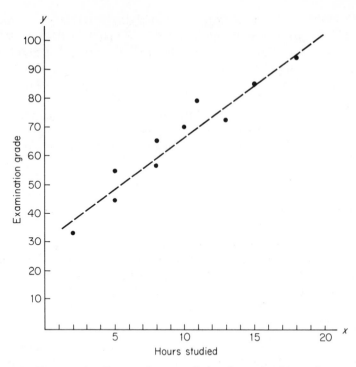

Figure 15.2. Data on hours studied and examination grades.

yield the *best possible predictions*. Logically speaking, there is no limit to the number of straight lines we can draw on a piece of paper; some of these are such obviously poor fits that they can be ruled out immediately, but many others will seem to provide a fairly good and close fit.* In order to single out *one* line as the one which "best" fits our data, we will have to state specifically what we mean here by "best"; in other words, we will have to provide a criterion on the basis of which we can decide which line is "best." If all the points actually fell on a straight line there would be no problem, but this is an extreme case we seldom, if ever, encounter in actual practice. In general, we have to be satisfied with lines which have certain desirable (though not perfect) characteristics.

---

* If the reader has ever had the opportunity to analyze paired data which were plotted as points on a piece of graph paper, he has probably felt the urge to take a ruler, juggle it around, and decide upon a line which, to the eye, presents a fairly good fit. There is no law which says that this cannot be done, but it certainly is not very "scientific." Another argument against *freehand curve fitting* is that it is largely subjective and, hence, there is no direct way of evaluating the "goodness" of subsequent predictions.

The criterion which, nowadays, is used almost exclusively for defining a "best" fit dates back to the early part of the nineteenth century and the French mathematician Adrien Legendre; it is known as *the criterion* (or *the method*) *of least squares.* As it will be used here, the least-squares criterion demands that the line which we fit to our data be such that *the sum of the squares of the vertical deviations (distances) from the points to the line is a minimum.* With regard to our numerical example, the method requires that the sum of the squares of the distances represented by the solid line segments of Figure 15.3 be as small as possible. The logic behind this approach may be explained as follows: considering, for example, the case where the student studied for 11 hours, we find that the actual grade which he received was 79. If we now read the grade which corresponds to $x = 11$ directly off the line of Figure 15.3, we find that the correspond-

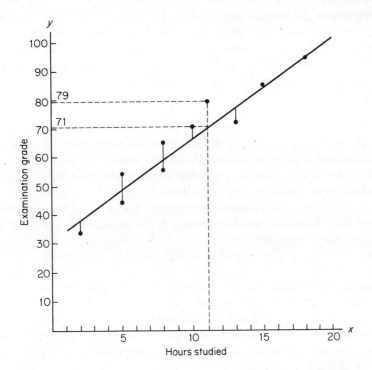

Figure 15.3. Line fitted to data on hours studied and examination grades.

ing "predicted" value is roughly 71, so that the *error* (namely, the difference between this predicted value and the value actually obtained) is $71 - 79 = -8$. There are ten such errors in our example corresponding to the ten *data points,* and the least-squares criterion demands that we *minimize the sum of their squares.* Observe that we do not minimize the sum of the deviations (distances) themselves—some of the differences between the observed y's and the corresponding values read off the line will be positive and others will be negative, so that their sum could be zero even though the errors are *numerically* very large.

As it does not matter which variable is called x and which variable is called y, *we shall agree to reserve y for the variable which is to be predicted in terms of the other.* If we wanted to predict x in terms of y, we would have to apply the method of least squares differently, minimizing the sum of the squares of the *horizontal deviations* from the line (see Exercise 6 on page 397).

To demonstrate how a *least-squares line* is fitted to a set of paired data, let us consider n pairs of numbers (x_1, y_1), (x_2, y_2), $\ldots$, (x_n, y_n), which might represent the heights and weights of n persons, I.Q.'s and examination grades of n students, measurements of the thrust and speed of n experimental rockets, the number of workers unemployed in two countries in n different years, and so on. Let us suppose, furthermore, that the line which we fit to these data has the equation

$$y' = a + bx$$

where the symbol y' is used to differentiate between observed values of y and the corresponding values calculated by means of the equation of the line. For each given value of x we thus have an observed value y and a calculated value y' obtained by substituting x into $y' = a + bx$.

Now, the least-squares criterion requires that we find the numerical values of the constants a and b appearing in the equation $y' = a + bx$ for which the sum of squares

$$\Sigma (y - y')^2$$

is as small as possible. In other words, we minimize the sum of the squares of the differences between the observed y's and the predicted y's (see Figure 15.4). We shall not go through the actual derivation (which requires either calculus or a process called "completing the square"), but simply state the result that $\Sigma (y - y')^2$ is minimized when a and b are solutions of the two equations

$$\Sigma y = na + b(\Sigma x)$$
$$\Sigma xy = a(\Sigma x) + b(\Sigma x^2)$$

Here n is the number of pairs of observations, Σx and Σy are, respectively, the sums of the given x's and y's, Σx^2 is the sum of the squares of

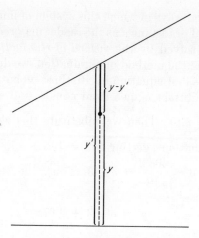

Figure 15.4. Least squares line.

the x's, and $\Sigma\, xy$ is the sum of the products obtained by multiplying each of the given x's by the corresponding observed value of y. The above equations, whose solution gives the desired least-squares values of a and b, are called the *normal equations*.

Returning now to our numerical example, let us fit a least-squares line to the given data. Copying the first two columns from the table on page 389 and performing the necessary calculations, we get

x	y	x^2	xy
8	56	64	448
5	44	25	220
11	79	121	869
13	72	169	936
10	70	100	700
5	54	25	270
18	94	324	1,692
15	85	225	1,275
2	33	4	66
8	65	64	520
95	652	1,121	6,996

Thus, we have $n = 10$, $\Sigma\, x = 95$, $\Sigma\, y = 652$, $\Sigma\, x^2 = 1{,}121$, and $\Sigma\, xy = 6{,}996$, and the next step is to substitute these quantities into the two normal equations. This gives

$$652 = 10a + 95b$$
$$6{,}996 = 95a + 1{,}121b$$

and there are several ways in which this system of linear equations can be solved for a and b. For instance, as the reader may recall from elementary algebra, it can be solved by the *method of elimination* or by the use of *determinants*. Using the method of elimination, we divide the expressions on both sides of the first equation by 10, those on both sides of the second equation by 95, subtract equals from equals, and get $8.44 = 2.3b$ and, hence, $b = \dfrac{8.44}{2.3} = 3.67$. Then we substitute this value into the first of the two normal equations, getting $652 = 10a + 95(3.67)$, $652 - 348.7 =$ $10a$, and, hence, $a = \dfrac{303.3}{10} = 30.33$. Thus, the equation of the *least squares line* becomes

$$y' = 30.33 + 3.67x$$

Since the reader may not be familiar with the method of elimination, or may have forgotten it, we shall solve the two normal equations *symbolically* for a and b, so that the values of a and b for any given set of paired data can be obtained by direct substitution. As the reader will be asked to verify in Exercise 13 on page 400, the resulting formulas are

$$a = \frac{(\Sigma\, y)(\Sigma\, x^2) - (\Sigma\, x)(\Sigma\, xy)}{n(\Sigma\, x^2) - (\Sigma\, x)^2}$$

$$b = \frac{n(\Sigma\, xy) - (\Sigma\, x)(\Sigma\, y)}{n(\Sigma\, x^2) - (\Sigma\, x)^2}$$

For our example, we would thus get

$$a = \frac{(652)(1{,}121) - (95)(6{,}996)}{10(1{,}121) - (95)^2} = 30.33$$

and

$$b = \frac{10(6{,}996) - (95)(652)}{10(1{,}121) - (95)^2} = 3.67$$

and this agrees with the result which we obtained before.

To simplify the calculations, we could first calculate b using the above formula, and then substitute the result into the first of the two normal equations to obtain a. Thus, an alternate formula for calculating a is given by

$$a = \frac{\Sigma\, y - b(\Sigma\, x)}{n}$$

For our example this would yield $a = \dfrac{652 - (3.67)(95)}{10} = 30.33$, and

this agrees with the result given above. A further simplification, which applies when the x's are *equally spaced*, is explained in Exercise 8 on page 398.

We can now use the equation of the least-squares line which we obtained for our example, namely,

$$y' = 30.33 + 3.67x$$

to predict, for example, the grade of a student who studies 11 hours for the examination. Substituting $x = 11$, we get

$$y' = 30.33 + 3.67(11) = 70.7$$

How such a prediction is to be interpreted and how we can judge its merits will be discussed in Section 15.5. Incidentally, the predicted grade could also have been obtained (approximately, at least) by plotting the line on graph paper and reading off the y which corresponds to $x = 11$. To plot the line, we have only to choose two arbitrary values of x (preferably not too close together), calculate the corresponding values of y', and draw a straight line through the corresponding points.

EXERCISES

1. The following data pertain to the number of years which ten applicants for a foreign service job studied French in high school or college and the grades they received in a proficiency test in that language:

Number of Years x	Grade in Test y
3	72
2	45
4	70
5	81
3	54
4	75
4	69
2	55
5	86
3	60

(a) Use the formulas on page 394 to calculate a and b for the equation of the least-squares line which will enable us to predict y in terms of x.

(b) Repeat part (a) by solving the two normal equations, and compare the results.

(c) Plot the least-squares line obtained in (a) or (b) and use it to *read off* the predicted grade of someone who has studied French in high school or college for four years.

(d) Use the equation obtained in (a) or (b) to predict the grade of someone who has studied French in high school or college for four years.

2. The following data pertain to the demand for a product (in thousands of units) and its price (in cents) charged in six different market areas:

Price x	15	18	13	16	12	20
Demand y	82	25	93	60	128	12

(a) Fit a least-squares line which will enable us to predict the demand for the product in terms of its price, and plot its graph.

(b) Predict the demand for the product in a market area where it is priced at 15 cents, using (i) the equation obtained in part (a), and (ii) the graph.

3. The following are data on the average annual yield of wheat (in bushels per acre) in a given county and the annual rainfall (in inches, measured from September through August):

Rainfall x	Yield of Wheat y
8.8	39.6
10.3	42.5
15.9	69.3
13.1	52.4
12.9	60.5
7.2	26.7
11.3	50.2
18.6	78.6

(a) Fit a least-squares line which will enable one to predict the yield of wheat in the given county in terms of the rainfall, and plot its graph.

(b) Use the equation obtained in (a) and also its graph to predict the average annual yield of wheat in the county when the annual rainfall is 9.0 inches.

4. The following data pertain to the advertising expenses (expressed as a percentage of total expenses) and the net operating profits (expressed as a percentage of total sales) in a sample of twelve sporting goods stores:

Advertising Expenses x	Profit y
1.5	3.1
0.8	1.9
2.6	4.2
1.0	2.3
0.6	1.2
2.8	4.9
1.2	2.8
0.9	2.1
0.4	1.4
1.3	2.4
1.2	2.4
2.0	3.8

Fit a least-squares line and use it to predict the profit of such a store (expressed as a percentage of total sales) when its advertising expenses are 1.2 per cent of its total expenses.

5. The following data pertain to the growth of a cactus graft under controlled environmental conditions:

Months after Grafting x	Height (Inches) y
1	0.8
2	2.4
3	4.0
4	5.1
5	7.3
6	9.4

Fit a least-squares line and use it to predict the cactus graft's height after 9 months. When a value is thus predicted beyond the range of x for which values are observed, this is referred to as an *extrapolation* and it involves certain risks, as is illustrated in Exercise 1 on page 408.

6. Suppose that in Exercise 4 we had wanted to predict, or estimate, how much should be spent on advertising so that there will be a net operat-

ing profit of 3 per cent of total sales. We could substitute $y = 3$ into the equation obtained in Exercise 4, but this would not be a prediction in the least-squares sense. To make predictions of x in terms of y as good as possible, we should minimize the sum of the squares of the errors in the *x-direction* (namely, the sum of the squares of the *horizontal deviations* from the line), and if we write the equation of the resulting line as $x = a + by$, the values of a and b may be obtained by means of the two formulas on page 394, with x substituted throughout for y, and vice versa. Find this kind of regression line for the data of Exercise 4, and use it to find the predicted value of x when $y = 3$.

7. Referring to Exercise 3 above, use the method of the preceding exercise to determine how much rainfall is needed so that there will be an average yield of 50.0 bushels per acre.

8. When the sum of the x's is zero, the calculation of a and b is greatly simplified; in fact, their formulas become

▲
$$a = \frac{\Sigma y}{n} \quad \text{and} \quad b = \frac{\Sigma xy}{\Sigma x^2}$$
▲

These formulas can also be used when the x's are *equally spaced*, that is, when the differences between successive values of x are all equal. In that case we use the above formulas after "coding" the x's by assigning them the values $\ldots, -3, -2, -1, 0, 1, 2, 3, \ldots$, when n is *odd*, or the values $\ldots, -5, -3, -1, 1, 3, 5, \ldots$, when n is *even*. Of course, the resulting equation of the least-squares line expresses y in terms of the coded x's, and we have to account for this when using the equation for purposes of prediction. For example, if we are given economic data for the years 1968, 1969, 1970, 1971, and 1972, and these years are coded as $x = -2, -1, 0, 1$, and 2, we must substitute $x = 4$ to obtain a prediction for the year 1974 and $x = 6$ to obtain a prediction for the year 1976.

(a) Use this technique to fit a least-squares line to the following data on a company's profits in millions of dollars:

Year x	Profit y
1965	5.5
1966	6.3
1967	7.1
1968	9.0
1969	9.9
1970	11.1
1971	11.7

(b) Use this technique to fit a least-squares line to the following data on Federal Reserve Bank credit outstanding in millions of dollars:

Year x	Credit Outstanding y
1962	33,218
1963	36,610
1964	39,873
1965	43,853
1966	46,864
1967	51,268
1968	56,610
1969	55,439

9. Use the kind of coding suggested in Exercise 8 to rework Exercise 5.

10. In the analysis of economic data, we sometimes describe the over-all pattern (trend, cycle, etc.) by means of a *moving average* rather than a specific kind of curve. A moving average is obtained by replacing each value in a series of equally spaced data (except for the first few and the last few) by the *mean* of itself and some of the values directly preceding it and directly following it. For instance, in a *three-year moving average* each annual figure is replaced by the mean of itself and the annual figures corresponding to the two adjacent years; in a *five-year moving average* each annual figure is replaced by the mean of itself and the annual figures corresponding to the two years which precede and the two years which follow it. Construct a three-year moving average for the following figures, representing the production of bituminous coal and lignite in the United States (in millions of short tons) for the years 1937 through 1969: 446, 349, 394, 460, 514, 583, 590, 620, 578, 534, 631, 600, 438, 516, 534, 467, 457, 392, 465, 501, 493, 410, 412, 416, 403, 422, 459, 487, 512, 534, 553, 545, and 556. Also plot a graph showing the moving average as well as the original data, drawing lines which connect the values corresponding to successive years.

11. Construct a five-year moving average for the data of Exercise 10 and draw a graph as suggested in that exercise.

12. The following figures represent the number of employees (in thousands) of the contract construction industry in the United States for the years 1931 through 1968: 1,214, 970, 809, 862, 912, 1,145, 1,112, 1,055, 1,150, 1,294, 1,790, 2,170, 1,567, 1,094, 1,132, 1,661, 1,982,

2,169, 2,165, 2,333, 2,603, 2,634, 2,623, 2,612, 2,802, 2,999, 2,923, 2,778, 2,960, 2,885, 2,816, 2,902, 2,963, 3,050, 3,186, 3,275, 3,203, and 3,259. Construct a three-year moving average and plot a graph as suggested in Exercise 10.

13.   To verify the formulas for a and b on page 394, multiply the expressions on both sides of the first of the two normal equations on page 392 by $\Sigma\, x$, multiply those of the second of the two normal equations by n, eliminate a by subtraction, and then solve for b. Then, substitute the expression obtained for b into the first of the two normal equations, and solve for a.

15.3 Non-Linear Regression*

So far we have studied only the problem of fitting straight lines to paired data. In this section, we shall first investigate two cases where the relationship between the variables under consideration is *not linear*, but the method of Section 15.2 can nevertheless be employed; then we shall give an example of *polynomial curve fitting* by fitting a *parabola* having the equation

$$y = a + b_1 x + b_2 x^2$$

to a given set of paired data.

It is common practice to plot paired data on various kinds of graph paper, to see whether there are scales for which the points will fall close to a straight line. Of course, when this is the case for ordinary graph paper, we go right ahead and use the method of Section 15.2. If it is the case when we use *semi-log paper* (with equal subdivisions for x and a logarithmic scale for y, as shown in Figure 15.6), this is an indication that an *exponential curve* will provide a good fit. The equation of such a curve is

$$y = a \cdot b^x$$

or in logarithmic form

$$\log y = \log a + x(\log b)$$

where log stands for logarithm to the base 10. Note that this last equation expresses a *linear relationship* between the variables x and $\log y$, and we can therefore use the method of Section 15.2 with $\log y$ substituted for y,

* The material in this section is somewhat more difficult and it may be omitted without loss of continuity.

log a substituted for a, and log b substituted for b, wherever these symbols occur. For instance, the two normal equations on page 392 become

$$\Sigma \log y = n(\log a) + (\log b)(\Sigma\, x)$$
$$\Sigma\, x(\log y) = (\log a)(\Sigma\, x) + (\log b)(\Sigma\, x^2)$$

and for a given set of data they can be solved for log a and log b and, hence, for a and b.

To illustrate this technique, let us fit an exponential curve to the following data on a company's net profits and the number of years it has been in business:

Years in Business x	Profit (Dollars) y
1	112,000
2	149,000
3	238,000
4	354,000
5	580,000
6	867,000

Plotting these data as in Figures 15.5 and 15.6, we find that the relationship between x and y is definitely not linear, but that the over-all pattern is pretty well "straightened out" when we use a logarithmic scale for y. Thus, it would seem justified to fit an exponential curve.

To obtain the sums needed for substitution into the two normal equa-

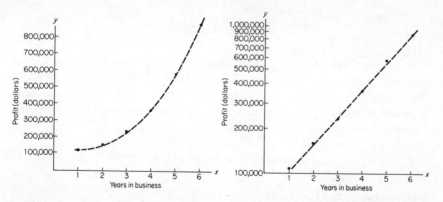

Figure 15.5. Data on net profit of company plotted on ordinary graph paper.

Figure 15.6. Data on net profit of company plotted on semi-log paper.

tions, we first determine the logarithms of the six y's with the use of Table X, and then perform the calculations shown in the following table:

x	y	$\log y$	$x \cdot \log y$	x^2
1	112,000	5.0492	5.0492	1
2	149,000	5.1732	10.3464	4
3	238,000	5.3766	16.1298	9
4	354,000	5.5490	22.1960	16
5	580,000	5.7634	28.8170	25
6	867,000	5.9380	35.6280	36
21		32.8494	118.1664	91

Substituting the column totals together with $n = 6$ into the two normal equations on page 401, we get

$$32.8494 = 6(\log a) + 21(\log b)$$
$$118.1664 = 21(\log a) + 91(\log b)$$

Dividing the expressions on both sides of the first equation by 6, those of the second equation by 21, and subtracting equals from equals, we get $0.1521 = 0.8333(\log b)$, $\log b = \dfrac{0.1521}{0.8333} = 0.1825$, and $b = 1.52$. Then, substituting the value obtained for $\log b$ into the first of the two normal equations, we get $32.8494 = 6(\log a) + 21(0.1825)$, $\log a = 4.8362$, and $a = 68,600$. Thus, the equation of the exponential curve which best describes the relationship between the company's net profits and the number of years it has been in business is given by

$$y' = 68,600(1.52)^x$$

where we again used the symbol y' to distinguish between the y-values on the curve and those which were actually observed. To estimate (predict) the company's net sales, say, for the eighth year it will be in business, it is easiest to substitute $x = 8$ into the logarithmic form of the exponential equation. This gives

$$\log y' = 4.8362 + 8(0.1825) = 6.2962$$

and, hence, $y' = \$1,980,000$.

If points representing paired data fall close to a straight line when plotted on *log-log paper* (that is, on paper having logarithmic scales for both x and y), this is an indication that an equation of the form

$$y = a \cdot x^b$$

will provide a good fit. In the logarithmic form, the equation of such a *power function* becomes

$$\log y = \log a + b(\log x)$$

and it should be noted that it expresses a *linear relationship* between $\log y$ and $\log x$. Thus, we can again use the method of Section 15.2—but this time with $\log x$ substituted for x, $\log y$ substituted for y, and $\log a$ substituted for a. The two normal equations become

$$\Sigma \log y = n(\log a) + b(\Sigma \log x)$$
$$\Sigma (\log x)(\log y) = (\log a)(\Sigma \log x) + b(\Sigma \log^2 x)$$

where $\Sigma (\log x)(\log y)$ is the sum of the products obtained by multiplying the logarithm of each observed value of x by the logarithm of the corresponding value of y, and $\Sigma \log^2 x$ is the sum of the *squares* of the logarithms of the x's. These two normal equations provide solutions for b and $\log a$, and, hence, for b and a. This work is very similar to that of the preceding illustration and we shall not give an example, but in Exercises 4 and 5 on pages 409 and 410 the reader will find data to which this method can be applied.

When the pattern of a set of paired data is such that the y-values *first increase and then decrease*, or *first decrease and then increase*, when the x's are arranged according to size, a *parabola* having the equation

$$y = a + b_1 x + b_2 x^2$$

may well provide a good fit. A curve like this can also be fitted by the method of least squares, and without any mathematical details, let us merely state that this leads to the following system of *normal equations:*

$$\Sigma y = na + b_1(\Sigma x) + b_2(\Sigma x^2)$$
$$\Sigma xy = a(\Sigma x) + b_1(\Sigma x^2) + b_2(\Sigma x^3)$$
$$\Sigma x^2 y = a(\Sigma x^2) + b_1(\Sigma x^3) + b_2(\Sigma x^4)$$

Here Σxy is the sum of the products obtained by multiplying each value of x by the corresponding value of y, $\Sigma x^2 y$ is the sum of the products obtained by multiplying the *square* of each value of x by the corresponding value of y, and Σx^2, Σx^3, and Σx^4 are, respectively, the sums of the second, third, and fourth powers of the x's.

To illustrate this technique, let us fit a parabola to the following data on the drying time of a varnish and the amount of a certain chemical that has been added:

Amount of Additive (Grams) x	Drying Time (Hours) y
1	7.2
2	6.7
3	4.7
4	3.7
5	4.7
6	4.2
7	5.2
8	5.7

As can be seen from Figure 15.7, where we have plotted the corresponding data points on ordinary graph paper, the y-values *first decrease and then increase,* and this suggests that a parabola may well give a good fit.

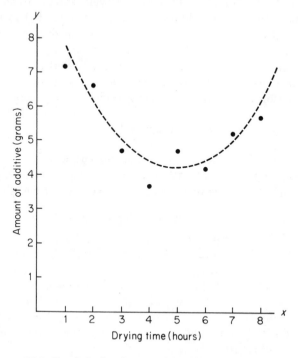

Figure 15.7. Parabola fitted to varnish-additive-drying-time example.

To obtain the sums needed for substitution into the three normal equations, we perform the calculations shown in the following table:

x	y	x^2	x^3	x^4	xy	x^2y
1	7.2	1	1	1	7.2	7.2
2	6.7	4	8	16	13.4	26.8
3	4.7	9	27	81	14.1	42.3
4	3.7	16	64	256	14.8	59.2
5	4.7	25	125	625	23.5	117.5
6	4.2	36	216	1296	25.2	151.2
7	5.2	49	343	2401	36.4	254.8
8	5.7	64	512	4096	45.6	364.8
36	42.1	204	1296	8772	180.2	1023.8

Then, substituting the column totals together with $n = 8$ into the three normal equations, we get

$$42.1 = 8a + 36b_1 + 204b_2$$
$$180.2 = 36a + 204b_1 + 1296b_2$$
$$1023.8 = 204a + 1296b_1 + 8772b_2$$

and the solution of this system of equations (rounded to one decimal) is $a = 9.2$, $b_1 = -2.0$, and $b_2 = 0.2$, as the reader will be asked to verify in Exercise 8 on page 411. Thus, the equation of the parabola is

$$y' = 9.2 - 2.0x + 0.2x^2$$

Having obtained this equation, let us use it to determine the drying time of the varnish when 6.5 grams of the additive is used, and also the amount of additive for which the drying time is the least. To answer the first of these two questions, we simply substitute $x = 6.5$ into the equation of the parabola, getting

$$y' = 9.2 - 2.0(6.5) + 0.2(6.5)^2 = 4.65 \text{ hours}$$

To answer the second question, we shall make use of the fact that a parabola of the form $y = a + b_1x + b_2x^2$ takes on its *minimum* (or *maximum*) value when $x = -\dfrac{b_1}{2b_2}$, and in our example this is $x = -\dfrac{-2.0}{2(0.2)} = 5.0$. The corresponding drying time is

$$y' = 9.2 - 2.0(5) + 0.2(5)^2 = 4.2 \text{ hours}$$

In case the reader is surprised that we obtained a minimum drying time of 4.2 hours while one of the observed values actually equaled 3.7, let us point out that the drying time is a random variable and that the parabola which we fitted to the data will have to be looked upon as a *regression curve*. In other words, we have shown that the *average* drying time is least when $x = 5$, and this is not contradicted by the fact that *one* of the observed drying times was actually less than this average. Of course, the answer to the first question, determining the drying time for a specific value of x, must also be interpreted as an average; that is, we showed in the example that the *average* drying time is 4.65 hours when 6.5 grams of the additive is being used.

15.4 Multiple Regression*

Although there are many problems in which one variable can be predicted quite accurately in terms of another, it stands to reason that predictions should improve if one considers additional relevant information. For instance, we should be able to make better predictions of a student's grade in a final examination if we consider not only the number of hours he studied for the examination, but also his I.Q. Similarly, we should be able to make better predictions of the attendance at a theater if we considered not only the quality of the performance, but also the size of the community, its wealth, and perhaps the number of competing attractions that are available to the public on the same night.

Many mathematical formulas can serve to express relationships between more than two variables, but most commonly used in statistics (partly for reasons of simplicity) are linear equations of the form

$$y = b_0 + b_1 x_1 + b_2 x_2 + b_3 x_3 + \ldots + b_k x_k$$

Here y is the variable which is to be predicted, x_1, x_2, x_3, $\ldots$, and x_k are the k known variables on which predictions are to be based, and b_0, b_1, b_2, b_3, $\ldots$, and b_k are numerical constants which have to be determined (usually by the method of least squares) on the basis of given data. To give an example, consider the following equation obtained in a study of the demand for beef and veal:

$$y = 3.489 - 0.090 x_1 + 0.064 x_2 + 0.019 x_3$$

where y stands for the total consumption of federally inspected beef and veal in millions of pounds, x_1 stands for the retail price of beef in cents per pound, x_2 stands for the retail price of pork in cents per pound, and x_3 stands for income as measured by a certain payroll index. Once we have

---

* The material in this section is somewhat more difficult and it may be omitted without loss of continuity.

obtained an equation like this, we can forecast the total consumption of federally inspected beef and veal on the basis of known (or estimated) values of x_1, x_2, and x_3.

The main problem of obtaining a linear equation in more than two variables which best describes a given set of data is that of finding numerical values for b_0, b_1, b_2, b_3, ..., and b_k. This is usually done by the method of least squares; that is, we minimize the sum of squares $\Sigma (y - y')^2$, where as before the y's are the observed values and the y''s are the values calculated by means of the linear equation. In principle, the problem of determining the values of b_0, b_1, b_2, b_3, ..., and b_k is no different from what we did before; however, the work becomes more tedious because the method of least squares yields as many normal equations as there are unknown constants b_0, b_1, b_2, b_3, ..., and b_k. For example, when there are two independent variables x_1 and x_2, and we want to fit a curve which has the equation $y = b_0 + b_1 x_1 + b_2 x_2$, we must solve the *three* normal equations

$$\Sigma y = n \cdot b_0 + b_1(\Sigma x_1) + b_2(\Sigma x_2)$$
$$\Sigma x_1 y = b_0(\Sigma x_1) + b_1(\Sigma x_1^2) + b_2(\Sigma x_1 x_2)$$
$$\Sigma x_2 y = b_0(\Sigma x_2) + b_1(\Sigma x_1 x_2) + b_2(\Sigma x_2^2)$$

Here $\Sigma x_1 y$ is the sum of the products obtained by multiplying each of the given values of x_1 by the corresponding y, $\Sigma x_1 x_2$ is the sum of the products obtained by multiplying each of the given values of x_1 by the corresponding value of x_2, and so on.

To illustrate this technique, let us consider the following data on the number of bedrooms, the number of baths, and the price of one-family homes recently sold in a community:

Number of Bedrooms x_1	Number of Baths x_2	Price (Dollars) y
3	2	23,800
2	1	19,300
4	3	28,800
2	1	19,200
3	2	24,700
2	2	19,900
5	3	33,400
4	2	27,900

What we want to find is an equation which will give us the *average price* of a house in terms of the number of bedrooms and baths, and to obtain the sums needed for substitution into the three normal equations, we perform the calculations shown in the following table:

x_1	x_2	y	$x_1 y$	$x_2 y$	x_1^2	$x_1 x_2$	x_2^2
3	2	23,800	71,400	47,600	9	6	4
2	1	19,300	38,600	19,300	4	2	1
4	3	28,800	115,200	86,400	16	12	9
2	1	19,200	38,400	19,200	4	2	1
3	2	24,700	74,100	49,400	9	6	4
2	2	19,900	39,800	39,800	4	4	4
5	3	33,400	167,000	100,200	25	15	9
4	2	27,900	111,600	55,800	16	8	4
25	16	197,000	656,100	417,700	87	55	36

Then, substituting the column totals together with $n = 8$ into the three normal equations, we get

$$197{,}000 = 8b_0 + 25b_1 + 16b_2$$
$$656{,}100 = 25b_0 + 87b_1 + 55b_2$$
$$417{,}700 = 16b_0 + 55b_1 + 36b_2$$

and the solution of this system of equations is $b_0 = 10{,}197$, $b_1 = 4{,}149$, and $b_2 = 731$, as the reader will be asked to verify in Exercise 11 on page 412. Thus, the equation is

$$y' = 10{,}197 + 4{,}149x_1 + 731x_2$$

which tells us that each extra bedroom costs on the average \$4,149 and each extra bath costs on the average \$731. Also, to estimate (predict) the *average price* of a three-bedroom house with two baths, we substitute $x_1 = 3$ and $x_2 = 2$, and get

$$y' = 10{,}197 + 4{,}149(3) + 731(2) = \$24{,}106$$

or approximately \$24,100.

EXERCISES

1. Referring to Exercise 5 on page 397, suppose that we are given data for two more months, so that the complete data are now

x	1	2	3	4	5	6	7	8
y	0.8	2.4	4.0	5.1	7.3	9.4	13.5	19.2

where x is the number of months after grafting and y is the height (in inches) of the cactus graft. Fit an exponential curve to these data and use it to predict the height of the graft after 9 months. Compare this with the result obtained earlier when a straight line was fitted to the data for the first six months. (*Hint:* $\log 0.8 = 9.9031 - 10 = -0.0969$.)

2. Fit an exponential curve to the following data on the percentage of the radial tires made by a certain manufacturer that are still usable after having been driven for the given number of miles:

Miles Driven (in thousands) x	Percentage Usable y
1	96.2
2	91.8
5	80.6
10	64.9
20	43.5
30	28.4
40	17.8

Also use the equation of the exponential curve (in the logarithmic form) to estimate what percentage of the tires can be expected to be usable after having been driven for 50,000 miles.

3. In 1956, 1958, 1960, 1962, 1964, 1966, and 1968, the number of non-immigrants admitted to the United States (in thousands) were, respectively, 686, 848, 1,141, 1,331, 1,745, 2,342, and 3,200. Coding these years as $-3, -2, -1, 0, 1, 2$, and 3, fit an exponential curve to these data, and use it to "predict" the number of nonimmigrants admitted to the United States in 1972.

4. Fit a power function to the following data on the demand for a certain product (in units of 1,000) and its price (in cents):

Price x	Demand y
5	62
10	80
15	107
20	140
25	175
30	231

Also estimate the demand for the product when its price is (a) 12 cents; (b) 24 cents.

5. The following data were obtained in a study of the relationship be-
    tween the volume of a gas (in cubic inches) and its pressure (in
    pounds per square inch), when the gas is compressed at a constant
    temperature:

Volume x	Pressure y
50	15.9
30	39.2
20	77.3
10	189.0
5	537.1

    Fit a power function to these data, and use it to estimate the pressure
    of this gas when it is compressed to a volume of 25 cubic inches.

6. A toy manufacturer experiments with the price of a new toy in five
    different market areas that are of equal size, and obtains the following
    results:

Price of the Toy (Dollars) x	Total Profit (1,000 Dollars) y
1.00	8.6
1.50	9.8
2.00	10.2
2.50	8.8
3.00	6.9

    Fit a parabola to these data and determine how the manufacturer
    should price the toy so as to maximize his estimated profit. (Carry
    four decimals while solving normal equations.)

7. When the x's are equally spaced, the work that is required to fit a
    parabola can be greatly reduced by using the same kind of *coding* as
    in Exercise 8 on page 398. Since this makes $\Sigma x = 0$ and also $\Sigma x^3 = 0$,
    the normal equations become

    $$\Sigma y = na + b_2(\Sigma x^2)$$
    $$\Sigma xy = b_1(\Sigma x^2)$$
    $$\Sigma x^2y = a(\Sigma x^2) + b_2(\Sigma x^4)$$

    and the values of b_1 can be obtained directly from the second equation.
    To obtain the values of a and b_2, we simultaneously solve the first
    and third of these normal equations, and all this is much less work
    than simultaneously solving three equations like those on page 403.

    (a) Use this method to fit a parabola to the data on nonimmigrants
        admitted to the United States given in Exercise 3. Use the
        equation of the parabola thus obtained to "predict" the number

of nonimmigrants admitted to the United States in 1972, and compare this with the corresponding "prediction" obtained in Exercise 3.

(b) Use this method to fit a parabola to the following data on the production of television sets (in thousands) in Japan for the years 1963 through 1969: 4,916, 5,273, 4,190, 5,652, 6,963, 9,001, and 12,118. Also use the equation of the parabola to predict the corresponding production figure for 1975.

8. Verify that the solution of the three normal equations on page 405 is $a = 9.2$, $b_1 = -2.0$, and $b_2 = 0.2$.

9. With reference to the illustration of Section 15.2, suppose that we are also given the I.Q.'s of the ten students, so that we now have the information shown in the following table:

Number of Hours Studied x_1	I.Q. x_2	Grade in Examination y
8	98	56
5	99	44
11	118	79
13	94	72
10	109	70
5	116	54
18	97	94
15	100	85
2	99	33
8	114	65

Fit an equation of the form $y = b_0 + b_1x_1 + b_2x_2$ to the given data, and use it to predict the grade of a student who has an I.Q. of 105 and studies 12 hours for the examination.

10. The following are data on the ages and incomes of five executives working for the same company, and the number of years they went to college:

Age x_1	Years College x_2	Income (Dollars) y
37	4	21,200
45	0	16,800
38	5	25,000
42	2	20,300
31	4	15,400

Fit an equation of the form $y = b_0 + b_1x_1 + b_2x_2$ to the given data, and use it to estimate how much *on the average* an executive working for this company will make if he is 40 years old and has had 4 years of college.

11.   Verify that the solution of the three normal equations on page 408 is $b_0 = 10{,}197$, $b_1 = 4{,}149$, and $b_2 = 731$. (There may be differences due to rounding.)

15.5 Regression Analysis

Earlier in this chapter, we used a least-squares line to predict that someone who studies for 11 hours for a given examination will get a grade of 70.7, but even if we interpret the line correctly as a *regression line*, and the prediction correctly as an *average* (or *expectation*), many questions remain. For instance,

(1)   "How *good* are the values we obtained for the constants a and b in the equation $y = a + bx$? After all, $a = 30.33$ and $b = 3.67$, the values we obtained on page 394, are only *estimates based on a sample*, and it stands to reason that if we took a different sample, data pertaining to different students who took the examination, the method of Section 15.2 would probably lead to different values of a and b."

(2)   "How *good* an estimate is 70.7 of the *true* average grade which students should get if they study 11 hours for the examination?"

(3)   "Can we construct limits (two numbers) for which we can assert with a probability of, say, 0.99 or 0.95 that they will contain the grade of a given student who has studied 11 hours for the examination?"

To be able to answer any of these questions, we shall have to make several assumptions. First, we shall assume that the *true mean* of the y's for a given value of x is given by an expression of the form $\alpha + \beta x$; in other words, we assume that the *true means* of the y's for given values of x (of the grades, in our example, for given numbers of hours studied) fall on the line

$$y = \alpha + \beta x$$

so that we can look upon the values obtained for a and b by the method of least squares as *estimates* of α and β. To distinguish between these two sets of values, we usually refer to α and β as the *(true) regression coefficients*, and to a and b as the *estimated regression coefficients*.

To clarify the idea of a *true* regression line, let us consider Figure 15.8,

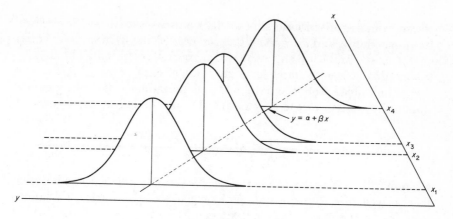

Figure 15.8. Distribution of y for given values of x.

in which we have drawn the distributions of y for several values of x. With reference to our example, these curves should be looked upon as the distributions of the grades of students who study, respectively, x_1 hours, x_2 hours, x_3 hours, and x_4 hours for the examination, and to complete the picture, the reader should be able to visualize similar distributions for all other values of x (within the range of values under consideration).

In *linear regression analysis* we assume that the x's are constants, *not* values of random variables, and that for each value of x the variable to be predicted has a certain distribution (as shown in Figure 15.8) whose mean is $\alpha + \beta x$. In *normal regression analysis* we assume, furthermore, that these distributions are all normal curves having the same standard deviation σ. In other words, the distributions pictured in Figure 15.8, as well as those we add mentally, are normal curves having the means $\alpha + \beta x$ and the standard deviation σ. (If the x's are also looked upon as values of random variables, the treatment of the data is referred to as *correlation analysis*, a subject we shall take up in Chapter 16.)

If we make all the assumptions of normal regression analysis, or meet them approximately in practice, questions like the first one on page 412 can be answered by making use of the statistics

$$ t = \frac{a - \alpha}{s_e \sqrt{\dfrac{1}{n} + \dfrac{n \cdot \bar{x}^2}{n(\Sigma\, x^2) - (\Sigma\, x)^2}}} $$

$$ t = \frac{b - \beta}{s_e} \sqrt{\frac{n(\Sigma\, x^2) - (\Sigma\, x)^2}{n}} $$

whose sampling distributions are t distributions with $n - 2$ degrees of freedom. Here α and β are the regression coefficients we want to estimate or test, and a and b are the estimated regression coefficients obtained by the method of least squares for a given set of data. Also, n, $\bar{x}$, $\Sigma\, x$, and $\Sigma\, x^2$ are obtained from the data, and s_e is an estimate of σ, the common standard deviation of the normal distributions of Figure 15.8, given by

$$s_e = \sqrt{\frac{\Sigma\,(y - y')^2}{n - 2}}$$

Customarily, s_e is referred to as the *standard error of estimate*, and it should be observed that s_e^2 is the sum of the squares of the vertical deviations from the least-squares line (for example, the sum of the squares of the lengths of the solid line segments of Figure 15.3) divided by $n - 2$. In actual practice, it is easier to calculate s_e by means of the short-cut formula

$$\blacktriangle \qquad s_e = \sqrt{\frac{\Sigma\, y^2 - a(\Sigma\, y) - b(\Sigma\, xy)}{n - 2}} \qquad \blacktriangle$$

and in Exercise 1 on page 416, the reader will be asked to verify that these two formulas for s_e will, indeed, give the same results.

To illustrate how the two t statistics are used in inferences concerning the regression coefficients α and β, suppose someone claims that $\beta = 4$ in our example, and that the data which we used in Section 15.2 had actually been obtained to check this claim. Note that β is the *slope* of the regression line, namely, the change (increase or decrease) in y which corresponds to an increase of one unit in x, and, hence, the above hypothesis asserts that *on the average* an extra hour of study will increase a student's grade by 4 points. Among the various quantities that are needed for substitution into the formula for the second of the two t statistics on page 413, we already know that $n = 10$, $\Sigma\, x = 95$, $\Sigma\, x^2 = 1{,}121$, and $b = 3.67$ from the work of Section 15.2. By assumption $\beta = 4$, and to calculate s_e, the only other quantity needed, we will first have to determine $\Sigma\, y^2$ from the original data on page 389. Since this sum is 45,688, we get

$$s_e = \sqrt{\frac{45{,}688 - 30.33(652) - 3.67(6{,}996)}{8}}$$

$$= 5.4$$

and, hence,

$$t = \frac{3.67 - 4}{5.4} \cdot \sqrt{\frac{10(1{,}121) - (95)^2}{10}}$$

$$= -0.90$$

Assuming that the null hypothesis $\beta = 4$ is to be tested against the two-sided alternative $\beta \neq 4$ at the level of significance $\alpha = 0.05$, we shall now have to compare $t = -0.90$ with $-t_{0.025}$ and $t_{0.025}$ for $n - 2 = 8$ degrees of freedom. Since $t_{0.025} = 2.306$ for 8 degrees of freedom (according to Table II), and the value we got falls *between* -2.306 and 2.306, we find that *the null hypothesis cannot be rejected*. In other words, we conclude that the difference between $b = 3.67$ and $\beta = 4$ may reasonably be attributed to chance.

Tests concerning the regression coefficient α are performed in the same way, except that we use the formula for the first, instead of the second, of the two t statistics on page 413. Note, however, that in most practical applications, the regression coefficient α is not of as much interest as the regression coefficient β, since α represents only the *y-intercept*, namely, the value of y which corresponds to $x = 0$. Thus, in our numerical example, $a = 30.33$ is the grade which we predict for a student who does not study at all for the examination. Let us also point out that if tests concerning α and β are performed with the use of the same data, problems will arise about the level of significance, since the estimates we get for α and β by the method of least squares are *not independent*. For instance, if we are dealing with positive data and it turns out, by chance, that the value we get for the slope b is *too high*, this will tend to make the y-intercept *too low*. To avoid this difficulty, *it is best not to perform tests concerning both regression coefficients on the basis of the same data*.

To construct confidence intervals for the regression coefficients α and β, we have only to substitute, respectively, the expressions for the two t statistics on page 413 into the middle term of the double inequality $-t_{\alpha/2} < t < t_{\alpha/2}$, and then manipulate these double inequalities algebraically so that α and β become the middle terms. Actually performing these steps, we obtain the *confidence limit formulas*

▲
$$a \pm t_{\alpha/2} \cdot s_e \sqrt{\frac{1}{n} + \frac{n \cdot \bar{x}^2}{n(\Sigma x^2) - (\Sigma x)^2}}$$
▲

and

▲
$$b \pm \frac{t_{\alpha/2} \cdot s_e}{\sqrt{\dfrac{n(\Sigma x^2) - (\Sigma x)^2}{n}}}$$
▲

where the degree of confidence is $1 - \alpha$ and $t_{\alpha/2}$ is the entry in Table II corresponding to $n - 2$ degrees of freedom. If both of these formulas are used for the same data, problems will arise about the degree of confidence, since these confidence limits for α and β are *not independent*.

To illustrate the calculation of such a confidence interval, let us find the one for β for the hours-studied-examination-grade example of Section 15.2. Since all of the quantities needed for substitution were already obtained in the preceding example, where we tested the null hypothesis that $\beta = 4$, the work is actually quite simple. Substituting the various quantities into the formula, we get

$$3.67 \pm \frac{2.306(5.4)}{\sqrt{\dfrac{10(1,121) - (95)^2}{10}}}$$

and, hence, the 0.95 confidence interval

$$2.83 < \beta < 4.51$$

Note that this is a 0.95 confidence interval for the average increase in a student's grade for each additional hour he studies for the examination.

To answer the second and third questions asked on page 412, we use methods that are very similar to the ones just discussed. In fact, they are based on two more t statistics, and they are discussed briefly in Exercises 8 and 9 below.

EXERCISES

1. Substitute $x = $ 8, 5, 11, 13, 10, 5, 18, 15, 2, and 8 into $y' = 30.33 + 3.67x$, the equation of the least-squares line obtained for the example of Section 15.2, and use these values of y' (together with the observed y's given on page 389) to calculate $\Sigma(y - y')^2$. Then, compare

    $$\sqrt{\frac{\Sigma(y - y')^2}{n - 2}}$$

    with $s_e = 5.4$, the value obtained for the standard error of estimate on page 414. (The difference should be small and due only to rounding.)

2. Referring to the illustration in the text, use the level of significance 0.05 to test the null hypothesis $\alpha = 35$ against the one-sided alternative $\alpha < 35$.

3. Referring to the illustration in the text, construct a 0.99 confidence interval for the regression coefficient α.

4. In each of the following, state *in words* what hypothesis is being tested, and perform the indicated tests:

(a) Referring to Exercise 1 on page 395, test the null hypothesis $\beta = 10$ against the alternative $\beta \neq 10$ at the level of significance 0.05.

(b) Referring to Exercise 3 on page 396, test the null hypothesis $\beta = 5$ against the alternative $\beta < 5$ at the level of significance 0.01.

(c) Referring to Exercise 4 on page 397, test the null hypotheses $\beta = 1.6$ against the alternative $\beta \neq 1.6$ at the level of significance 0.05.

(d) Referring to Exercise 5 on page 397, test the null hypothesis $\beta = 1.0$ against the alternative $\beta > 1.0$ at the level of significance 0.01.

5. In each of the following, state *in words* what hypothesis is being tested, and perform the indicated tests:

   (a) Referring to Exercise 1 on page 395, test the null hypothesis $\alpha = 30$ against the alternative $\alpha \neq 30$ at the level of significance 0.05.

   (b) Referring to Exercise 4 on page 397, test the null hypothesis $\alpha = 0.5$ against the alternative $\alpha > 0.5$ at the level of significance 0.05.

6. For each of the following, construct a confidence interval for β at the indicated degree of confidence:

   (a) 0.95, for Exercise 1 on page 395;

   (b) 0.99, for Exercise 2 on page 396;

   (c) 0.95, for Exercise 3 on page 396;

   (d) 0.99, for Exercise 4 on page 397;

   (e) 0.95, for Exercise 5 on page 397.

7. For each of the following, construct a confidence interval for α at the indicated degree of confidence:

   (a) 0.95, for Exercise 1 on page 395;

   (b) 0.99, for Exercise 4 on page 397.

8. (*Confidence limits for the true mean of y for a given value of x.*) To answer the second of the three questions on page 412, we can use the following $1 - \alpha$ confidence limits for $\alpha + \beta x_0$, where x_0 is the value of x for which we want to estimate the mean of the y's:

$$\blacktriangle \qquad (a + bx_0) \pm t_{\alpha/2} \cdot s_e \sqrt{\frac{1}{n} + \frac{n(x_0 - \bar{x})^2}{n(\Sigma\, x^2) - (\Sigma\, x)^2}} \qquad \blacktriangle$$

As before, $t_{\alpha/2}$ is to be obtained from Table II; the number of degrees of freedom is $n - 2$. To illustrate, let us refer again to the hours-studied-examination-grade example, and let us find a 0.95 confidence interval for the average grade students should get if they study

$x_0 = 11$ hours for the examination. Making use of the previous results that $a + bx_0 = 30.33 + 3.67(11) = 70.7$ and $s_e = 5.4$, we get

$$70.7 \pm 2.306(5.4) \sqrt{\frac{1}{10} + \frac{10(11 - 9.5)^2}{10(1,121) - (95)^2}}$$

$$70.7 \pm 12.45 \sqrt{0.11}$$

and, hence, the confidence interval $66.6 - 74.8$.

(a) Referring to Exercise 1 on page 395, use this method to find a a 0.95 confidence interval for the true average grade applicants for the job should get on the test if they have studied French (in high school or college) for 4 years.

(b) Referring to Exercise 3 on page 396, use this method to find a 0.99 confidence interval for the true average yield of wheat in the given county where there are 12 inches of rain.

(c) Referring to Exercise 4 on page 397, use this method to find a 0.95 confidence interval for the true net operating profits (expressed as a percentage of total sales), which managers of such sporting goods stores can expect if their advertising expenses are 2 per cent of their total expenses

9. (*Limits of Prediction*) The third question on page 412 differs from the other two insofar as it does *not* concern an estimate of a population parameter; *instead it pertains to the prediction of a single future observation*. A set of limits for which we can assert (with the probability $1 - \alpha$) that they will contain such an observation are called *limits of prediction;* if the y-value is to be observed when $x = x_0$, appropriate limits of prediction are given by

$$\blacktriangle \qquad (a + bx_0) \pm t_{\alpha/2} \cdot s_e \sqrt{1 + \frac{1}{n} + \frac{n(x_0 - \bar{x})^2}{n(\Sigma x^2) - (\Sigma x)^2}} \qquad \blacktriangle$$

Again, $t_{\alpha/2}$ must be obtained from Table II and the number of degrees of freedom is $n - 2$. To illustrate, let us refer again to the hours-studied-examination-grade example, and let us find 0.95 limits of prediction for the grade a student should get if he studies $x_0 = 11$ hours for the examination. Noting that the only difference between the above limits and the ones of Exercise 8 is that we add 1 to the quantity inside the square-root sign, we can immediately write the limits of prediction as

$$70.7 \pm 12.45 \sqrt{1.11}$$

and, hence, as $57.6 - 83.8$.

(a) Referring to Exercise 1 on page 395, find 0.95 limits of prediction for the grade an applicant for the job will get on the test if he has studied French in high school or college for 4 years.

(b) Referring to Exercise 3 on page 396, use this method to find 0.99 limits of prediction for the yield of wheat in the given county when there were 15 inches of rain.

(c) Referring to Exercise 5 on page 397, find 0.95 limits of prediction for the height of such a graft $x_0 = 4$ months after grafting.

BIBLIOGRAPHY

Methods of deciding which kind of curve to fit to a given set of paired data may be found in books on numerical analysis and more advanced texts in statistics. A detailed treatment of problems of regression, including multiple regression, may be found in

EZEKIEL, M., and FOX, K. A., *Methods of Correlation and Regression Analysis*, 3rd ed. New York: John Wiley & Sons, Inc., 1959.

DRAPER, N. R., and SMITH, H., *Applied Regression Analysis*. New York: John Wiley & Sons, Inc., 1966.

16

Correlation

16.1 The Coefficient of Correlation

Now that we have learned how to fit a least-squares line to paired data, let us see how we might describe, or measure, *how well such a line actually fits*. Of course, we can get a fair idea by inspection, say, by looking at a diagram like that of Figure 15.3, but to show how we can be more objective, let us take another look at the original data given in the table on page 389. As can be seen from this table, there are considerable differences among the y's, with the smallest being 33 and the largest being 94. It can also be seen, though, that the grade of 33 was obtained by a student who studied 2 hours for the examination, while the grade of 94 was obtained by a student who studied 18 hours, and this suggests that the differences among the y's may well be due (at least, in part) to the fact that the students did not all study the same amount of time. *This raises the question as to how much of the total variation of the y's can be attributed to the relationship with x, and how much of it can be attributed to all other factors, including chance.* With reference to our example, we would thus want to know what part of the total variation of the grades can be accounted for by the differences in the amount of time which the students studied for the examination, and what part can be attributed to all other factors (including the students' intelligence, their previous training, their physical and mental health on the day of the examination, . . ., and perhaps their having studied *by chance* exactly what was asked in the test).

Essentially, we are thus faced with a problem of *analysis of variance*, and the quantity which we shall want to analyze is $\Sigma (y - \bar{y})^2$, the *total sum of squares* of the y's (in our example, the grades), which is simply the *variance* of the y's multiplied by $n - 1$. Since $\Sigma y = 652$ in our example (according to the work on page 393), we find that $\bar{y} = \dfrac{652}{10} = 65.2$, and, hence, that

$$
\begin{aligned}
\Sigma (y - \bar{y})^2 &= (56 - 65.2)^2 + (44 - 65.2)^2 + (79 - 65.2)^2 + (72 - 65.2)^2 \\
&\quad + (70 - 65.2)^2 + (54 - 65.2)^2 + (94 - 65.2)^2 \\
&\quad + (85 - 65.2)^2 + (33 - 65.2)^2 + (65 - 65.2)^2 \\
&= 3{,}177.6
\end{aligned}
$$

If the amount of time which the students studied for the examination were the *only* thing that affected their grades, the points would all have fallen on a straight line (assuming, of course, that the relationship *is* linear). However, the fact that they do *not* fall on a straight line, as is apparent from Figure 15.3, indicates that there are other factors at play. The extent to which the points fluctuate above and below the line, namely, *the variation caused by these other factors,* is usually measured by the sum of the squares of the vertical deviations from the points to the line (see Figure 15.3), that is, by the quantity

$$
\Sigma (y - y')^2
$$

on which we based the least-squares criterion of Section 15.2.

To calculate this sum of squares for our numerical example, where the equation of the least-squares line was $y' = 30.33 + 3.67x$, we first have to substitute into this equation the ten given values of x, calculate the corresponding values of y', and then add the squares of the differences between the respective values of y and y'. Since the reader was already asked to perform these calculations in Exercise 1 on page 416, let us merely state the result that $\Sigma (y - y')^2 = 233.9$. Thus, we can say that the difference

$$
\begin{aligned}
\Sigma (y - \bar{y})^2 - \Sigma (y - y')^2 &= 3{,}177.6 - 233.9 \\
&= 2{,}943.7
\end{aligned}
$$

measures the variation of the y's that can be attributed to the relationship with x, and that

$$
\frac{\Sigma (y - \bar{y})^2 - \Sigma (y - y')^2}{\Sigma (y - \bar{y})^2} = \frac{2{,}943.7}{3{,}177.6} = 0.93
$$

is the *proportion of the total variation of the y's that can be attributed to the relationship with x.* This quantity is sometimes called the *coefficient* *of determination* and it is denoted r^2; its square root, r, is called the *coefficient of correlation*. When taking the square root, the sign of r is chosen so that it is the same as that of the coefficient b in the equation of the least-squares line, and for our numerical example we thus get

$$r = \sqrt{0.93} = 0.96$$

which is positive since b equalled $+3.67$.

What do we mean when we say that $r = 0.96$ for the given example? We mean that $100 \cdot r^2 = 93$ per cent of the total variation of the grades is accounted for by the differences in the amount of time which the students studied for the examination, and this suggests a pretty strong dependence. In general, if the dependence between x and y is *strong*, $\Sigma (y - y')^2$ will be small compared to $\Sigma (y - \bar{y})^2$, and r will be close to $+1$ or -1; indeed, when r *equals* $+1$ or -1, this is indicative of the fact that $\Sigma (y - y')^2 = 0$, which means that *all the points actually fall on a straight line*. On the other hand, if the dependence between x and y is *weak*, $\Sigma (y - y')^2$ will be almost as large as $\Sigma (y - \bar{y})^2$, and r will be close to 0. When $r = 0$, we say that there is *no correlation*.

Let us also point out that since *part of the variation of the y's cannot exceed their total variation*, $\Sigma (y - y')^2$ cannot exceed $\Sigma (y - \bar{y})^2$, and it follows that r cannot take on a value greater than $+1$ or less than -1.

The statistic r which we have introduced is, undoubtedly, the most widely used measure of the *strength of the linear relationship between two variables*. It indicates the goodness of the fit of a line fitted by the method of least squares, and this, in turn, tells us whether or not it is reasonable to say that there exists a linear relationship (correlation) between x and y.

Actually, the method which we used to introduce r serves mainly to *define* the coefficient of correlation, and in actual practice it is seldom, if ever, used. A great deal of work can be saved by using instead the short-cut formula*

$$r = \frac{n(\Sigma xy) - (\Sigma x)(\Sigma y)}{\sqrt{n(\Sigma x^2) - (\Sigma x)^2} \sqrt{n(\Sigma y^2) - (\Sigma y)^2}}$$

---

* This short-cut formula follows directly from an alternate, though equivalent, definition of the coefficient of correlation in terms of the *sample covariance*

$$s_{xy} = \frac{\Sigma (x - \bar{x})(y - \bar{y})}{n - 1}$$

In this formula, we add the products obtained by multiplying the deviation of each x from $\bar{x}$ by the deviation of the corresponding y from $\bar{y}$, divide by $n - 1$, and in this way we are literally measuring how the values of x and y *vary together*. If the relationship between the x's and the y's is such that large values of x tend to go with large values of y, and small values of x with small values of y, the deviations $x - \bar{x}$ and $y - \bar{y}$

A derivation of this formula is referred to in the Bibliography at the end of this chapter, and in Exercise 1 on page 430, the reader will be asked to calculate r *both ways* for a given set of data and compare the results; except for possible errors due to rounding, the answers should be the same.

Although the short-cut formula may look rather complicated, it is quite easy to use. To find r we have only to determine the five sums $\Sigma\ x$, $\Sigma\ x^2$, $\Sigma\ y$, $\Sigma\ y^2$, and $\Sigma\ xy$, and substitute them into the formula together with n, the number of pairs of observations. To illustrate this, let us refer again to the numerical example of Section 15.2, where we fitted a least-squares line to the hours-studied-examination-grades data. Having found on page 393 that $n = 10$, $\Sigma\ x = 95$, $\Sigma\ y = 652$, $\Sigma\ x^2 = 1{,}121$, and $\Sigma\ xy = 6{,}996$, and on page 414 that $\Sigma\ y^2 = 45{,}688$, we thus get

$$r = \frac{10(6{,}996) - (95)(652)}{\sqrt{10(1{,}121) - (95)^2}\ \sqrt{10(45{,}688) - (652)^2}} = 0.96$$

and this agrees, as it should, with the result obtained on page 422.

To give another example, let us calculate r as a measure of the "degree of association" between x, the age of a certain make two-door sedan, and y, its price as advertised in a Phoenix newspaper. Writing the original data in the first two columns, we obtain the necessary sums as shown in the following table:

x (Years)	y (Dollars)	x^2	y^2	xy
4	895	16	801,025	3,580
10	125	100	15,625	1,250
2	1,395	4	1,946,025	2,790
1	1,795	1	3,222,025	1,795
3	1,245	9	1,550,025	3,735
4	695	16	483,025	2,780
24	6,150	146	8,017,750	15,930

are apt to be *both positive* or *both negative,* so that most of the products $(x - \bar{x})(y - \bar{y})$ and, hence, the covariance, will be *positive.* On the other hand, if the relationship between the x's and the y's is such that large values of x tend to go with small values of y and vice versa, the deviations $x - \bar{x}$ and $y - \bar{y}$ are apt to be of *opposite sign,* so that most of the products $(x - \bar{x})(y - \bar{y})$ and, hence, the covariance, will be *negative.* Using this concept of the sample covariance, we can define the coefficient of correlation by means of the formula

$$r = \frac{s_{xy}}{s_x \cdot s_y}$$

where s_x and s_y are, respectively, the standard deviations of the x's and the y's.

(When calculations like these are performed with a desk calculator, the various totals can be accumulated directly without writing down the individual products and squares.) Substituting $n = 6$ and the five sums into the computing formula for r, we get

$$r = \frac{6(15,930) - (24)(6,150)}{\sqrt{6(146) - (24)^2} \, \sqrt{6(8,017,750) - (6,150)^2}} = -0.94$$

The fact that r was positive in the first example and negative in the second agrees with the rule on page 422, according to which the sign of r should be the same as that of the coefficient b (namely, the *slope*) of the least-squares line which we fit to the data. In the example of Section 15.2, the least-squares line had the positive slope $b = 3.67$, and if we plot the data pertaining to the ages and the prices of the cars as in Figure 16.1, it can be seen that the line which we fitted to the data

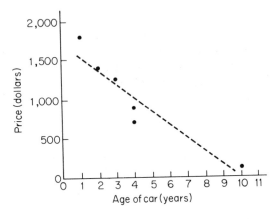

Figure 16.1. Least-squares line with negative slope.

has a *downward slope* (going from left to right), so that b will be *negative*. *Note that the sign of r is, thus, automatically determined if we use the short-cut formula on page* 422, and this should not come as a surprise in view of what we said in the footnote on that page. Geometrically speaking, the idea of a *positive correlation* and that of a *negative correlation* are illustrated in the first two diagrams of Figure 16.2, and the third diagram illustrates the case where the least-squares line is horizontal, $r = 0$, and there is *no correlation*.

Since r does not depend on the scales of x and y, its calculation can often be simplified by adding a suitable positive or negative number to each x, to each y, or to both, or by multiplying each x, each y, or both

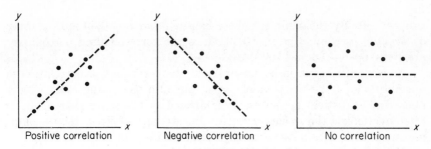

Figure 16.2. Types of correlation.

by arbitrary positive constants (see Exercises 5 and 7 on page 431). What numbers to add or subtract and by what numbers to multiply will have to depend on the nature of the data. The important thing to keep in mind is that the purpose of all this is to simplify the arithmetic as much as possible. When dealing with very large sets of data, the calculation of r can also be simplified by first grouping the data into a *two-way table* as is illustrated in Section 16.6, the Technical Note at the end of this chapter. However, grouping does entail some loss of information and there is no real advantage to it if we can use some kind of computer.

16.2 The Interpretation of r

When $r = 0$, $+1$, or -1, there is generally no difficulty about the interpretation of the coefficient of correlation; when it is 0, the points are so scattered and the fit of the least-squares line is so poor, that it does not aid in the prediction of y; when it is $+1$ or -1, all the points actually lie on a straight line, and it stands to reason that we should be able to make excellent predictions by using the equation of the line. However, when it comes to other values of r, one must be careful not to believe, say, that a correlation of $r = 0.80$ is "twice as good" or "twice as strong" as a correlation of $r = 0.40$, or that a correlation of 0.60 is "three times as strong" as a correlation of $r = 0.20$. To understand the fallacy of these two comparisons, we have only to refer to the definition of the coefficient of correlation on page 421, according to which r^2 gives the *proportion* (or $100r^2$ the *percentage*) *of the total variation of the y's which can be attributed to the relationship with x*. Thus, if $r = 0.80$ in a given example, then $100(0.80)^2 = 64$ per cent of the total variation of the y's is accounted for by the relationship with x, but if $r = 0.40$, the corresponding percentage is only $100(0.40)^2 = 16$. In the sense of "percentage of the variation of the y's which can be attributed to the relationship with x" we can thus say that $r = 0.80$ is *four times as strong a correlation* as $r = 0.40$. By the same token we can say that $r = 0.60$ is *nine times as strong a correlation* as $r = 0.20$, since the corresponding percentages are $100(0.60)^2 = 36$ and $100(0.20)^2 = 4$.

As a word of caution let us add that the coefficient of correlation is not only one of the most widely used, but also one of the most widely abused of statistical measures. It is abused in the sense that (1) it is often overlooked that r measures only the strength of *linear* relationships and that (2) it does not necessarily imply a cause-effect relationship.

If r is calculated indiscriminately, for instance, for the data of Figure 16.3, a value of r close to 0 does not imply that the two variables are not related. The dashed curve of Figure 16.3 provides an excellent fit even

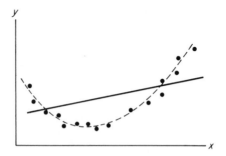

Figure 16.3. Nonlinear relationship.

though the straight line does not. *Let us remember, therefore, that r measures only the strength of linear relationships.*

The fallacy of interpreting high values of r as implying *cause-effect relationships* is best explained with a few examples. One such example, which is frequently used as an illustration, is the high positive correlation one obtains for data pertaining to teachers' salaries and the consumption of liquor over the years. This is obviously not a cause-effect relationship; it results from the fact that both variables are effects of a common cause—the over-all standard of living. Another classical example is the strong positive correlation which has been obtained for the number of storks seen nesting in English villages and the number of childbirths recorded in the same communities. In this case it is the *size* of the villages which affects both, the number of childbirths and the number of storks.

16.3 Correlation Analysis

It is sometimes overlooked that when r is calculated on the basis of sample data, we may get a strong (positive or negative) correlation purely by chance, even though there is actually no relationship whatsoever between the two variables under investigation. To illustrate this point, suppose we take a pair of dice, one red and one green, we roll them five times, and we obtain the following results:

Red Die x	Green Die y
4	5
2	2
4	6
2	1
6	4

If we calculate r for these data, we get the surprisingly high value $r = 0.66$, and this raises the question whether anything is wrong with the assumption that there should be no relationship at all. *After all, one die does not know what the other is doing.* In order to answer this question, we shall have to investigate whether this high value of r might be attributed entirely to chance.

When a correlation coefficient is calculated on the basis of sample data, as in the above example, the value we obtain for r is only an *estimate* of a corresponding parameter, the *population correlation coefficient*, which we refer to as ρ (the Greek letter *rho*). *Rho* measures for populations what r measures for samples. To test the null hypothesis of no correlation, namely, the hypothesis that $\rho = 0$, we shall have to make several assumptions about the distribution of the random variables whose values we observe. In *normal correlation analysis* we make the assumptions of *normal regression analysis* (see page 413), except that now the x's are not looked upon as constants but as values of a random variable having itself a normal distribution. If all these assumptions are met, it can be shown that if the null hypothesis $\rho = 0$ is true, then

▲
$$t = \frac{r \sqrt{n - 2}}{\sqrt{1 - r^2}}$$
▲

has the t distribution with $n - 2$ degrees of freedom. Thus, to check whether the value of r which we obtained above for the two dice is significant, we substitute $r = 0.66$ and $n = 5$, getting

$$t = \frac{(0.66) \sqrt{5 - 2}}{\sqrt{1 - (0.66)^2}} = 1.52$$

and since this does not exceed 3.182, the value of $t_{0.025}$ for $5 - 2 = 3$ degrees of freedom, we find that *the null hypothesis of no correlation cannot be rejected* at the level of significance $\alpha = 0.05$. In other words, we conclude that the value of r which we obtained for the two dice is not significant.

Since the method we have discussed can be used only when the null hypothesis is $\rho = 0$, let us indicate an alternate technique, based on an approximation, which can be used more generally to test the null hypothesis $\rho = \rho_0$ or to construct confidence intervals for ρ. It is based on the *Fisher Z transformation* of the correlation coefficient r, namely, a change of scale from r to Z, which is given by the equation

$$Z = \frac{1}{2}\cdot\log_e\left(\frac{1 + r}{1 - r}\right)$$

(Here, $\log_e$ denotes "logarithm to the base e," where e is an irrational number close to 2.718; actually, we shall not have to work with these logarithms, since we shall perform the change of scale by means of a special table.) The Fisher Z transformation is named after R. A. Fisher, a prominent statistician, who showed that for any value of ρ, the distribution of Z is approximately a normal distribution with the mean μ_Z, which is the Z-value obtained when $r = \rho$, and the standard deviation*

$$\sigma_Z = \frac{1}{\sqrt{n - 3}}$$

Hence, the statistic

▲ $$z = \frac{Z - \mu_Z}{1/\sqrt{n - 3}} = (Z - \mu_Z)\sqrt{n - 3}$$ ▲

has approximately the standard normal distribution. As we have already indicated, this alternate method is facilitated by the use of Table VII at the end of the book, which gives the values of Z corresponding to $r = 0.00, 0.01, 0.02, 0.03, \ldots$, and 0.99. Note that only positive values are given in the table, for if r is negative, we simply look up $-r$ and take the *negative* of the corresponding Z.

To illustrate this technique, let us refer again to the hours-studied-examination-grades example, where we had $n = 10$ and $r = 0.96$, and let us test the null hypothesis $\rho = 0.80$ against the alternative hypothesis $\rho > 0.80$ at the level of significance $\alpha = 0.05$. Since the Z-values corre-

---

* Symbolically,

$$\mu_Z = \frac{1}{2}\cdot\log_e\left(\frac{1 + \rho}{1 - \rho}\right)$$

sponding to 0.96 and 0.80 are, respectively, 1.946 and 1.099 according to Table VII, we find that substitution into the above formula for z yields

$$z = (1.946 - 1.099) \sqrt{10 - 3} = 2.24$$

Since this exceeds $z_{0.05} = 1.64$ (see page 276), we find that *the null hypothesis will have to be rejected;* in other words, we conclude that the correlation between the number of hours which the students study for the examination and their grades is *stronger than 0.80.*

To construct a confidence interval for ρ, we first construct a confidence interval for μ_Z, and then convert to r and ρ by means of Table VII. The confidence interval for μ_Z may be obtained by substituting

$$z = (Z - \mu_Z) \sqrt{n - 3}$$

for the middle term of the double inequality $-z_{\alpha/2} < z < z_{\alpha/2}$, and then manipulating the terms algebraically so that the middle term is μ_Z. This leads to the following $1 - \alpha$ *confidence interval for μ_Z:*

▲
$$Z - \frac{z_{\alpha/2}}{\sqrt{n - 3}} < \mu_Z < Z + \frac{z_{\alpha/2}}{\sqrt{n - 3}}$$
▲

To illustrate its use, suppose that for a given set of paired data (pertaining, say, to the history and sociology grades of 30 students) we got $r = 0.70$. Looking up the value of Z which corresponds to $r = 0.70$ in Table VII, we get 0.867, and substituting this value together with $n = 30$ and $z_{\alpha/2} = 1.96$ into the above formula, we get the 0.95 confidence interval

$$0.867 - \frac{1.96}{\sqrt{27}} < \mu_Z < 0.867 + \frac{1.96}{\sqrt{27}}$$

or

$$0.490 < \mu_Z < 1.244$$

Finally, looking up the values of r which come closest to Z-values of 0.490 and 1.244 in Table VII, we get the 0.95 confidence interval

$$0.45 < \rho < 0.85$$

for the "true" strength of the linear relationship between the grades of students in the two given subjects. Note that the width of this interval serves to illustrate the fact that *correlation coefficients based on relatively small samples are generally not very reliable;* that is, they are subject to considerable chance fluctuations.

To give an example where we run into negative values of r and Z, suppose that for a given set of data $n = 40$ and $r = 0.20$. Since the Z-value which corresponds to $r = 0.20$ is 0.203 according to Table VII, we first get the 0.95 confidence interval

$$0.203 - \frac{1.96}{\sqrt{37}} < \mu_Z < 0.203 + \frac{1.96}{\sqrt{37}}$$

or

$$-0.119 < \mu_Z < 0.525$$

Then, looking up the values of r which come closest to Z-values of 0.119 and 0.525 in Table VII, we finally get the 0.95 confidence interval

$$-0.12 < \rho < 0.48$$

EXERCISES

1. Referring to Exercise 4 on page 397, calculate the coefficient of correlation between advertising expenses and profit
   (a) by first determining $\Sigma (y - \bar{y})^2$ and $\Sigma (y - y')^2$ and then finding the square root of the proportion of the variation of the y's that can be attributed to the relationship with x;
   (b) by using the short-cut formula on page 422.
   Compare the results and also use the t test to test the null hypothesis of no correlation at the level of significance $\alpha = 0.01$.

2. The following are the number of mistakes which twelve waitresses made during the first hour after coming to work and also during the last hour before going home:

First Hour x	Last Hour y
3	5
2	1
4	4
5	6
4	3
1	4
2	3
5	2
0	3
3	2
2	2
4	5

Calculate r and test the null hypothesis of no correlation at the level of significance $\alpha = 0.05$.

3. Calculate r for the data of Exercise 5 on page 397 and test whether it is significant at the level of significance $\alpha = 0.05$.

4. If we actually calculated r for each of the following sets of data, should we be surprised to get $r = 1$ and $r = -1$, respectively? Explain.

	x	y			x	y
(a)				(b)		
	19	7			13	5
	25	11			20	2

5. Calculate r for the data of Exercise 1 on page 395 and repeat the calculations after subtracting 3 from each x and 65 from each y. *The results should be the same.* Also test the null hypothesis of no correlation at the level of significance $\alpha = 0.05$.

6. Referring to Exercise 10 on page 411, calculate r for each of the following pairs of variables:
   (a) income and age;
   (b) income and years college;
   (c) age and years college.
   In each case test the null hypothesis of no correlation at the level of significance $\alpha = 0.05$.

7. Calculate r for the data of Exercise 3 on page 396 after making suitable simplifications in the data. Also test the null hypothesis of no correlation at the level of significance $\alpha = 0.05$.

8. State in each case whether you would expect to obtain a positive correlation, a negative correlation, or no correlation:
   (a) The I.Q.'s of husbands and wives.
   (b) The number of hours which a swimmer practices and his time in a 100-meter race.
   (c) A person's shoe size and the amount of income tax he has to pay.
   (d) Total rainfall in New York City during July and August and the attendance at the Mets' home games during these months.
   (e) Hair color and height.
   (f) Pollen count and the sale of anti-allergy drugs.
   (g) The amount of rubber on a tire and the number of miles it has been driven.
   (h) Shirt size and sense of humor.
   (i) A company's dividends and earnings per share.
   (j) January temperatures in the North Central states and tourist business in Arizona.

9. In a study of the relationship between the amount of irrigation applied to the soil and the corresponding amount of cotton harvested, an

agronomist obtained $r = 0.45$ for $n = 15$ test plots. Test the null hypothesis of no correlation at the level of significance $\alpha = 0.05$, using
  (a) the t test described on page 427;
  (b) the Z transformation.

10. In a study of the relationship between the ages and prices of second-hand typewriters, a business student obtained $r = -0.57$ on the basis of the ages and prices of 25 second-hand typewriters. Test the null hypothesis of no correlation at the level of significance $\alpha = 0.01$, using
  (a) the t test described on page 427;
  (b) the Z transformation.

11. In a study of the relationship between the grades which his students get in the mid-term and final examinations, a professor obtained $r = 0.78$ on the basis of the grades of $n = 50$ of his students. Test the null hypothesis $\rho = 0.70$ against the alternative hypothesis $\rho > 0.70$ at the level of significance $\alpha = 0.05$.

12. In a study of the relationship between the annual production of cotton and citrus in Arizona, data for $n = 12$ years yielded the correlation coefficient $r = -0.56$. Test the null hypothesis $\rho = -0.50$ against the alternative hypothesis $\rho \neq -0.50$ at the level of significance $\alpha = 0.05$.

13. In a study of the relationship between a person's reaction time to a visual stimulus before and after he has been administered a certain drug, a group of psychologists obtained $r = 0.44$ on the basis of data for $n = 18$ persons. Test the null hypothesis $\rho = 0.30$ against the alternative hypothesis $\rho > 0.30$ at the level of significance $\alpha = 0.05$.

14. In a study of the relationship between the supply and the demand for certain food items, an economist obtained $r = -0.62$ for $n = 24$ of these items. Test the null hypothesis $\rho = -0.20$ against the alternative hypothesis $\rho \neq -0.20$ at the level of significance $\alpha = 0.05$.

15. Construct 0.95 confidence intervals for ρ when the necessary assumptions are met and
  (a) $r = 0.75$ and $n = 15$;
  (b) $r = -0.57$ and $n = 32$;
  (c) $r = 0.39$ and $n = 100$;
  (d) $r = 0.16$ and $n = 25$.

16. Construct 0.99 confidence intervals for ρ when the necessary assumptions are met and
  (a) $r = 0.82$ and $n = 19$;
  (b) $r = -0.44$ and $n = 26$;
  (c) $r = 0.12$ and $n = 35$;
  (d) $r = -0.20$ and $n = 13$.

17. When $\rho = 0$ and n is *large*, the sampling distribution of r can be approximated with a normal distribution having the mean 0 and the standard deviation

$$\sigma_r = \frac{1}{\sqrt{n-1}}$$

What is the *numerically* smallest value of r that is significant at the 0.05 level of significance, when
  (a) $n = 100$;   (c) $n = 500$;
  (b) $n = 200$;   (d) $n = 1,000$?

16.4 Rank Correlation

Since the assumptions on which the methods of Section 16.3 are based are rather stringent, it is often desirable to use a nonparametric alternative which can be applied under much more general conditions. To test the null hypothesis $\rho = 0$, we often use a nonparametric test based on the *rank-correlation coefficient* (often called *Spearman's rank-correlation coefficient*), which is essentially the coefficient of correlation for the ranks of the x's and the y's within the respective samples. As we shall see, the rank-correlation coefficient has the added advantage that it is usually much easier to determine than the coefficient of correlation and, hence, provides a "quick and easy" substitute.

To illustrate the calculation of the rank-correlation coefficient, let us refer again to the data on page 389, which we have used to demonstrate most of the regression techniques. Copying the x's and the y's from page 389, ranking the x's among themselves (giving Rank 1 to the largest value, Rank 2 to the second largest value, ..., and Rank 10 to the smallest value), and similarly ranking the y's, we get the values shown in the first four columns of the following table:

Number of Hours Studied x	Grade in Examination y	Rank of x	Rank of y	d	d^2
8	56	6.5	7	-0.5	0.25
5	44	8.5	9	-0.5	0.25
11	79	4	3	1.0	1.00
13	72	3	4	-1.0	1.00
10	70	5	5	0.0	0.00
5	54	8.5	8	0.5	0.25
18	94	1	1	0.0	0.00
15	85	2	2	0.0	0.00
2	33	10	10	0.0	0.00
8	65	6.5	6	0.5	0.25
					3.00

Note that *when there are ties, we assign to each of the tied observations the ranks which they jointly occupy.* Since, among the x's, the 6th and the 7th largest values were tied, we gave each the rank $\dfrac{6+7}{2} = 6.5$, and since the 8th and 9th largest values were also tied, we gave each the rank $\dfrac{8+9}{2} = 8.5$. Had three values been tied, say, the 7th, 8th, and 9th, we would have assigned each the rank $\dfrac{7+8+9}{3} = 8$.

Proceeding from here, we could calculate r for the two sets of ranks using the short-cut computing formula on page 422, but it is generally much easier to use the formula

▲
$$r' = 1 - \frac{6(\Sigma\, d^2)}{n(n^2 - 1)}$$
▲

which actually *defines* the *rank-correlation coefficient.* Here n is the number of pairs of observations and the d's are the *differences* between the ranks of the corresponding x's and y's. When there are no ties, r' will actually *equal* r calculated for the two sets of ranks, but when there are ties, there may be a small (but usually negligible) difference.

Returning to the table, we find that the d's are given in the fifth column, their squares in the sixth column, and that $\Sigma\, d^2 = 3.00$. Since $n = 10$ in this example, we thus get

$$r' = 1 - \frac{6 \cdot 3}{10(10^2 - 1)} = 0.98$$

Note that the difference between $r' = 0.98$ and $r = 0.96$ (which we obtained in Section 16.1) is quite small.

To test whether a rank-correlation coefficient is significant, we make use of the fact that under the null hypothesis that there is no relationship between the x's and the y's (in fact, that they are *randomly matched*), the sampling distribution of r' has the mean 0 and the standard deviation*

$$\sigma_{r'} = \frac{1}{\sqrt{n - 1}}$$

Since, furthermore, this sampling distribution can be approximated with a normal distribution even for relatively small values of n, we base the

---

* There exists a correction for $\sigma_{r'}$ that accounts for ties in rank; it is seldom used, however, unless the number of ties is large. A reference to this correction may be found in the Bibliography at the end of this chapter.

test of the hypothesis on the statistic

$$z = \frac{r' - 0}{1/\sqrt{n - 1}} = r'\sqrt{n - 1}$$

which has approximately the standard normal distribution. For our example, we get

$$z = 0.98\sqrt{10 - 1} = 2.94$$

and since this exceeds $z_{\alpha/2} = z_{0.005} = 2.58$, we can say that there is a *significant relationship* (between the number of hours a student studies for the examination and the grade which he gets) at the level of significance $\alpha = 0.01$.

16.5 Multiple and Partial Correlation

In the beginning of this chapter we defined the correlation coefficient as a measure of the goodness of the fit of a least-squares line to a set of paired data. If predictions are to be made with a multiple linear regression equation of the form

$$y' = b_0 + b_1 x_1 + b_2 x_2 + \ldots + b_k x_k$$

as in Section 15.4, we define the *multiple correlation coefficient* in the same way in which we originally defined r in the beginning of this chapter. We take the square root of

$$\frac{\Sigma (y - \bar{y})^2 - \Sigma (y - y')^2}{\Sigma (y - \bar{y})^2}$$

which is the *proportion of the total variation of the y's that can be attributed to the relationship with the x's*, and the only difference is that we now calculate y' by means of the *multiple* regression equation instead of the equation $y' = a + bx$.

To illustrate, let us refer to the example on page 408, where we obtained the regression equation $y' = 10,197 + 4,149x_1 + 731x_2$ to estimate the price of a house in terms of the number of bedrooms, x_1, and the number of baths, x_2. Without going through any details, let us merely point out that for the given data $\Sigma (y - y')^2 = 686,719$ and $\Sigma (y - \bar{y})^2 = 185,955,000$, so that

$$r^2 = \frac{185,955,000 - 686,719}{185,955,000} = 0.9963$$

and, hence, $r = 0.997$. This indicates that we should be able to make excellent predictions of the price of a house in terms of the number of bedrooms and the number of baths. Actually, though, the example is not very instructive because for x_1 and y alone the correlation coefficient is already $r = 0.996$, so that virtually nothing is gained by considering also the number of baths. As the reader will find, the situation is quite different in Exercise 9 on page 439, where *two variables together will yield much better predictions than either variable alone.*

When we discussed the problem of *correlation and causation*, we showed that a high correlation between two variables can be due entirely to their dependence on a third variable. We illustrated this with the examples of birth registrations and storks, and teachers' salaries and the consumption of liquor. To give another example, let us consider the variables x_1, the weekly amount of hot chocolate sold by a refreshment stand at a summer resort, and x_2, the weekly number of tourists staying at the resort. Let us suppose, also, that on the basis of appropriate data on these variables we obtain a correlation coefficient of $r = -0.30$. This result is surprising; surely, we would expect higher sales of hot chocolate when there are more tourists and vice versa, and hence a *positive* correlation.

If we investigate this situation more closely, we surmise that the negative correlation of -0.30 may well be accounted for by the fact that sales of hot chocolate as well as the number of tourists staying at the resort are related to a third variable x_3, the average weekly temperature at the resort. If the temperature is high, there will be many tourists, but they will prefer cold drinks to hot chocolate; if the temperature is low, there will be fewer tourists, and they will prefer hot chocolate to cold drinks. Thus, let us suppose that further data yielded a correlation coefficient of $r = -0.70$ for x_1 and x_3 (the sale of hot chocolate and temperature), and a correlation coefficient of $r = 0.80$ for x_2 and x_3 (the number of tourists and temperature). These values seem reasonable since low sales of hot chocolate should go with high temperatures and vice versa, while the number of tourists should be high when the temperature is high, and low when the temperature is low.

To study the *actual* effect of the number of tourists on the sale of hot chocolate, we should investigate the relationship between x_1 and x_2 *when all other factors, primarily temperature, are held fixed.* As it is seldom, if ever, possible to control things to such an extent, it has been found that a statistic called the *partial correlation coefficient* does a fair job of eliminating the effects of other variables. If we write the "ordinary" correlation coefficient for x_1 and x_2 as r_{12}, that for x_1 and x_3 as r_{13}, and that for x_2 and x_3 as r_{23}, the *partial correlation coefficient for x_1 and x_2 with x_3 fixed* is given by

$$r_{12.3} = \frac{r_{12} - r_{13} \cdot r_{23}}{\sqrt{1 - r_{13}^2} \ \sqrt{1 - r_{23}^2}}$$

Substituting the numerical values of our example, we get

$$r_{12.3} = \frac{(-0.30) - (-0.70)(0.80)}{\sqrt{1 - (-0.70)^2} \sqrt{1 - (0.80)^2}} = 0.61$$

and this shows that, as we should have expected, there is a *positive* correlation between sales of hot chocolate and the number of tourists visiting the resort *when the temperature is fixed;* that is, when the effect of variations in temperature is eliminated.

We have given this example primarily to illustrate what is meant by *partial correlation;* at the same time the example has also served to emphasize again that "ordinary" correlation coefficients can lead to very misleading conclusions unless they are interpreted with great care. The formula we have given for $r_{12.3}$ provides a measure of the strength of the correlation between two variables when a third variable is held fixed, but it should be observed that the theory on which it is based can be generalized so that it applies also when more than one variable is held fixed.

EXERCISES

1. Calculate r' for the data of Exercise 2 on page 396 and test the null hypothesis of no correlation at the level of significance $\alpha = 0.05$.

2. Calculate r' for the data of Exercise 3 on page 396 and test the null hypothesis of no correlation at the level of significance $\alpha = 0.05$.

3. Calculate r' for the data of Exercise 4 on page 397 and test the null hypothesis of no correlation at the level of significance $\alpha = 0.01$.

4. Calculate r' for the following data representing the English grades, x, and the psychology grades, y, of 18 college freshmen:

x	91	67	68	50	46	71	67	71	83
y	82	64	81	68	60	68	68	82	71
x	71	86	94	75	59	50	71	91	55
y	66	83	87	77	61	52	68	87	64

Also test for significance at $\alpha = 0.05$.

5. The following are the total 1969 and 1971 attendance figures for the 12 baseball teams in the National League:

	1969	1971
Atlanta	1,458,320	1,006,420
Chicago	1,674,993	1,653,007
Cincinnati	987,991	1,501,122
Houston	1,442,995	1,261,589
Los Angeles	1,784,527	2,064,594
Montreal	1,212,608	1,290,963
New York	2,175,373	2,266,680
Philadelphia	519,414	1,511,233
Pittsburgh	769,368	1,501,132
St. Louis	1,682,783	1,604,671
San Diego	512,970	549,085
San Francisco	873,603	1,122,599

Calculate r' for these data.

6. The following shows how a panel of nutrition experts and a panel of housewives ranked fifteen breakfast foods:

Breakfast food	Nutrition experts	Housewives
I	3	5
II	7	4
III	11	8
IV	9	14
V	1	2
VI	4	6
VII	10	12
VIII	8	7
IX	5	1
X	13	15
XI	12	9
XII	2	3
XIII	15	10
XIV	6	11
XV	14	13

Calculate r' as a measure of the consistency of the two rankings.

7. The following are the rankings which two critics gave twelve television programs:

	Critic I	Critic II
Program A	3	2
Program B	6	8
Program C	12	10
Program D	1	4
Program E	7	6
Program F	5	5
Program G	10	12
Program H	2	1
Program I	11	9
Program J	4	3
Program K	9	11
Program L	8	7

Calculate r' as a measure of the *consistency* of the two rankings.

8. The following are the rankings given to the works of ten artists by three judges:

Judge A	6	4	2	5	9	3	1	8	10	7
Judge B	2	5	4	8	10	1	6	9	7	3
Judge C	7	3	1	2	10	6	4	9	8	5

Calculate r' for each pair of rankings and decide
   (a) which pair of rankings is the most consistent;
   (b) which pair of rankings is the least consistent.

9. Use the least-squares equation obtained in Exercise 10 on page 411 to calculate y' for each of the five data points, determine the two sums of squares $\Sigma (y - \bar{y})^2$ and $\Sigma (y - y')^2$, and use them to find the value of the multiple correlation coefficient. Compare this value of the multiple correlation coefficient with the values of r obtained for age and income, and for years college and income, in Exercise 6 on page 431.

10. Referring to the example in the text, calculate the partial correlation coefficient for x_1 and x_3 (sales of hot chocolate and temperature) when x_2 (number of tourists) is held fixed. (*Hint:* interchange 2 and 3 wherever these numbers appear in the subscripts of the formula for the partial correlation coefficient on page 436.)

11. Referring to the results of Exercise 6 on page 431, find
   (a) the partial correlation coefficient for income and age when "years college" is held fixed;
   (b) the partial correlation coefficient for "years college" and income when age is held fixed.

12. With reference to the data of Exercise 9 on page 411, find
   (a) the coefficient of correlation for the I.Q.'s and the examination grades;
   (b) the partial correlation coefficient for the I.Q.'s and the examination grades when the number of hours studied is held fixed.

16.6 Technical Note (The Calculation of r for Grouped Data)

To illustrate the steps needed to group paired data and to calculate the coefficient of correlation on the basis of the resulting *two-way frequency table*, let us consider the following data obtained while checking the *reliability* of an achievement test in spelling, which may eventually be given to a great many students. Such a test is said to be reliable, if there would be relatively small chance variations among the scores of a student who repeatedly took the test. Rather than have students repeat the test, however, they are graded separately for the even-numbered problems and the odd-numbered problems, and if the test is reliable, the correlation between the pairs of scores should be very high. Thus, let us consider the following data representing the scores of 25 students on the even-numbered problems, x, and the odd-numbered problems, y:

x	37	32	45	27	43	37	43	36	43	30	48	35	44
y	42	34	42	28	41	39	48	31	37	34	46	35	42

x	45	32	42	38	36	47	38	40	38	33	39	46
y	37	31	42	33	37	44	39	43	41	38	36	40

The problems we must face when grouping paired data are almost identical with the ones we met in the construction of ordinary frequency distributions. *We must decide how many classes to use for each variable and from where to where each class is to go.* Choosing the five classes 25–29, 30–34, 35–39, 40–44, and 45–49 for the x's as well as the y's, we obtain the following *two-way table*:

x

	25–29	30–34	35–39	40–44	45–49
25–29					
30–34					
y 35–39					
40–44					
45–49					

The next step is to *tally* the data; for instance, $x = 37$ and $y = 42$ goes into the cell belonging to the fourth row and the third column, and $x = 32$ and $y = 34$ goes into the cell belonging to the second row and the second column. After counting the number of cases falling into each cell, we obtain the following *two-way frequency distribution:*

x

	25–29	30–34	35–39	40–44	45–49
25–29	1				
30–34		3	2		
y 35–39		1	5	1	1
40–44			2	4	3
45–49				1	1

Graphically, such a two-way distribution may be represented by a *three-dimensional histogram* like the one shown in Figure 16.4. Here the heights of the blocks represent the frequencies of the cells on which they stand just as the heights of the rectangles of an ordinary histogram represent the individual class frequencies.

In order to calculate r from a two-way frequency table, we shall have to assume (as in the case of the mean and the standard deviation) that all measurements within a class (cell) are located at the class marks; for instance, we assume that the three "data points" which belong to the second row and the second column are all $x = 32$ and $y = 32$, and that the two "data points" which belong to the fourth row and the third column

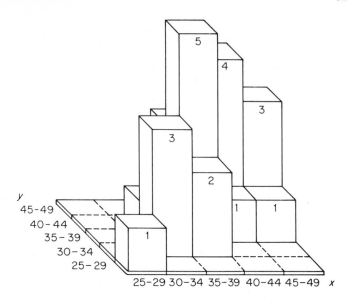

Figure 16.4. Three-dimensional histogram.

are all $x = 37$ and $y = 42$. If the respective intervals for x and y are all equal, as they are in our example, we perform the same kind of *coding* (change of scale) as in Chapters 3 and 4; that is, we replace the x-scale with a u-scale and the y-scale with a v-scale, numbering successive class marks $\ldots, -3, -2, -1, 0, 1, 2, 3, \ldots$ As before, the choice of the zero of the coded scales is arbitrary, but assigning it in each case to the middle class, we get

		-2	-1	*u-Scale* 0	1	2
	-2	1				
	-1		3	2		
v-Scale	0		1	5	1	1
	1			2	4	3
	2				1	1

As we pointed out on page 424, the coefficient of correlation is not affected if we add a constant to all the x's or all the y's or if we multiply all the x's or all the y's by positive constants. *This implies that if we calculate*

r *in terms of the u's and v's, we will get the same result as if we had used the* *class marks in the original scales.* Thus, substituting u's and v's for the x's and y's in the computing formula for r and allowing for the fact that we are now dealing with grouped data, *the formula for r for grouped data* becomes

$$\blacktriangle \qquad r = \frac{n(\Sigma\ uvf) - (\Sigma\ uf_u)(\Sigma\ vf_v)}{\sqrt{n(\Sigma\ u^2 f_u) - (\Sigma\ uf_u)^2}\ \sqrt{n(\Sigma\ v^2 f_v) - (\Sigma\ vf_v)^2}} \qquad \blacktriangle$$

The quantities which are needed for the calculation of r for grouped data may be obtained by arranging the calculations as in the following table.

					(1) v	(2) f_v	(3) vf_v	(4) $v^2 f_v$	(5) uvf
1					-2	1	-2	4	4
	3	2			-1	5	-5	5	3
	1	5	1	1	0	8	0	0	0
		2	4	3	1	9	9	9	10
			1	1	2	2	4	8	6

							(1)	(2)	(3)	(4)
(6)	u	-2	-1	0	1	2		25	6	26	23
(7)	f_u	1	4	9	6	5	25				
(8)	uf_u	-2	-4	0	6	10	10				
(9)	$u^2 f_u$	4	4	0	6	20	34				
(10)	uvf	4	3	0	6	10	23				

To simplify the over-all appearance of this table, we put the u- and v-scales at the bottom and at the right-hand side, labeling them row (6) and column (1), respectively. Column (2) contains the frequencies f_v, the number of times each v occurs, and they are obtained by adding the frequencies in the respective rows. Column (3) contains the products vf_v, which are obtained by multiplying the corresponding entries of columns (1) and (2), and column (4) contains the products $v^2 f_v$, which may be obtained by either squaring each entry of column (1) and multiplying by the corresponding entry of column (2) *or* by multiplying the corresponding entries of columns (1) and (3). The numbers shown in rows (7), (8), and (9) are obtained by performing the identical operations on the u's and f_u's.

The totals of columns (2), (3), and (4) provide n, $\Sigma\, vf_v$, and $\Sigma\, v^2 f_v$, while the totals of rows (7), (8), and (9) provide n, $\Sigma\, uf_u$, and $\Sigma\, u^2 f_u$. To calculate r we still lack $\Sigma\, uvf$, which stands for the sum of the quantities obtained by individually multiplying each cell frequency by the u and the v of the row and column to which it belongs. To simplify these calculations, we shall work on each row separately, first multiplying each cell frequency by the corresponding u and then multiplying the *sum* of these products by the corresponding v. For the *second row* we thus get

$$[3(-1) + 2(0)](-1) = 3$$

for the *fourth row* we get

$$[2(0) + 4(1) + 3(2)](1) = 10$$

and these are the corresponding entries shown in column (5). Interchanging the roles of u and v, we can also calculate $\Sigma\, uvf$ by working separately on each column, multiplying each cell frequency by the corresponding v, and the multiplying the *sum* of these products by the corresponding u. For the *fourth column* we thus get

$$[1(0) + 4(1) + 1(2)](1) = 6$$

and this is the corresponding entry shown in row (10). Comparison of the totals of column (5) and row (10) provides a check on the calculations.

If we now substitute the totals of the appropriate rows and columns of the *correlation table* (as such a table is sometimes called) into the formula for r, we finally get

$$r = \frac{25(23) - (10)(6)}{\sqrt{25(34) - (10)^2}\ \sqrt{25(26) - (6)^2}}$$
$$= 0.76$$

and it would seem reasonable to conclude that the spelling test under consideration is *quite reliable*.

EXERCISES

1. In a study of the growth of saguaro cacti, an experiment was performed to determine how well the height of such cacti can be estimated from aerial photographs. The following are the heights of 36 saguaros (in inches) estimated from aerial photographs, x, and measured on the ground, y:

x	y	x	y
118	103	163	163
166	160	124	137
141	143	171	173
164	187	165	112
150	111	123	132
151	134	142	151
133	121	144	148
122	143	130	117
165	141	135	165
168	149	139	112
109	125	161	121
153	128	170	189
135	101	148	156
158	136	136	158
104	117	174	182
183	121	186	161
173	156	194	153
125	130	181	183

Group these data into a two-way table having the classes 100–119, 120–139, 140–159, 160–179, and 180–199 for x as well as y. Then calculate r from this table and test the null hypothesis of no correlation at the level of significance $\alpha = 0.05$.

2. The following are the number of minutes it took 30 students to learn two lists of French verbs, one in the morning and one in the late afternoon:

Morning x	Afternoon y	Morning x	Afternoon y
15	16	18	21
21	28	25	23
17	22	13	19
23	23	20	24
23	17	16	26
12	17	24	25
28	25	18	27
23	26	22	28
16	24	27	25
25	29	19	18
22	21	23	22
21	20	14	22
17	15	26	26
21	29	24	21
22	18	19	23

Group these data into a two-way table having the classes 10–14, 15–19, 20–24, and 25–29 for x as well as y, calculate r, and test the null hypothesis of no correlation at the level of significance $\alpha = 0.01$.

3. Calculate r for the data grouped into the following table, where x and y stand, respectively, for height (in inches) and weight (in pounds):

y (*weights*) \ x (*heights*)	60–62	63–65	66–68	69–71	72–74	75–77
110–129	2	3	1			
130–149		1	4	1		
150–169		1	3	5	1	
170–189			2	6	3	1
190–209			1	5	4	1
210–229				1	3	
230–249					1	1

4. Calculate r from the following correlation table, with x representing grades in an English test and y representing grades in a statistics examination:

y-Scale \ *x-Scale*	1–20	21–40	41–60	61–80	81–100
1–20	1	2	1		
21–40		4	3	1	
41–60		1	5	6	1
61–80			2	3	3
81–100			1	2	4

BIBLIOGRAPHY

More detailed information about multiple and partial correlation may be found in the book by M. Ezekiel and K. A. Fox listed on page 419. The book by S. Siegel, listed on page 385, gives the correction in the formula for σ_r' that is needed when there are many ties. A derivation of the computing formula for r may be found in

RICHARDSON, C. H., *An Introduction to Statistical Analysis.* New York: Harcourt, Brace & World, Inc., 1944.

17

Planning Surveys
and Experiments

17.1 Introduction

Throughout this book it was assumed that the samples with which we dealt were random and that the experiments to which we referred in exercises and illustrations were such that valid generalizations could be made. In this chapter we shall discuss briefly how to *improve* on simple random sampling, where possible, and what to do when simple random sampling is physically impossible or too expensive. So far as the planning of experiments is concerned, we shall further develop some of the ideas introduced in Chapter 13 in connection with the analysis of variance.

17.2 Sample Designs

A sample design is a definite plan, completely determined before any data are collected, for obtaining a sample from a given population. Thus, the plan to take a simple random sample of size 10 from among the 220 members of a Medical Association by using a table of random numbers in a prescribed way constitutes a sample design. Generally speaking, there are always many ways in which a sample can be taken from a given population; some of these are quite simple, while others are relatively involved. Sometimes two or more procedures are combined in sampling different parts of the same population. In the remainder of this section, we shall briefly mention some of the most important kinds of sample designs;

detailed treatments of this subject are referred to in the Bibliography at the end of the chapter.

17.3 Systematic Sampling

There are many situations in which the most practical way of sampling is to select, say, every 20th voucher in a file, every 10th name on a list, every 50th piece coming off an assembly line, and so on. Sampling of this sort is referred to as *systematic sampling*, and an element of randomness can be introduced into this kind of sampling by using random numbers or some gambling device to pick the unit with which to start. Although a systematic sample may not be a (simple) random sample in accordance with our definition, it is often reasonable to treat systematic samples as if they were random samples. Whether or not this is justified depends entirely on the structure (order) of the list, or arrangement, from which the sample is obtained. In some instances, systematic sampling actually provides an improvement over simple random sampling inasmuch as the sample is "spread more evenly" over the entire population.

The danger in systematic sampling lies in the possible presence of *hidden periodicities*. For instance, if we inspect every 50th piece made by a machine, our result would be quite biased if (owing to a regularly recurring failure) it so happened that every 25th piece had blemishes. Also, a systematic sample might yield biased results if we interviewed the residents of every 12th house along a certain route and it so happened that each of these houses was a corner house on a double lot.

17.4 Stratified Sampling

As we saw in the basic chapters on statistical inference, the "goodness" of a generalization (that is, the reliability of an estimate or the risk of erroneously accepting or rejecting a hypothesis) depends both on the size of the sample and the variability of the population. If cost were no factor, we could always make the inference as "good" as we want by choosing samples which are sufficiently large, but since this is seldom the case, we must look for other ways of reducing the risks to which we are exposed.

When a population can be subdivided into a number of subpopulations, or *strata*, each of which is relatively *uniform or homogeneous*, inferences about the parameters of the population can generally be improved by what is called *stratified sampling*. In this kind of sampling, we divide the population into a number of nonoverlapping homogeneous subpopulations, to which we then allocate certain portions of the total sample. If the items selected from each of the strata constitute a simple random sample, the entire procedure (first stratification and then simple random sampling) is called *stratified (simple) random sampling.* Although the concept of this kind of sampling is simple, a number of problems immediately arise: How many strata should be formed? What should be the sample size for the different strata? How are the samples within the strata to be selected? Stratification does not guarantee good results, but if properly executed, stratified sampling generally leads to a higher degree of precision, that is, to improved generalizations about the whole population.

To illustrate the general idea behind stratified sampling, let us consider the following oversimplified, though concrete, example. Suppose we want to estimate the mean weight of 4 persons on the basis of a sample of size 2; the (supposedly unknown) weights of the 4 persons are, respectively, 115, 135, 185, and 205 pounds, so that the population mean we are trying to estimate is

$$\mu = \frac{115 + 135 + 185 + 205}{4} = 160 \text{ pounds}$$

If we take a simple random sample of size 2 from this population, it can easily be seen that the sample mean may be as small as 125 pounds or as large as 195 pounds. In fact, the 6 possible samples of size 2 that can be taken from this population have means of 125, 150, 160, 160, 170, and 195 pounds, so that

$$\sigma_{\bar{x}}^2 = [(125 - 160)^2 + (150 - 160)^2 + (160 - 160)^2$$
$$+ (160 - 160)^2 + (170 - 160)^2 + (195 - 160)^2] \cdot \frac{1}{6}$$
$$= 441.7$$

and $\sigma_{\bar{x}}$ is approximately 21.0.

Now suppose that we make use of the fact that among the four persons there are 2 men and 2 women and that we *stratify* our sample by randomly selecting one of the two women and one of the two men. Assuming that the two smaller weights are those of the two women, we now find that the sample mean varies on the much smaller interval from 150 to 170. In fact, the four possible stratified samples have means of 150, 160, 160, and 170, so that

$$\sigma_{\bar{x}}^2 = [(150 - 160)^2 + (160 - 160)^2 + (160 - 160)^2 + (170 - 160)^2] \cdot \frac{1}{4}$$
$$= 50$$

and $\sigma_{\bar{x}}$ is approximately 7.1. *This illustrates how, by stratifying the sample, we were able to reduce the standard error of the mean, $\sigma_{\bar{x}}$, from 21.0 to 7.1.*

Essentially, the goal of stratification is to form strata in such a way that there is some relationship between being in a particular stratum and the answer sought in the statistical study, and that within the individual strata there is as much homogeneity (uniformity) as possible. Note that in our example there was such a connection between sex and weight and that there was much less variability in weight within the two strata than there was within the entire population. Note also that we used what is called *proportional allocation,* which means that the sizes of the individual samples were proportional to the sizes of the respective strata. In general, if we divide a population of size N into k strata of size $N_1, N_2, \ldots,$ and $N_k,$ and take a sample of size n_1 from the first stratum, a sample of size n_2 from the second stratum, $\ldots,$ and a sample of size n_k from the kth stratum, we say that the *allocation is proportional* if

$$\frac{n_1}{N_1} = \frac{n_2}{N_2} = \ldots = \frac{n_k}{N_k}$$

or, at least, if these ratios are as nearly equal as possible. Note that in the example of the preceding paragraph, we had $N_1 = 2, N_2 = 2, n_1 = 1,$ and $n_2 = 1,$ so that

$$\frac{n_1}{N_1} = \frac{1}{2} \quad \text{and} \quad \frac{n_2}{N_2} = \frac{1}{2}$$

and the allocation is, indeed, proportional.

To consider another example, suppose that a sample of size $n = 60$ is to be taken from a population of size $N = 4,000,$ which is divided into three strata of size $N_1 = 2,000, N_2 = 1,200,$ and $N_3 = 800,$ respectively. Since the first stratum constitutes half the population, we allocate to it half the sample, and this makes $n_1 = 30;$ correspondingly, the second and third strata constitute, respectively, 30 per cent and 20 per cent of the population, and we let $n_2 = (0.30)60 = 18$ and $n_3 = (0.20)60 = 12.$

After we have taken a sample of size n_1 from the first stratum, a sample of size n_2 from the second stratum, $\ldots,$ and the means of the respective samples are $\bar{x}_1, \bar{x}_2, \ldots,$ and, $\bar{x}_k,$ we estimate the mean of the whole population as

$$\blacktriangle \qquad \bar{x}_w = \frac{N_1\bar{x}_1 + N_2\bar{x}_2 + \ldots + N_k\bar{x}_k}{N_1 + N_2 + \ldots + N_k} \qquad \blacktriangle$$

This is a *weighted mean* of the individual $\bar{x}$'s, and the weights are the sizes of the corresponding strata. Actually, the above formula applies regardless of how we allocate parts of the total sample to the strata, but *when allocation is proportional, the result is simply the mean of all the data combined* (see Exercise 7 on page 456), and it is in this sense that proportional allocation is said to be *self-weighting*.

In proportional allocation we account for the relative importance (that is, size) of each stratum, but we do not account for possible differences in their *variability*. Clearly, if one stratum has a greater variability than another, it will take a large sample to estimate its mean with the same precision, and to account for this we choose the n's so that

$$\frac{n_1}{N_1 \sigma_1} = \frac{n_2}{N_2 \sigma_2} = \ldots = \frac{n_k}{N_k \sigma_k}$$

or, at least, so that these ratios are as nearly equal as possible, where σ_1, σ_2, ..., and σ_k are the standard deviations of the individual strata. In this method of allocation, called *optimum allocation*, the more variable strata will contribute relatively more items to the total sample than in proportional allocation, and it can be shown (see Exercise 8 on page 456) that if the total sample is of size n, then n_1, n_2, ..., and n_k, are given by the formula

$$\blacktriangle \qquad n_i = \frac{n \cdot N_i \sigma_i}{N_1 \sigma_1 + N_2 \sigma_2 + \ldots + N_k \sigma_k} \qquad \blacktriangle$$

for $i = 1, 2, \ldots$, and k. To illustrate this method of allocation, let us refer again to the example on page 451, and let us suppose that the standard deviations of the three strata are $\sigma_1 = 8$, $\sigma_2 = 15$, and $\sigma_3 = 32$. Since we had $N_1 = 2,000$, $N_2 = 1,200$, and $N_3 = 800$, and the total sample size was $n = 60$, we now get

$$n_1 = \frac{60 \cdot 2,000(8)}{2,000(8) + 1,200(15) + 800(32)}$$

which gives $n_1 = 16$, and similarly we find that $n_2 = 18$ and $n_3 = 26$. Note that because of the large variability of the third stratum, n_3 is now larger than n_1 and n_2.

So far we have mentioned only stratification with respect to one variable, or characteristic, but populations are often stratified with respect to *several* characteristics, say, with respect to income, education, age, and geographical location, as in a survey designed to determine public opinion concerning a political issue. We might thus allocate part of the total sample to persons with low incomes, who are not college graduates, in the 25–

45 year age bracket, and live in the Northeast; another portion to persons with high incomes, who are college graduates, in the over-65 age bracket, and live in the Southwest; and so forth. This process is called *cross-stratification*, and it is widely used, particularly in public opinion polls and market surveys.

17.5 Cluster Sampling

To illustrate another important kind of sampling, suppose that a large foundation wants to study the changing patterns of family expenditures for recreation in the Detroit area. In attempting to complete schedules for a sample of 800 families, the foundation runs into a number of difficulties; simple random sampling is practically impossible since suitable lists are not available and the cost of contacting families scattered over a wide area (sometimes with 2 or 3 call-backs for the not-at-homes) can be very high. One way in which a sample can be taken in this example is to divide the total area of interest into a number of smaller, nonoverlapping areas, say, city blocks. A number of these blocks can then be chosen with the use of random numbers, and all (or samples of) the families residing in these blocks are included in the sample. This kind of sampling is referred to as *cluster sampling;* that is, the total population is divided into a number of relatively small subdivisions (which are themselves *clusters* of still smaller units) and then some of these subdivisions, or clusters, are randomly selected for inclusion in the over-all sample. If the clusters are geographic subdivisions, as in our example, this kind of sampling is also called *area sampling*.

Although estimates based on cluster samples are generally not as reliable as estimates based on simple random samples of the same size [see Exercise 4(c) on page 455], they are usually more reliable *per unit cost*. Referring again to the survey of family expenditures for recreation in the Detroit area, it is easy to see that it may well be possible to obtain a cluster sample several times the size of a simple random sample at the same expense. (It is much cheaper to visit and interview families living close together in clusters than families selected at random over a wide area.)

In practice, several of the methods we have discussed may well be used in the same survey. For instance, if government statisticians wanted to study the attitude of elementary school teachers toward certain federal programs, they might first stratify the country by states or some other geographic subdivision. To obtain a sample from each stratum they might then use cluster sampling, subdividing each stratum, say, into school districts, and finally they might use simple random sampling or systematic sampling within each cluster.

17.6 Quota Sampling

In many applications of stratified sampling the selection of individuals within the various strata has traditionally been *nonrandom*. Instead, interviewers have been given quotas to be filled from the various strata, without too many restrictions as to *how* they are to be filled. For instance, in determining attitude toward increased foreign aid expenditures, an interviewer working a certain area for an opinion research organization might be told to interview 5 retail merchants of German origin who own homes, 12 wage-earners of Anglo-Saxon origin who live in rented apartments, 2 retired persons who live in trailers, etc., with the actual selection of the individuals being left to the interviewer's discretion. This is a convenient, relatively inexpensive, and often necessary procedure, but as it is usually executed, the resulting quota samples do not have the essential features of random samples. Interviewers naturally tend to select individuals who are most readily available, that is, persons who work in the same building, shop in the same store, or perhaps reside in the same general area. Quota samples are, thus, essentially *judgment samples*, and they generally do not lend themselves to any sort of formal statistical evaluation.

EXERCISES

1. The following are the number of hospitals in 20 large cities: 7, 16, 6, 21, 17, 10, 23, 8, 7, 18, 81, 9, 26, 5, 7, 9, 11, 114, 13, and 32.
   (a) List the four possible systematic samples of size 5 that can be taken from this list by starting with one of the first four numbers and then taking each fourth number on the list.
   (b) Calculate the means of the four samples obtained in part (a) and verify that *their* mean equals the average (mean) number of hospitals in the given 20 cities.

2. The following are the number of commercial FM radio stations in operation in 1968 in the fifty states (listed in alphabetic order): 38, 3, 12, 25, 134, 22, 16, 3, 70, 45, 3, 5, 84, 67, 28, 19, 50, 31, 10, 30, 35, 67, 24, 22, 23, 4, 13, 7, 10, 23, 14, 82, 65, 5, 99, 31, 14, 99, 6, 31, 2, 52, 107, 7, 1, 48, 31, 20, 63, and 1. List the five possible systematic samples of size 10 that can be taken from this list by starting with one of the first five numbers and then taking each fifth number on the list.

3. With reference to Exercise 2, calculate the standard deviation of the means of the five systematic samples, and compare it with the value of the standard error of the mean for random samples of size 10 from the given set of 50 numbers, which is $\sigma_{\bar{x}} = 9.0$.

4. To generalize the example on page 450, suppose that in a group of 6 athletes there are 3 tennis players whose weights are 145 lb, 155 lb, and 165 lb, and 3 football players whose weights are 215 lb, 225 lb, and 235 lb.

   (a) List the 15 possible random samples of size 2 from this population of weights, calculate their means, and show that $\sigma_{\bar{x}} = 22.7$.

   (b) List the 9 possible stratified random samples of size 2 which may be obtained by choosing the weight of one of the tennis players and the weight of one of the football players, calculate their means, and show that $\sigma_{\bar{x}} = 5.8$.

   (c) Suppose that the 6 athletes are divided into clusters according to their sport, each cluster is assigned the probability 0.50, and a random sample of size 2 is taken from the cluster which has been chosen. List the 6 possible samples that can thus be obtained, calculate their means, and show that $\sigma_{\bar{x}} = 35.2$.

5. Show that if a sample of size n is to be proportionally allocated to the strata of a population of size N, the sample size for the ith stratum is given by

$$\blacktriangle \qquad n_i = n \cdot \frac{N_i}{N} \qquad \blacktriangle$$

or as near to this quantity as possible. Use this formula to solve the following problems:

   (a) If a random sample of size $n = 40$ is to be taken from a population of size $N = 2,000$, which consists of four strata for which $N_1 = 500$, $N_2 = 1,200$, $N_3 = 200$, and $N_4 = 100$, how large a sample should be taken from each stratum?

   (b) If a random sample of size $n = 200$ is to be taken from a population of size $N = 50,000$, which consists of five strata for which $N_1 = 20,000$, $N_2 = 15,000$, $N_3 = 5,000$, $N_4 = 8,000$, and $N_5 = 2,000$, how large a sample should be taken from each stratum?

6. The records of a casualty insurance company shows that among 3,800 claims filed against the company over a period of time, 2,600 were minor claims (under \$200), while the other 1,200 were major claims (\$200 or more). To estimate the average size of these claims, the company takes a one per cent sample, proportionally allocated to the two strata, and obtains the following results (rounded to the nearest dollar):

   Minor claims: 42, 115, 63, 78, 45, 148, 195, 66, 18, 73, 55, 89, 170, 41, 92, 103, 22, 138, 49, 62, 88, 113, 29, 71, 58, 83

   Major claims: 246, 355, 872, 649, 253, 338, 491, 860, 755, 502, 488, 311

(a) Find the means of these two samples and then determine their weighted mean, using as weights the respective sizes of the two strata.

(b) Verify that the result of part (a) equals the ordinary mean of the 38 claims; that is, verify for this example that proportional allocation is *self-weighting*.

7. Verify symbolically that for proportional allocation the weighted mean given by the formula on page 451 equals the ordinary mean of the sample values obtained for all the strata. Make use of the formula of Exercise 5 and the fact that $n_i \cdot \bar{x}_i$ equals the sum of the values obtained for the ith stratum.

8. Verify that if the n_i are calculated according to the formula on page 452 for optimum allocation, the ratios $\dfrac{n_i}{N_i \sigma_i}$ for $i = 1, 2, \ldots,$ and k are all equal to the same quantity and the sum of the n_i is equal to n.

9. A population is divided into two strata so that $N_1 = 10{,}000$, $N_2 = 30{,}000$, $\sigma_1 = 45$, and $\sigma_2 = 60$. How should a sample of size $n = 100$ be allocated to the two strata if we use
    (a) proportional allocation;
    (b) optimum allocation?

10. A population is divided into four strata so that $N_1 = 2{,}000$, $N_2 = 6{,}000$, $N_3 = 10{,}000$, $N_4 = 4{,}000$, $\sigma_1 = 25$, $\sigma_2 = 20$, $\sigma_3 = 15$, and $\sigma_4 = 30$. How should a sample of size $n = 440$ be allocated to the four strata if we use
    (a) proportional allocation;
    (b) optimum allocation?

11. Suppose we want to estimate the mean weight of 6 persons on the basis of a sample of size 3. The supposedly unknown weights of the 6 persons are, respectively, 135, 141, 159, 165, 171, and 267 pounds, so that the population mean which we want to estimate is $\mu = 173$. If, furthermore, the first four of these weights are those of women and the other two are weights of men, and we stratify according to sex, show that
    (a) proportional allocation leads to $n_1 = 2$ and $n_2 = 1$;
    (b) optimum allocation leads to $n_1 = 1$ and $n_2 = 2$.
    Also show that
    (c) if the allocation is proportional, there are 12 possible samples for which $\sigma_{\bar{x}} = 16.7$ (provided the selection within each stratum is random);
    (d) if we use optimum allocation, there are only 4 possible samples for which $\sigma_{\bar{x}_w} = 8.2$ (provided the selection within each stratum is random.)
    (*Hint:* the quantity $\sigma_{\bar{x}_w}$ is the standard error of the weighted means determined according to the formula on page 189.)

17.7 The Design of Experiments

Anyone who has at one time been engaged in applied research, or even thought or read about it, should be able to understand how difficult it can be to plan an experiment so that it will serve the purpose for which it is designed. All too often, it happens that *an experiment purported to test one thing tests another or that an experiment which is not properly designed will not serve any useful purpose at all.* For instance, in the performance of statistical tests, we may conclude that one gasoline is better than another, that the students in one school get better instruction than those in another, or that one rocket is more accurate than another, completely overlooking the fact that the "poorer" gasoline may have been tested in an older car having a less efficient engine, that the students in the second school may have done poorly in the test (on which the judgment was based) because they happened to be thinking about a big football game scheduled for the same day, and that the person who fired the second rocket may have been a poorer marksman than the one who fired the first.

There are essentially two ways of avoiding situations like those described. One is to perform a rigorously *controlled experiment* in which all variables are held fixed except for the one with which we are concerned. For instance, if we wanted to compare the mileage yields of three kinds of gasoline, all test runs could be made with the same car (which is carefully inspected after each run), with the same driver, and over identical routes. In that case, if there *is* a significant difference in the average mileage yields of the three gasolines, we would know that it is *not* due to differences in cars, drivers, or routes. On the positive side, this may tell us that one of the gasolines performs better than the other two *provided it is used in a certain kind of car, by a given driver, and over a given route.* Although this kind of information may be useful under very special conditions (say, if a resort hotel has only one car and one driver, and always uses the same route to pick up its guests at the airport), it really does not tell us very much. Indeed, it would be dangerous and risky to infer that similar results would be obtained with different cars, different drivers, and over different routes (say, in freeway driving instead of crawling in heavy city traffic). Generally speaking, experiments which are "overcontrolled" like this will not provide us with the kind of information we really need.

The other way of handling problems of this kind is to *design* the experiment in such a way that we can not only compare the merits of the three gasolines under more general conditions, but that we can also test whether the other factors really affect their performance. To illustrate how this might be done, suppose that the test runs are performed in two cars, a low-priced car L and a high-priced car H, by two drivers, a good driver Mr. G and a poor driver Mr. P, and over two routes, a rural route R and a freeway route F. Suppose, furthermore, that each test run is performed with a gallon of the respective gasoline, A, B, or C, and that the experiment consists of the following 24 test runs:

Test Run	Gasoline	Car	Driver	Route
1	B	L	P	F
2	A	H	P	F
3	C	H	G	R
4	B	H	P	F
5	A	L	P	F
6	C	L	P	R
7	B	H	G	R
8	C	L	G	R
9	A	L	G	R
10	B	H	G	F
11	B	L	G	R
12	B	H	P	R
13	C	L	G	F
14	B	L	G	F
15	A	L	G	F
16	C	H	P	F
17	C	H	G	F
18	A	H	P	R
19	A	H	G	R
20	C	H	P	R
21	C	L	P	F
22	B	L	P	R
23	A	L	P	R
24	A	H	G	F

This means that the first test run is performed with gasoline B in the low-priced car, by the poor driver, over the freeway route; the second test run is performed with gasoline A in the high-priced car, by the poor driver, over the freeway route; the third test run is performed with gasoline C in the high-priced car, by the good driver, over the rural route; . . . ; and the twenty-fourth test run is performed with gasoline A in the high-priced car, by the good driver, over the freeway route. It is customary to refer to this kind of scheme as a *completely balanced design*—each gasoline is used the same number of times (once, in this case) with each possible combination of cars, drivers, and routes.

Another important feature of the above scheme (which may not be apparent) is that we protected ourselves by *randomization*. First we wrote down the 24 possible ways in which we can select one of the three gasolines, one of the two cars, one of the two drivers, and one of the two routes, and then we *randomly* selected the order in which the test runs are to be performed. If we did not randomize the experiment in this way, extraneous factors might conceivably upset the results; for instance, if we used gasoline A in the first eight test runs, gasoline B is the next eight, and gasoline C in the last eight, the results might be affected by deterioration of the equipment, increasing driver fatigue, or differences in traffic conditions along the chosen routes. Similarly, we might have asked for trouble if we had performed the first twelve test runs with car L and the others with car H, or if we had performed the first twelve test runs with Mr. G and the others with Mr. P.

Another important consideration in the design of an experiment is that of *replication* (or *repetition*), which we already mentioned in Chapter 13. Any time we want to decide whether an observed difference between sample means is significant or whether a sample mean differs significantly from an assumed value, we must have some estimate of *chance variation*, or as it is usually called in complicated experiments like the one we have just described, the *experimental error*. Such an estimate is usually obtained by repeating all or part of the experimental scheme, and in our example we might thus perform 48 test runs, 2 of each of the possible combinations listed above. Whatever differences there are between the corresponding pairs may then be attributed to chance (unless there is another reason as in Exercises 9 and 10 on page 359).

The purpose of our example has been to introduce some of the basic ideas of experimental design, and it is important to realize that the 24 test runs would not only enable us to decide whether there really is a difference between the gasolines, but also whether their performance is affected by differences in cars, differences in drivers, and differences in driving conditions. The actual analysis of a *four-factor experiment* like the one just described is fairly complicated and it will not be discussed in this text; conceptually, though, its analysis is a straightforward generalization of that of the *one-factor* and *two-factor experiments* which we studied in Chapter 13.

17.8 Latin Squares

The greatest problem with completely balanced designs is that they may require a very large number of observations. For instance, if we wanted to compare the breaking strength of six kinds of linen thread, and the measurements are made by five different technicians with four different instruments, a completely balanced design would require that each kind of thread be measured by each technician with each instrument, and hence, it would require $6 \cdot 5 \cdot 4 = 120$ measurements. To give another example, suppose that a market research organization wants to study the potential market for a new breakfast food, and that it wants to try it out in four different cities, with four different kinds of packaging, and with four different advertising campaigns. In other words, the organization wants to determine not only whether there is a difference in the demand for the product in the four cities, but also whether the demand is affected by the differences in packaging and/or the differences in advertising. Although a completely balanced design would require that the product be tried in $4 \cdot 4 \cdot 4 = 64$ market areas (representing all possible combinations of cities, packaging, and advertising), it is of interest to note that with proper planning 16 market areas would suffice. To illustrate, let us refer to the four cities as 1, 2, 3, and 4, the four kinds of packaging as I, II, III, and IV, and the four kinds of advertising as A, B, C, and D. Then, let us consider the following arrangement, called a *Latin square:*

Packaging

	I	II	III	IV
1	A	B	C	D
2	B	C	D	A
3	C	D	A	B
4	D	A	B	C

Cities

In general, a Latin square is a square array of the letters A, B, C, D, ..., of the Roman alphabet, which is such that *each letter occurs once and only once in each row and in each column.*

The above Latin square, looked upon as an experimental design, suggests that advertising A be used in city 1 with packaging I, in city 2 with packaging IV, in city 3 with packaging III, and in city 4 with packaging II; that advertising B be used in city 1 with packaging II, in city 2 with packaging I, in city 3 with packaging IV, and in city 4 with packaging III; and so on. Note that *each kind of advertising is thus used once in each city and once for each kind of packaging, each kind of packaging is used once in each city and once with each kind of advertising, and that each city is used once for each kind of packaging and once for each kind of advertising.*

The analysis of an *r-by-r Latin square* (namely, a Latin square with r rows and r columns) is very similar to the two-way analysis of variance of Section 13.4. The formula for the *total sum of squares* is as on page 343 with $n = k = r$, namely,

▲
$$SST = \sum_{i=1}^{r} \sum_{j=1}^{r} x_{ij}^2 - \frac{1}{r^2} \cdot T_{..}^2.$$
▲

and what we referred to on pages 343 and 350 as the treatment and block sum of squares are now the *row and column sums of squares*

▲
$$SSR = \frac{1}{r} \cdot \sum_{i=1}^{r} T_{i\cdot}^2 - \frac{1}{r^2} \cdot T_{..}^2.$$
▲

and

▲
$$SSC = \frac{1}{r} \cdot \sum_{j=1}^{r} T_{\cdot j}^2 - \frac{1}{r^2} \cdot T_{..}^2.$$
▲

The symbols x_{ij}, $T_{..}$, $T_{i\cdot}$, and $T_{\cdot j}$ are defined as in Chapter 13, and it should be noted that SSR measures the variation of the observations x_{ij} which is due to differences in the demand for the product in the four cities, while SSC measures the variation of the x_{ij} which is due to the different kinds of packaging. So far, the work has been like that of Section 13.4, but the importance of a Latin square design is that we can also attribute part of the total variation to the third variable represented by the letters A, B, C, ..., which in our example consists of the four kinds of advertising. Referring to the four kinds of advertising as *treatments*, we can calculate the *treatment sum of squares* by means of the formula

▲
$$SS(Tr) = \frac{1}{r} \cdot (T_A^2 + T_B^2 + T_C^2 + \ldots) - \frac{1}{r^2} \cdot T_{..}^2.$$
▲

where T_A is the total of the measurements or observations corresponding to treatment A, T_B is the total of the measurements or observations corresponding to treatment B, and so on. Finally, the *error sum of squares* is obtained by subtraction, namely, by means of the equation

▲
$$SSE = SST - SSR - SSC - SS(Tr)$$
▲

and the analysis of variance is performed as is indicated in the following table:

Source of Variation	Degrees of Freedom	Sum of Squares	Mean Square	F
Rows	$r - 1$	SSR	$MSR = \dfrac{SSR}{r - 1}$	$\dfrac{MSR}{MSE}$
Columns	$r - 1$	SSC	$MSC = \dfrac{SSC}{r - 1}$	$\dfrac{MSC}{MSE}$
Treatments	$r - 1$	SS(Tr)	$MS(Tr) = \dfrac{SS(Tr)}{r - 1}$	$\dfrac{MS(Tr)}{MSE}$
Error	$(r - 1)(r - 2)$	SSE	$MSE = \dfrac{SSE}{(r - 1)(r - 2)}$	
Total	$r^2 - 1$	SST		

The numerator and denominator degrees of freedom for all three of the F tests (for row, columns, and treatments) are $r - 1$ and $(r - 1)(r - 2)$, respectively.

To illustrate the analysis of a Latin square, suppose that in the breakfast-food example the market research organization actually obtained the data shown in the following table, where the figures are one week's sales in \$10,000:

Packaging

		I	II	III	IV
	1	A 48	B 38	C 42	D 53
	2	B 39	C 43	D 50	A 54
Cities	3	C 42	D 50	A 47	B 44
	4	D 46	A 48	B 46	C 52

As can easily be checked, the row totals are, respectively, 181, 186, 183, and 192; the column totals are, respectively, 175, 179, 185, and 203; the treatment totals for A, B, C, and D are, respectively, 197, 167, 179, and 199; and the grand total is 742. Thus, since $r = 4$, we get

$$SST = (48^2 + 38^2 + \ldots + 52^2) - \frac{1}{16} \cdot 742^2$$

$$= 345.75$$

$$SSR = \frac{1}{4}(181^2 + 186^2 + 183^2 + 192^2) - \frac{1}{16} \cdot 742^2$$

$$= 17.25$$

$$SSC = \frac{1}{4}(175^2 + 179^2 + 185^2 + 203^2) - \frac{1}{16} \cdot 742^2$$

$$= 114.75$$

$$SS(Tr) = \frac{1}{4}(197^2 + 167^2 + 179^2 + 199^2) - \frac{1}{16} \cdot 742^2$$

$$= 174.75$$

and hence,

$$SSE = 345.75 - 17.25 - 114.75 - 174.75$$
$$= 39$$

and the analysis of variance table becomes

Source of varation	Degrees of freedom	Sum of squares	Mean square	F
Rows	3	17.25	$\frac{17.25}{3} = 5.75$	$\frac{5.75}{6.5} = 0.9$
Columns	3	114.75	$\frac{114.75}{3} = 38.25$	$\frac{38.25}{6.5} = 5.9$
Treatments	3	174.75	$\frac{174.75}{3} = 58.25$	$\frac{58.25}{6.5} = 9.0$
Error	6	39	$\frac{39}{6} = 6.5$	
Total	15	345.75		

Using the level of significance $\alpha = 0.05$, we find that for $4 - 1 = 3$ and $(4 - 1)(4 - 2) = 6$ degrees of freedom $F_{0.05} = 4.76$ and, hence that the differences due to packaging (columns) and advertising (treatments) *are significant*, while those due to the different cities (rows) *are not significant*.

We have included the material of this section mainly to impress upon the reader how a suitably designed experiment can yield a wealth of information on the basis of relatively few data. Of course, one should not expect too much from small samples, for as we saw in earlier chapters, they tend to lead to wide confidence intervals or may make it difficult to reject a null hypothesis unless it is "way off." This is reflected by the fact that in the table on page 462 there are only $(r - 1)(r - 2)$ degrees of freedom for the experimental error, which means that for $r = 3$ and $r = 4$, for example, there is very little information about the variation that is attributed to chance. So far as the construction of Latin squares is concerned, this is a problem of pure mathematics, and in most practical work they are looked up in tables like the one referred to in the Bibliography at the end of this chapter.

17.9 Some Further Considerations

There exist a great variety of experimental designs besides the ones we have discussed in this chapter and in Chapter 13, and they serve an equally great variety of special purposes. Among the more widely used designs are the *incomplete block designs*, which are characterized by the feature that each treatment is not represented in each block. The need for such designs arises, for example, when we want to compare 13 kinds of tires but cannot put all of them on a test car at the same time, or when a taste tester has to compare 10 kinds of wine, but cannot compare more than three of the wines at a time   With regard to the first example, where only 4 of the 13 tires can be tested at the same time, we might use the incomplete block design shown at the top of page 465, which has the important property that *each kind of tire (numbered from 1 through 13) is used together with each other kind of tire once and only once during the course of the experiment*. There are 13 test runs, or blocks, and since each kind of tire appears together with each other kind of tire once within the same block, the design is referred to as a *balanced incomplete block design*. The fact that each kind of tire appears together with each other kind of tire once within the same block facilities the statistical analysis, for it assures that we have the same amount of information for comparing each pair of tires.

In the wine-tasting experiment we cannot construct a design which is such that each wine is compared directly with each other wine *once and only once* during the course of the experiment. This follows from the fact that each wine has to be compared directly with 9 other wines, in each comparison a wine is compared directly with two others, and 9 is not

Number of Test Run	Kinds of Tires			
1	1	2	4	10
2	2	3	5	11
3	3	4	6	12
4	4	5	7	13
5	5	6	8	1
6	6	7	9	2
7	7	8	10	3
8	8	9	11	4
9	9	10	12	5
10	10	11	13	6
11	11	12	1	7
12	12	13	2	8
13	13	1	3	9

divisible by 2. To get around this difficulty, we simply compare each wine with each other wine *twice*, as in the following *balanced incomplete block design*, where the wines are numbered from 1 through 10:

Comparison	Wines	Comparison	Wines	Comparison	Wines
1	1, 2, 3	11	1, 2, 4	21	1, 3, 5
2	2, 5, 8	12	2, 3, 6	22	2, 6, 7
3	3, 4, 7	13	3, 4, 8	23	3, 8, 9
4	1, 4, 6	14	4, 5, 9	24	2, 4, 10
5	5, 7, 8	15	1, 5, 7	25	3, 5, 6
6	4, 6, 9	16	6, 8, 9	26	1, 6, 8
7	1, 7, 9	17	3, 7, 10	27	2, 7, 9
8	2, 8, 10	18	1, 8, 10	28	4, 7, 8
9	3, 9, 10	19	2, 5, 9	29	1, 9, 10
10	5, 6, 10	20	6, 7, 10	30	4, 5, 10

The actual analysis of balanced incomplete block designs is fairly complicated and we shall not go into it here, as it has been our purpose only to
demonstrate what *can* be done in the important, and ever-growing part of
statistics, which comes under the heading of "the design of experiments."

EXERCISES

1. An agronomist wants to compare the yield of 18 varieties of corn, and
   at the same time study the effects of 5 different fertilizers and 2 methods of irrigation. How many test plots will we have to plant for a
   completely balanced design with only one observation of each kind?

2. Suppose that in the illustration of Section 13.5 we consider not only
   four machine operators and three machines, but the production of
   five different parts in three different shifts. How many observations
   would have to be made for a completely balanced design with three
   replications?

3. A manufacturer of pharmaceuticals wants to market a new cold
   remedy which is actually a combination of four drugs, but first he
   wants to experiment with two different dosages for each drug. If A_L
   and A_H denote the low and high dosage of drug A, B_L and B_H the low
   and high dosage of drug B, C_L and C_H the low and high dosage of
   drug C, and D_L and D_H the low and high dosage of drug D, list the
   16 preparations he has to test if each dosage of each drug is to be used
   once in combination with each dosage of the other drugs.

4. A clothing manufacturer wants to test a new sewing machine using
   three kinds of needles, a, b, and c, and three kinds of thread, I, II,
   and III.
   (a) List the nine combinations of needles and threads which will
       have to be used for a completely balanced design.
   (b) Use the Latin square

A	B	C
B	C	A
C	A	B

   to indicate how three of the nine combinations of needles and
   threads of part (a) might be assigned to three seamstresses,
   Mrs. A, Mrs. B, and Mrs. C, so that each of the seamstresses
   works once with each kind of needle and once with each kind
   of thread.

5. Making use of the fact that each of the letters must occur once and only once in each row and each column, complete the following Latin square:

A	D		
	B		
		C	B
			D

6. Among the nine persons interviewed by a poll, three are Easterners, three are Southerners, and three are Westerners. By profession, three of them are teachers, three are lawyers, and three are doctors, and no two of the same profession come from the same part of the United States. Also, three of them are Democrats, three are Republicans, and three are Independents, and no two of the same political sympathy are of the same profession or come from the same part of the United States. If one of the teachers is an Easterner and an Independent, another teacher is a Southerner and a Republican, and one of the lawyers is a Southerner and a Democrat, what are the political sympathies of the doctor who is a Westerner? (*Hint:* refer to a 3-by-3 Latin square; this exercise is a simplified version of a famous problem posed by R. A. Fisher in his classical work *The Design of Experiments.*)

7. The figures in the following 3-by-3 Latin square are the grades in a French test obtained by nine students of ethnic background E_1, E_2, or or E_3, of age A_1, A_2, or A_3, who were taught by teacher A, teacher B, or teacher C:

	E_1	E_2	E_3
A_1	A 73	B 84	C 67
A_2	B 93	C 77	A 84
A_3	C 68	A 81	B 91

Use the level of significance $\alpha = 0.05$ to test
  (a) the null hypothesis that differences in ethnic background have no effect on a student's grade in the test;
  (b) the null hypothesis that differences in age have no effect on a student's grade in the test;
  (c) the null hypothesis that having a different teacher has no effect on a student's grade in the test.

8. The figures in the following 5-by-5 Latin square are the number of minutes engines E_1, E_2, E_3, E_4, and E_5, tuned up by mechanics M_1, M_2, M_3, M_4, and M_5, ran with a gallon of fuel A, B, C, D, or E:

	E_1	E_2	E_3	E_4	E_5
M_1	A 31	B 24	C 20	D 20	E 18
M_2	B 21	C 27	D 23	E 25	A 31
M_3	C 21	D 27	E 25	A 29	B 21
M_4	D 21	E 25	A 33	B 25	C 22
M_5	E 21	A 37	B 24	C 24	D 20

Use the level of significance $\alpha = 0.01$ to test
  (a) the null hypothesis that there is no difference in the performance of the five engines;
  (b) the null hypothesis that the persons who tuned up these engines have no effect on their performance;
  (c) the null hypothesis that the engines perform equally well with each of the fuels.

9. To test their ability to make decisions under pressure, the 9 senior executives of a company are to be interviewed by each of 4 psychologists. As it takes a psychologist a full day to interview three of the executives, the schedule for the interviews is arranged as follows, where the nine executives are denoted A, B, C, D, E, F, G, H, and I:

Day	Psychologist	Executives		
March 2	I	B	C	?
March 3	I	E	F	G
March 4	I	H	I	A
March 5	II	C	?	H
March 6	II	B	F	A
March 9	II	D	E	?
March 10	III	D	G	A
March 11	III	C	F	?
March 12	III	B	E	H
March 13	IV	B	?	I
March 16	IV	C	?	A
March 17	IV	D	F	H

Replace the six question marks with the appropriate letters, given that each of the 9 executives is to be interviewed together with each of the other executives once and only once on the same day. Note that this will make the arrangement a balanced incomplete block design, which may be important because each executive is thus tested together with each other executive once under *identical conditions*.

10. A company has seven vice-presidents who are assigned to its various administrative committees as shown in the following table:

Committee	Vice-Presidents
Investments	Davis, Frost, Green, Allen
Advertising	Baker, Eaton, Green, Allen
Sales	Baker, Cooke, Frost, Allen
Personnel	Baker, Cooke, Davis, Green
Insurance	Cooke, Eaton, Frost, Green
Purchasing	Baker, Davis, Eaton, Frost
Research	Cooke, Davis, Eaton, Allen

(a) Verify that this arrangement is a balanced incomplete block design. (In a situation like this the balance of the arrangement may not seem too important, and yet it may well prevent a clique from taking over the operation of the company.)

(b) If Davis, Baker, and Cooke are (in that order) appointed chairmen of the first three committees, how will the chairmen of the other four committees have to be chosen so that each of the seven vice-presidents is chairman of one of the committees? How many different solutions are there to this problem?

11. A newspaper regularly prints the columns of seven writers, but has room for only three in each edition. Complete the following schedule, in which the writers are numbered from 1 through 7, so that each

writer's columns appear three times per week, and a column of each writer appears together with a column of each other writer once per week:

	Writers		
Monday	1	2	3
Tuesday	4		
Wednesday	1	4	5
Thursday	2		
Friday	1	6	7
Saturday	5		
Sunday	2	4	6

(*Hint:* begin with Tuesday or Thursday)

BIBLIOGRAPHY

Detailed descriptions of various kinds of sample designs may be found in

COCHRAN, W. G., *Sampling Techniques*, 2nd ed. New York: John Wiley & Sons, Inc., 1963.

DEMING, W. E., *Sample Design in Business Research*. New York: John Wiley & Sons, Inc., 1960.

STUART, A., *Basic Ideas of Scientific Sampling*. New York: Hafner Publishing Company, 1962.

The following are some detailed treatments of problems relating to the design of experiments

COCHRAN, W. G., and Cox, G. M., *Experimental Design*, 2nd ed. New York: John Wiley & Sons, Inc., 1957.

FINNEY, D. J., *An Introduction to Experimental Design*. Chicago: University of Chicago Press, 1960.

FISHER, R. A., *The Design of Experiments*, 8th ed. Edinburgh: Oliver & Boyd Ltd., 1966.

HICKS, C. R., *Fundamental Concepts in the Design of Experiments*. New York: Holt, Rinehart & Winston, Inc., 1964.

LI, C. C., *Introduction to Experimental Statistics*. New York: McGraw-Hill, Inc., 1964.

SNEDECOR, G. W., and COCHRAN, W. G., *Statistical Methods*. Ames, Iowa: Iowa State University Press, 1967.

A table of Latin squares of size $r = 3, 4, 5, ..$, and 12 may be found in the above-mentioned book by W. G. Cochran and G. M. Cox.

Statistical Tables

   I   **The Standard Normal Distribution**

  II   **The t Distribution**

 III   **The Chi-Square Distribution**

 IV   **The F Distribution**

  V   **Confidence Intervals for Proportions**

 VI   **The Binomial Distribution**

VII   **Values of $Z = \dfrac{1}{2} \cdot \log_e \left(\dfrac{1 + r}{1 - r} \right)$**

VIII   **Binomial Coefficients**

 IX   **Random Numbers**

  X   **Logarithms**

 XI   **Values of e^{-x}**

XII   **Squares and Square Roots**

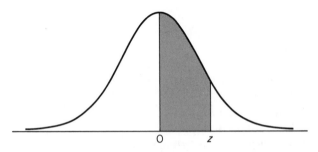

The entries in Table I are the probabilities that a random variable having the standard normal distribution assumes a value between 0 and z; they are given by the area under the curve shaded in the figure shown above.

Table I

THE STANDARD NORMAL DISTRIBUTION

z	.00	.01	.02	.03	.04	.05	.06	.07	.08	.09
0.0	.0000	.0040	.0080	.0120	.0160	.0199	.0239	.0279	.0319	.0359
0.1	.0398	.0438	.0478	.0517	.0557	.0596	.0636	.0675	.0714	.0753
0.2	.0793	.0832	.0871	.0910	.0948	.0987	.1026	.1064	.1103	.1141
0.3	.1179	.1217	.1255	.1293	.1331	.1368	.1406	.1443	.1480	.1517
0.4	.1554	.1591	.1628	.1664	.1700	.1736	.1772	.1808	.1844	.1879
0.5	.1915	.1950	.1985	.2019	.2054	.2088	.2123	.2157	.2190	.2224
0.6	.2257	.2291	.2324	.2357	.2389	.2422	.2454	.2486	.2517	.2549
0.7	.2580	.2611	.2642	.2673	.2704	.2734	.2764	.2794	.2823	.2852
0.8	.2881	.2910	.2939	.2967	.2995	.3023	.3051	.3078	.3106	.3133
0.9	.3159	.3186	.3212	.3238	.3264	.3289	.3315	.3340	.3365	.3389
1.0	.3413	.3438	.3461	.3485	.3508	.3531	.3554	.3577	.3599	.3621
1.1	.3643	.3665	.3686	.3708	.3729	.3749	.3770	.3790	.3810	.3830
1.2	.3849	.3869	.3888	.3907	.3925	.3944	.3962	.3980	.3997	.4015
1.3	.4032	.4049	.4066	.4082	.4099	.4115	.4131	.4147	.4162	.4177
1.4	.4192	.4207	.4222	.4236	.4251	.4265	.4279	.4292	.4306	.4319
1.5	.4332	.4345	.4357	.4370	.4382	.4394	.4406	.4418	.4429	.4441
1.6	.4452	.4463	.4474	.4484	.4495	.4505	.4515	.4525	.4535	.4545
1.7	.4554	.4564	.4573	.4582	.4591	.4599	.4608	.4616	.4625	.4633
1.8	.4641	.4649	.4656	.4664	.4671	.4678	.4686	.4693	.4699	.4706
1.9	.4713	.4719	.4726	.4732	.4738	.4744	.4750	.4756	.4761	.4767
2.0	.4772	.4778	.4783	.4788	.4793	.4798	.4803	.4808	.4812	.4817
2.1	.4821	.4826	.4830	.4834	.4838	.4842	.4846	.4850	.4854	.4857
2.2	.4861	.4864	.4868	.4871	.4875	.4878	.4881	.4884	.4887	.4890
2.3	.4893	.4896	.4898	.4901	.4904	.4906	.4909	.4911	.4913	.4916
2.4	.4918	.4920	.4922	.4925	.4927	.4929	.4931	.4932	.4934	.4936
2.5	.4938	.4940	.4941	.4943	.4945	.4946	.4948	.4949	.4951	.4952
2.6	.4953	.4955	.4956	.4957	.4959	.4960	.4961	.4962	.4963	.4964
2.7	.4965	.4966	.4967	.4968	.4969	.4970	.4971	.4972	.4973	.4974
2.8	.4974	.4975	.4976	.4977	.4977	.4978	.4979	.4979	.4980	.4981
2.9	.4981	.4982	.4982	.4983	.4984	.4984	.4985	.4985	.4986	.4986
3.0	.4987	.4987	.4987	.4988	.4988	.4989	.4989	.4989	.4990	.4990

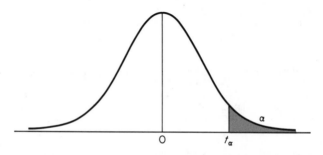

The entries in Table II are values for which the area to their right under the t distribution with given degrees of freedom (the area shaded in the figure shown above) is equal to α.

Table II

THE t DISTRIBUTION (VALUES OF t_α)*

d.f.	$t_{.100}$	$t_{.050}$	$t_{.025}$	$t_{.010}$	$t_{.005}$	d.f.
1	3.078	6.314	12.706	31.821	63.657	1
2	1.886	2.920	4.303	6.965	9.925	2
3	1.638	2.353	3.182	4.541	5.841	3
4	1.533	2.132	2.776	3.747	4.604	4
5	1.476	2.015	2.571	3.365	4.032	5
6	1.440	1.943	2.447	3.143	3.707	6
7	1.415	1.895	2.365	2.998	3.499	7
8	1.397	1.860	2.306	2.896	3.355	8
9	1.383	1.833	2.262	2.821	3.250	9
10	1.372	1.812	2.228	2.764	3.169	10
11	1.363	1.796	2.201	2.718	3.106	11
12	1.356	1.782	2.179	2.681	3.055	12
13	1.350	1.771	2.160	2.650	3.012	13
14	1.345	1.761	2.145	2.624	2.977	14
15	1.341	1.753	2.131	2.602	2.947	15
16	1.337	1.746	2.120	2.583	2.921	16
17	1.333	1.740	2.110	2.567	2.898	17
18	1.330	1.734	2.101	2.552	2.878	18
19	1.328	1.729	2.093	2.539	2.861	19
20	1.325	1.725	2.086	2.528	2.845	20
21	1.323	1.721	2.080	2.518	2.831	21
22	1.321	1.717	2.074	2.508	2.819	22
23	1.319	1.714	2.069	2.500	2.807	23
24	1.318	1.711	2.064	2.492	2.797	24
25	1.316	1.708	2.060	2.485	2.787	25
26	1.315	1.706	2.056	2.479	2.779	26
27	1.314	1.703	2.052	2.473	2.771	27
28	1.313	1.701	2.048	2.467	2.763	28
29	1.311	1.699	2.045	2.462	2.756	29
inf.	1.282	1.645	1.960	2.326	2.576	inf.

* This table is abridged from Table IV of R. A. Fisher, *Statistical Methods for Research Workers*, published by Oliver and Boyd, Ltd., Edinburgh, by permission of the author's literary executor and publishers.

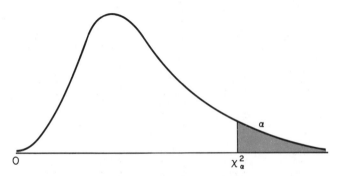

The entries in Table III are values for which the area to their right under the chi-square distribution with given degrees of freedom (the area shaded in the figure shown above) is equal to α.

Table III

THE CHI-SQUARE DISTRIBUTION (VALUES OF χ_α^2)*

d.f.	$\chi_{.995}^2$	$\chi_{.99}^2$	$\chi_{.975}^2$	$\chi_{.95}^2$	$\chi_{.05}^2$	$\chi_{.025}^2$	$\chi_{.01}^2$	$\chi_{.005}^2$	d.f.
1	.0000393	.000157	.000982	.00393	3.841	5.024	6.635	7.879	1
2	.0100	.0201	.0506	.103	5.991	7.378	9.210	10.597	2
3	.0717	.115	.216	.352	7.815	9.348	11.345	12.838	3
4	.207	.297	.484	.711	9.488	11.143	13.277	14.860	4
5	.412	.554	.831	1.145	11.070	12.832	15.086	16.750	5
6	.676	.872	1.237	1.635	12.592	14.449	16.812	18.548	6
7	.989	1.239	1.690	2.167	14.067	16.013	18.475	20.278	7
8	1.344	1.646	2.180	2.733	15.507	17.535	20.090	21.955	8
9	1.735	2.088	2.700	3.325	16.919	19.023	21.666	23.589	9
10	2.156	2.558	3.247	3.940	18.307	20.483	23.209	25.188	10
11	2.603	3.053	3.816	4.575	19.675	21.920	24.725	26.757	11
12	3.074	3.571	4.404	5.226	21.026	23.337	26.217	28.300	12
13	3.565	4.107	5.009	5.892	22.362	24.736	27.688	29.819	13
14	4.075	4.660	5.629	6.571	23.685	26.119	29.141	31.319	14
15	4.601	5.229	6.262	7.261	24.996	27.488	30.578	32.801	15
16	5.142	5.812	6.908	7.962	26.296	28.845	32.000	34.267	16
17	5.697	6.408	7.564	8.672	27.587	30.191	33.409	35.718	17
18	6.265	7.015	8.231	9.390	28.869	31.526	34.805	37.156	18
19	6.844	7.633	8.907	10.117	30.144	32.852	36.191	38.582	19
20	7.434	8.260	9.591	10.851	31.410	34.170	37.566	39.997	20
21	8.034	8.897	10.283	11.591	32.671	35.479	38.932	41.401	21
22	8.643	9.542	10.982	12.338	33.924	36.781	40.289	42.796	22
23	9.260	10.196	11.689	13.091	35.172	38.076	41.638	44.181	23
24	9.886	10.856	12.401	13.848	36.415	39.364	42.980	45.558	24
25	10.520	11.524	13.120	14.611	37.652	40.646	44.314	46.928	25
26	11.160	12.198	13.844	15.379	38.885	41.923	45.642	48.290	26
27	11.808	12.879	14.573	16.151	40.113	43.194	46.963	49.645	27
28	12.461	13.565	15.308	16.928	41.337	44.461	48.278	50.993	28
29	13.121	14.256	16.047	17.708	42.557	45.722	49.588	52.336	29
30	13.787	14.953	16.791	18.493	43.773	46.979	50.892	53.672	30

* This table is based on Table 8 of *Biometrika Tables for Statisticians*, *Volume I*, by permission of the *Biometrika* trustees.

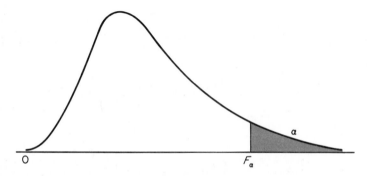

The entries in Tables IVa and IVb are values for which the area to their right under the F distribution with given degrees of freedom (the area shaded in the figure shown above) is equal to α.

Table IVa

THE F DISTRIBUTION (VALUES OF $F_{.05}$)*

Degrees of freedom for numerator

	1	2	3	4	5	6	7	8	9	10	12	15	20	24	30	40	60	120	∞
1	161	200	216	225	230	234	237	239	241	242	244	246	248	249	250	251	252	253	254
2	18.5	19.0	19.2	19.2	19.3	19.3	19.4	19.4	19.4	19.4	19.4	19.4	19.4	19.5	19.5	19.5	19.5	19.5	19.5
3	10.1	9.55	9.28	9.12	9.01	8.94	8.89	8.85	8.81	8.79	8.74	8.70	8.66	8.64	8.62	8.59	8.57	8.55	8.53
4	7.71	6.94	6.59	6.39	6.26	6.16	6.09	6.04	6.00	5.96	5.91	5.86	5.80	5.77	5.75	5.72	5.69	5.66	5.63
5	6.61	5.79	5.41	5.19	5.05	4.95	4.88	4.82	4.77	4.74	4.68	4.62	4.56	4.53	4.50	4.46	4.43	4.40	4.37
6	5.99	5.14	4.76	4.53	4.39	4.28	4.21	4.15	4.10	4.06	4.00	3.94	3.87	3.84	3.81	3.77	3.74	3.70	3.67
7	5.59	4.74	4.35	4.12	3.97	3.87	3.79	3.73	3.68	3.64	3.57	3.51	3.44	3.41	3.38	3.34	3.30	3.27	3.23
8	5.32	4.46	4.07	3.84	3.69	3.58	3.50	3.44	3.39	3.35	3.28	3.22	3.15	3.12	3.08	3.04	3.01	2.97	2.93
9	5.12	4.26	3.86	3.63	3.48	3.37	3.29	3.23	3.18	3.14	3.07	3.01	2.94	2.90	2.86	2.83	2.79	2.75	2.71
10	4.96	4.10	3.71	3.48	3.33	3.22	3.14	3.07	3.02	2.98	2.91	2.85	2.77	2.74	2.70	2.66	2.62	2.58	2.54
11	4.84	3.98	3.59	3.36	3.20	3.09	3.01	2.95	2.90	2.85	2.79	2.72	2.65	2.61	2.57	2.53	2.49	2.45	2.40
12	4.75	3.89	3.49	3.26	3.11	3.00	2.91	2.85	2.80	2.75	2.69	2.62	2.54	2.51	2.47	2.43	2.38	2.34	2.30
13	4.67	3.81	3.41	3.18	3.03	2.92	2.83	2.77	2.71	2.67	2.60	2.53	2.46	2.42	2.38	2.34	2.30	2.25	2.21
14	4.60	3.74	3.34	3.11	2.96	2.85	2.76	2.70	2.65	2.60	2.53	2.46	2.39	2.35	2.31	2.27	2.22	2.18	2.13
15	4.54	3.68	3.29	3.06	2.90	2.79	2.71	2.64	2.59	2.54	2.48	2.40	2.33	2.29	2.25	2.20	2.16	2.11	2.07
16	4.49	3.63	3.24	3.01	2.85	2.74	2.66	2.59	2.54	2.49	2.42	2.35	2.28	2.24	2.19	2.15	2.11	2.06	2.01
17	4.45	3.59	3.20	2.96	2.81	2.70	2.61	2.55	2.49	2.45	2.38	2.31	2.23	2.19	2.15	2.10	2.06	2.01	1.96
18	4.41	3.55	3.16	2.93	2.77	2.66	2.58	2.51	2.46	2.41	2.34	2.27	2.19	2.15	2.11	2.06	2.02	1.97	1.92
19	4.38	3.52	3.13	2.90	2.74	2.63	2.54	2.48	2.42	2.38	2.31	2.23	2.16	2.11	2.07	2.03	1.98	1.93	1.88
20	4.35	3.49	3.10	2.87	2.71	2.60	2.51	2.45	2.39	2.35	2.28	2.20	2.12	2.08	2.04	1.99	1.95	1.90	1.84
21	4.32	3.47	3.07	2.84	2.68	2.57	2.49	2.42	2.37	2.32	2.25	2.18	2.10	2.05	2.01	1.96	1.92	1.87	1.81
22	4.30	3.44	3.05	2.82	2.66	2.55	2.46	2.40	2.34	2.30	2.23	2.15	2.07	2.03	1.98	1.94	1.89	1.84	1.78
23	4.28	3.42	3.03	2.80	2.64	2.53	2.44	2.37	2.32	2.27	2.20	2.13	2.05	2.01	1.96	1.91	1.86	1.81	1.76
24	4.26	3.40	3.01	2.78	2.62	2.51	2.42	2.36	2.30	2.25	2.18	2.11	2.03	1.98	1.94	1.89	1.84	1.79	1.73
25	4.24	3.39	2.99	2.76	2.60	2.49	2.40	2.34	2.28	2.24	2.16	2.09	2.01	1.96	1.92	1.87	1.82	1.77	1.71
30	4.17	3.32	2.92	2.69	2.53	2.42	2.33	2.27	2.21	2.16	2.09	2.01	1.93	1.89	1.84	1.79	1.74	1.68	1.62
40	4.08	3.23	2.84	2.61	2.45	2.34	2.25	2.18	2.12	2.08	2.00	1.92	1.84	1.79	1.74	1.69	1.64	1.58	1.51
60	4.00	3.15	2.76	2.53	2.37	2.25	2.17	2.10	2.04	1.99	1.92	1.84	1.75	1.70	1.65	1.59	1.53	1.47	1.39
120	3.92	3.07	2.68	2.45	2.29	2.18	2.09	2.02	1.96	1.91	1.83	1.75	1.66	1.61	1.55	1.50	1.43	1.35	1.25
∞	3.84	3.00	2.60	2.37	2.21	2.10	2.01	1.94	1.88	1.83	1.75	1.67	1.57	1.52	1.46	1.39	1.32	1.22	1.00

Degrees of freedom for denominator

* This table is reproduced from M. Merrington and C. M. Thompson, "Tables of percentage points of the inverted beta (F) distribution," *Biometrika*, vol. 33 (1943), by permission of the *Biometrika* trustees.

Table IVb

THE F DISTRIBUTION (VALUES OF $F_{.01}$)*

Degrees of freedom for numerator

df (denom)	1	2	3	4	5	6	7	8	9	10	12	15	20	24	30	40	60	120	∞
1	4,052	5,000	5,403	5,625	5,764	5,859	5,928	5,982	6,023	6,056	6,106	6,157	6,209	6,235	6,261	6,287	6,313	6,339	6,366
2	98.5	99.0	99.2	99.2	99.3	99.3	99.4	99.4	99.4	99.4	99.4	99.4	99.4	99.5	99.5	99.5	99.5	99.5	99.5
3	34.1	30.8	29.5	28.7	28.2	27.9	27.7	27.5	27.3	27.2	27.1	26.9	26.7	26.6	26.5	26.4	26.3	26.2	26.1
4	21.2	18.0	16.7	16.0	15.5	15.2	15.0	14.8	14.7	14.5	14.4	14.2	14.0	13.9	13.8	13.7	13.7	13.6	13.5
5	16.3	13.3	12.1	11.4	11.0	10.7	10.5	10.3	10.2	10.1	9.89	9.72	9.55	9.47	9.38	9.29	9.20	9.11	9.02
6	13.7	10.9	9.78	9.15	8.75	8.47	8.26	8.10	7.98	7.87	7.72	7.56	7.40	7.31	7.23	7.14	7.06	6.97	6.88
7	12.2	9.55	8.45	7.85	7.46	7.19	6.99	6.84	6.72	6.62	6.47	6.31	6.16	6.07	5.99	5.91	5.82	5.74	5.65
8	11.3	8.65	7.59	7.01	6.63	6.37	6.18	6.03	5.91	5.81	5.67	5.52	5.36	5.28	5.20	5.12	5.03	4.95	4.86
9	10.6	8.02	6.99	6.42	6.06	5.80	5.61	5.47	5.35	5.26	5.11	4.96	4.81	4.73	4.65	4.57	4.48	4.40	4.31
10	10.0	7.56	6.55	5.99	5.64	5.39	5.20	5.06	4.94	4.85	4.71	4.56	4.41	4.33	4.25	4.17	4.08	4.00	3.91
11	9.65	7.21	6.22	5.67	5.32	5.07	4.89	4.74	4.63	4.54	4.40	4.25	4.10	4.02	3.94	3.86	3.78	3.69	3.60
12	9.33	6.93	5.95	5.41	5.06	4.82	4.64	4.50	4.39	4.30	4.16	4.01	3.86	3.78	3.70	3.62	3.54	3.45	3.36
13	9.07	6.70	5.74	5.21	4.86	4.62	4.44	4.30	4.19	4.10	3.96	3.82	3.66	3.59	3.51	3.43	3.34	3.25	3.17
14	8.86	6.51	5.56	5.04	4.70	4.46	4.28	4.14	4.03	3.94	3.80	3.66	3.51	3.43	3.35	3.27	3.18	3.09	3.00
15	8.68	6.36	5.42	4.89	4.56	4.32	4.14	4.00	3.89	3.80	3.67	3.52	3.37	3.29	3.21	3.13	3.05	2.96	2.87
16	8.53	6.23	5.29	4.77	4.44	4.20	4.03	3.89	3.78	3.69	3.55	3.41	3.26	3.18	3.10	3.02	2.93	2.84	2.75
17	8.40	6.11	5.19	4.67	4.34	4.10	3.93	3.79	3.68	3.59	3.46	3.31	3.16	3.08	3.00	2.92	2.83	2.75	2.65
18	8.29	6.01	5.09	4.58	4.25	4.01	3.84	3.71	3.60	3.51	3.37	3.23	3.08	3.00	2.92	2.84	2.75	2.66	2.57
19	8.19	5.93	5.01	4.50	4.17	3.94	3.77	3.63	3.52	3.43	3.30	3.15	3.00	2.92	2.84	2.76	2.67	2.58	2.49
20	8.10	5.85	4.94	4.43	4.10	3.87	3.70	3.56	3.46	3.37	3.23	3.09	2.94	2.86	2.78	2.69	2.61	2.52	2.42
21	8.02	5.78	4.87	4.37	4.04	3.81	3.64	3.51	3.40	3.31	3.17	3.03	2.88	2.80	2.72	2.64	2.55	2.46	2.36
22	7.95	5.72	4.82	4.31	3.99	3.76	3.59	3.45	3.35	3.26	3.12	2.98	2.83	2.75	2.67	2.58	2.50	2.40	2.31
23	7.88	5.66	4.76	4.26	3.94	3.71	3.54	3.41	3.30	3.21	3.07	2.93	2.78	2.70	2.62	2.54	2.45	2.35	2.26
24	7.82	5.61	4.72	4.22	3.90	3.67	3.50	3.36	3.26	3.17	3.03	2.89	2.74	2.66	2.58	2.49	2.40	2.31	2.21
25	7.77	5.57	4.68	4.18	3.86	3.63	3.46	3.32	3.22	3.13	2.99	2.85	2.70	2.62	2.53	2.45	2.36	2.27	2.17
30	7.56	5.39	4.51	4.02	3.70	3.47	3.30	3.17	3.07	2.98	2.84	2.70	2.55	2.47	2.39	2.30	2.21	2.11	2.01
40	7.31	5.18	4.31	3.83	3.51	3.29	3.12	2.99	2.89	2.80	2.66	2.52	2.37	2.29	2.20	2.11	2.02	1.92	1.80
60	7.08	4.98	4.13	3.65	3.34	3.12	2.95	2.82	2.72	2.63	2.50	2.35	2.20	2.12	2.03	1.94	1.84	1.73	1.60
120	6.85	4.79	3.95	3.48	3.17	2.96	2.79	2.66	2.56	2.47	2.34	2.19	2.03	1.95	1.86	1.76	1.66	1.53	1.38
∞	6.63	4.61	3.78	3.32	3.02	2.80	2.64	2.51	2.41	2.32	2.18	2.04	1.88	1.79	1.70	1.59	1.47	1.32	1.00

Degrees of freedom for denominator

* This table is reproduced from M. Merrington and C. M. Thompson, "Tables of percentage points of the inverted beta (F) distribution," *Biometrika*, vol. 33 (1943), by permission of the *Biometrika* trustees.

Table Va

0.95 CONFIDENCE INTERVALS FOR PROPORTIONS*

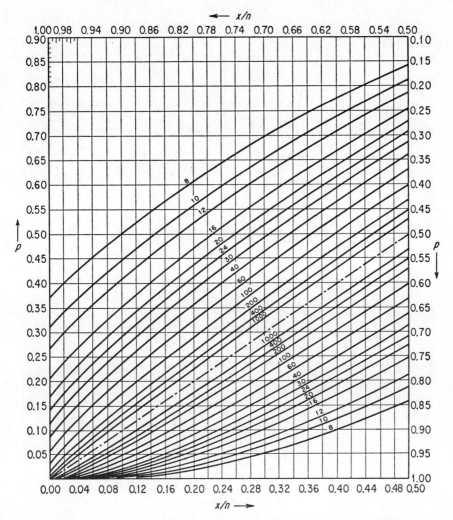

* This table is reproduced from Table 41 of the *Biometrika Tables for Statisticians*, Vol. I (New York: Cambridge University Press, 1954) by permission of the *Biometrika* trustees.

Table Vb

0.99 CONFIDENCE INTERVALS FOR PROPORTIONS*

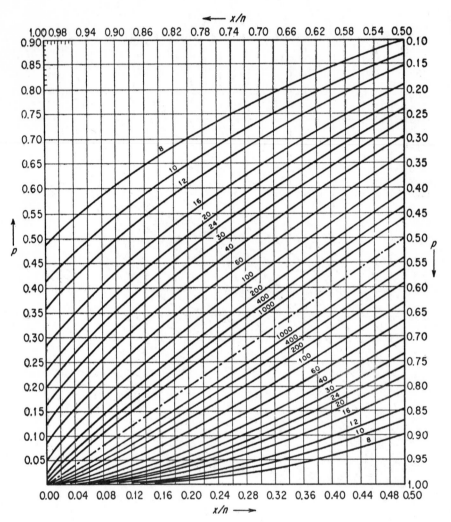

* This table is reproduced from Table 41 of the *Biometrika Tables for Statisticians*, Vol. I (New York: Cambridge University Press, 1954) by permission of the *Biometrika* trustees.

Table VI

BINOMIAL PROBABILITIES*

n	x						p					
		0.05	0.1	0.2	0.3	0.4	0.5	0.6	0.7	0.8	0.9	0.95
2	0	0.902	0.810	0.640	0.490	0.360	0.250	0.160	0.090	0.040	0.010	0.002
	1	0.095	0.180	0.320	0.420	0.480	0.500	0.480	0.420	0.320	0.180	0.095
	2	0.002	0.010	0.040	0.090	0.160	0.250	0.360	0.490	0.640	0.810	0.902
3	0	0.857	0.729	0.512	0.343	0.216	0.125	0.064	0.027	0.008	0.001	
	1	0.135	0.243	0.384	0.441	0.432	0.375	0.288	0.189	0.096	0.027	0.007
	2	0.007	0.027	0.096	0.189	0.288	0.375	0.432	0.441	0.384	0.243	0.135
	3		0.001	0.008	0.027	0.064	0.125	0.216	0.343	0.512	0.729	0.857
4	0	0.815	0.656	0.410	0.240	0.130	0.062	0.026	0.008	0.002		
	1	0.171	0.292	0.410	0.412	0.346	0.250	0.154	0.076	0.026	0.004	
	2	0.014	0.049	0.154	0.265	0.346	0.375	0.346	0.265	0.154	0.049	0.014
	3		0.004	0.026	0.076	0.154	0.250	0.346	0.412	0.410	0.292	0.171
	4			0.002	0.008	0.026	0.062	0.130	0.240	0.410	0.656	0.815
5	0	0.774	0.590	0.328	0.168	0.078	0.031	0.010	0.002			
	1	0.204	0.328	0.410	0.360	0.259	0.156	0.077	0.028	0.006		
	2	0.021	0.073	0.205	0.309	0.346	0.312	0.230	0.132	0.051	0.008	0.001
	3	0.001	0.008	0.051	0.132	0.230	0.312	0.346	0.309	0.205	0.073	0.021
	4			0.006	0.028	0.077	0.156	0.259	0.360	0.410	0.328	0.204
	5				0.002	0.010	0.031	0.078	0.168	0.328	0.590	0.774
6	0	0.735	0.531	0.262	0.118	0.047	0.016	0.004	0.001			
	1	0.232	0.354	0.393	0.303	0.187	0.094	0.037	0.010	0.002		
	2	0.031	0.098	0.246	0.324	0.311	0.234	0.138	0.060	0.015	0.001	
	3	0.002	0.015	0.082	0.185	0.276	0.312	0.276	0.185	0.082	0.015	0.002
	4		0.001	0.015	0.060	0.138	0.234	0.311	0.324	0.246	0.098	0.031
	5			0.002	0.010	0.037	0.094	0.187	0.303	0.393	0.354	0.232
	6				0.001	0.004	0.016	0.047	0.118	0.262	0.531	0.735
7	0	0.698	0.478	0.210	0.082	0.028	0.008	0.002				
	1	0.257	0.372	0.367	0.247	0.131	0.055	0.017	0.004			
	2	0.041	0.124	0.275	0.318	0.261	0.164	0.077	0.025	0.004		
	3	0.004	0.023	0.115	0.227	0.290	0.273	0.194	0.097	0.029	0.003	
	4		0.003	0.029	0.097	0.194	0.273	0.290	0.227	0.115	0.023	0.004

* All values omitted in this table are 0.0005 or less.

Table VI

BINOMIAL PROBABILITIES *(Continued)*

						p						
n	x	0.05	0.1	0.2	0.3	0.4	0.5	0.6	0.7	0.8	0.9	0.95
7	5			0.004	0.025	0.077	0.164	0.261	0.318	0.275	0.124	0.041
	6				0.004	0.017	0.055	0.131	0.247	0.367	0.372	0.257
	7					0.002	0.008	0.028	0.082	0.210	0.478	0.698
8	0	0.663	0.430	0.168	0.058	0.017	0.004	0.001				
	1	0.279	0.383	0.336	0.198	0.090	0.031	0.008	0.001			
	2	0.051	0.149	0.294	0.296	0.209	0.109	0.041	0.010	0.001		
	3	0.005	0.033	0.147	0.254	0.279	0.219	0.124	0.047	0.009		
	4		0.005	0.046	0.136	0.232	0.273	0.232	0.136	0.046	0.005	
	5			0.009	0.047	0.124	0.219	0.279	0.254	0.147	0.033	0.005
	6			0.001	0.010	0.041	0.109	0.209	0.296	0.294	0.149	0.051
	7				0.001	0.008	0.031	0.090	0.198	0.336	0.383	0.279
	8					0.001	0.004	0.017	0.058	0.168	0.430	0.663
9	0	0.630	0.387	0.134	0.040	0.010	0.002					
	1	0.299	0.387	0.302	0.156	0.060	0.018	0.004				
	2	0.063	0.172	0.302	0.267	0.161	0.070	0.021	0.004			
	3	0.008	0.045	0.176	0.267	0.251	0.164	0.074	0.021	0.003		
	4	0.001	0.007	0.066	0.172	0.251	0.246	0.167	0.074	0.017	0.001	
	5		0.001	0.017	0.074	0.167	0.246	0.251	0.172	0.066	0.007	0.001
	6			0.003	0.021	0.074	0.164	0.251	0.267	0.176	0.045	0.008
	7				0.004	0.021	0.070	0.161	0.267	0.302	0.172	0.063
	8					0.004	0.018	0.060	0.156	0.302	0.387	0.299
	9						0.002	0.010	0.040	0.134	0.387	0.630
10	0	0.599	0.349	0.107	0.028	0.006	0.001					
	1	0.315	0.387	0.268	0.121	0.040	0.010	0.002				
	2	0.075	0.194	0.302	0.233	0.121	0.044	0.011	0.001			
	3	0.010	0.057	0.201	0.267	0.215	0.117	0.042	0.009	0.001		
	4	0.001	0.011	0.088	0.200	0.251	0.205	0.111	0.037	0.006		
	5		0.001	0.026	0.103	0.201	0.246	0.201	0.103	0.026	0.001	
	6			0.006	0.037	0.111	0.205	0.251	0.200	0.088	0.011	0.001
	7			0.001	0.009	0.042	0.117	0.215	0.267	0.201	0.057	0.010
	8				0.001	0.011	0.044	0.121	0.233	0.302	0.194	0.075
	9					0.002	0.010	0.040	0.121	0.268	0.387	0.315
	10						0.001	0.006	0.028	0.107	0.349	0.599

Table VI

BINOMIAL PROBABILITIES *(Continued)*

						p						
n	x	0.05	0.1	0.2	0.3	0.4	0.5	0.6	0.7	0.8	0.9	0.95
11	0	0.569	0.314	0.086	0.020	0.004						
	1	0.329	0.384	0.236	0.093	0.027	0.005	0.001				
	2	0.087	0.213	0.295	0.200	0.089	0.027	0.005	0.001			
	3	0.014	0.071	0.221	0.257	0.177	0.081	0.023	0.004			
	4	0.001	0.016	0.111	0.220	0.236	0.161	0.070	0.017	0.002		
	5		0.002	0.039	0.132	0.221	0.226	0.147	0.057	0.010		
	6			0.010	0.057	0.147	0.226	0.221	0.132	0.039	0.002	
	7			0.002	0.017	0.070	0.161	0.236	0.220	0.111	0.016	0.001
	8				0.004	0.023	0.081	0.177	0.257	0.221	0.071	0.014
	9				0.001	0.005	0.027	0.089	0.200	0.295	0.213	0.087
	10					0.001	0.005	0.027	0.093	0.236	0.384	0.329
	11							0.004	0.020	0.086	0.314	0.569
12	0	0.540	0.282	0.069	0.014	0.002						
	1	0.341	0.377	0.206	0.071	0.017	0.003					
	2	0.099	0.230	0.283	0.168	0.064	0.016	0.002				
	3	0.017	0.085	0.236	0.240	0.142	0.054	0.012	0.001			
	4	0.002	0.021	0.133	0.231	0.213	0.121	0.042	0.008	0.001		
	5		0.004	0.053	0.158	0.227	0.193	0.101	0.029	0.003		
	6			0.016	0.079	0.177	0.226	0.177	0.079	0.016		
	7			0.003	0.029	0.101	0.193	0.227	0.158	0.053	0.004	
	8			0.001	0.008	0.042	0.121	0.213	0.231	0.133	0.021	0.002
	9				0.001	0.012	0.054	0.142	0.240	0.236	0.085	0.017
	10					0.002	0.016	0.064	0.168	0.283	0.230	0.099
	11						0.003	0.017	0.071	0.206	0.377	0.341
	12							0.002	0.014	0.069	0.282	0.540
13	0	0.513	0.254	0.055	0.010	0.001						
	1	0.351	0.367	0.179	0.054	0.011	0.002					
	2	0.111	0.245	0.268	0.139	0.045	0.010	0.001				
	3	0.021	0.100	0.246	0.218	0.111	0.035	0.006	0.001			
	4	0.003	0.028	0.154	0.234	0.184	0.087	0.024	0.003			
	5		0.006	0.069	0.180	0.221	0.157	0.066	0.014	0.001		
	6		0.001	0.023	0.103	0.197	0.209	0.131	0.044	0.006		
	7			0.006	0.044	0.131	0.209	0.197	0.103	0.023	0.001	

Table VI

BINOMIAL PROBABILITIES *(Continued)*

							p					
n	x	0.05	0.1	0.2	0.3	0.4	0.5	0.6	0.7	0.8	0.9	0.95
13	8			0.001	0.014	0.066	0.157	0.221	0.180	0.069	0.006	
	9				0.003	0.024	0.087	0.184	0.234	0.154	0.028	0.003
	10				0.001	0.006	0.035	0.111	0.218	0.246	0.100	0.021
	11					0.001	0.010	0.045	0.139	0.268	0.245	0.111
	12						0.002	0.011	0.054	0.179	0.367	0.351
	13							0.001	0.010	0.055	0.254	0.513
14	0	0.488	0.229	0.044	0.007	0.001						
	1	0.359	0.356	0.154	0.041	0.007	0.001					
	2	0.123	0.257	0.250	0.113	0.032	0.006	0.001				
	3	0.026	0.114	0.250	0.194	0.085	0.022	0.003				
	4	0.004	0.035	0.172	0.229	0.155	0.061	0.014	0.001			
	5		0.008	0.086	0.196	0.207	0.122	0.041	0.007			
	6		0.001	0.032	0.126	0.207	0.183	0.092	0.023	0.002		
	7			0.009	0.062	0.157	0.209	0.157	0.062	0.009		
	8			0.002	0.023	0.092	0.183	0.207	0.126	0.032	0.001	
	9				0.007	0.041	0.122	0.207	0.196	0.086	0.008	
	10				0.001	0.014	0.061	0.155	0.229	0.172	0.035	0.004
	11					0.003	0.022	0.085	0.194	0.250	0.114	0.026
	12					0.001	0.006	0.032	0.113	0.250	0.257	0.123
	13						0.001	0.007	0.041	0.154	0.356	0.359
	14							0.001	0.007	0.044	0.229	0.488
15	0	0.463	0.206	0.035	0.005							
	1	0.366	0.343	0.132	0.031	0.005						
	2	0.135	0.267	0.231	0.092	0.022	0.003					
	3	0.031	0.129	0.250	0.170	0.063	0.014	0.002				
	4	0.005	0.043	0.188	0.219	0.127	0.042	0.007	0.001			
	5	0.001	0.010	0.103	0.206	0.186	0.092	0.024	0.003			
	6		0.002	0.043	0.147	0.207	0.153	0.061	0.012	0.001		
	7			0.014	0.081	0.177	0.196	0.118	0.035	0.003		
	8			0.003	0.035	0.118	0.196	0.177	0.081	0.014		
	9			0.001	0.012	0.061	0.153	0.207	0.147	0.043	0.002	
	10				0.003	0.024	0.092	0.186	0.206	0.103	0.010	0.001
	11				0.001	0.007	0.042	0.127	0.219	0.188	0.043	0.005
	12					0.002	0.014	0.063	0.170	0.250	0.129	0.031
	13						0.003	0.022	0.092	0.231	0.267	0.135
	14							0.005	0.031	0.132	0.343	0.366
	15								0.005	0.035	0.206	0.463

Table VII

$$\text{VALUES OF } Z = \tfrac{1}{2}\cdot\log_e\left(\frac{1+r}{1-r}\right)$$

r	.00	.01	.02	.03	.04	.05	.06	.07	.08	.09
0.0	0.000	0.010	0.020	0.030	0.040	0.050	0.060	0.070	0.080	0.090
0.1	0.100	0.110	0.121	0.131	0.141	0.151	0.161	0.172	0.182	0.192
0.2	0.203	0.213	0.224	0.234	0.245	0.255	0.266	0.277	0.288	0.299
0.3	0.310	0.321	0.332	0.343	0.354	0.365	0.377	0.388	0.400	0.412
0.4	0.424	0.436	0.448	0.460	0.472	0.485	0.497	0.510	0.523	0.536
0.5	0.549	0.563	0.576	0.590	0.604	0.618	0.633	0.648	0.662	0.678
0.6	0.693	0.709	0.725	0.741	0.758	0.775	0.793	0.811	0.829	0.848
0.7	0.867	0.887	0.908	0.929	0.950	0.973	0.996	1.020	1.045	1.071
0.8	1.099	1.127	1.157	1.188	1.221	1.256	1.293	1.333	1.376	1.422
0.9	1.472	1.528	1.589	1.658	1.738	1.832	1.946	2.092	2.298	2.647

For negative values of r put a minus sign in front of the corresponding Z's, and vice versa.

Table VIII

BINOMIAL COEFFICIENTS

n	$\binom{n}{0}$	$\binom{n}{1}$	$\binom{n}{2}$	$\binom{n}{3}$	$\binom{n}{4}$	$\binom{n}{5}$	$\binom{n}{6}$	$\binom{n}{7}$	$\binom{n}{8}$	$\binom{n}{9}$	$\binom{n}{10}$
0	1										
1	1	1									
2	1	2	1								
3	1	3	3	1							
4	1	4	6	4	1						
5	1	5	10	10	5	1					
6	1	6	15	20	15	6	1				
7	1	7	21	35	35	21	7	1			
8	1	8	28	56	70	56	28	8	1		
9	1	9	36	84	126	126	84	36	9	1	
10	1	10	45	120	210	252	210	120	45	10	1
11	1	11	55	165	330	462	462	330	165	55	11
12	1	12	66	220	495	792	924	792	495	220	66
13	1	13	78	286	715	1287	1716	1716	1287	715	286
14	1	14	91	364	1001	2002	3003	3432	3003	2002	1001
15	1	15	105	455	1365	3003	5005	6435	6435	5005	3003
16	1	16	120	560	1820	4368	8008	11440	12870	11440	8008
17	1	17	136	680	2380	6188	12376	19448	24310	24310	19448
18	1	18	153	816	3060	8568	18564	31824	43758	48620	43758
19	1	19	171	969	3876	11628	27132	50388	75582	92378	92378
20	1	20	190	1140	4845	15504	38760	77520	125970	167960	184756

For $r > 10$ it may be necessary to make use of the identity $\binom{n}{r} = \binom{n}{n-r}$.

Table IX

RANDOM NUMBERS*

04433	80674	24520	18222	10610	05794	37515
60298	47829	72648	37414	75755	04717	29899
67884	59651	67533	68123	17730	95862	08034
89512	32155	51906	61662	64130	16688	37275
32653	01895	12506	88535	36553	23757	34209
95913	15405	13772	76638	48423	25018	99041
55864	21694	13122	44115	01601	50541	00147
35334	49810	91601	40617	72876	33967	73830
57729	32196	76487	11622	96297	24160	09903
86648	13697	63677	70119	94739	25875	38829
30574	47609	07967	32422	76791	39725	53711
81307	43694	83580	79974	45929	85113	72268
02410	54905	79007	54939	21410	86980	91772
18969	75274	52233	62319	08598	09066	95288
87863	82384	66860	62297	80198	19347	73234
68397	71708	15438	62311	72844	60203	46412
28529	54447	58729	10854	99058	18260	38765
44285	06372	15867	70418	57012	72122	36634
86299	83430	33571	23309	57040	29285	67870
84842	68668	90894	61658	15001	94055	36308
56970	83609	52098	04184	54967	72938	56834
83125	71257	60490	44369	66130	72936	69848
55503	52423	02464	26141	68779	66388	75242
47019	76273	33203	29608	54553	25971	69573
84828	32592	79526	29554	84580	37859	28504
68921	08141	79227	05748	51276	57143	31926
36458	96045	30424	98420	72925	40729	22337
95752	59445	36847	87729	81679	59126	59437
26768	47323	58454	56958	20575	76746	49878
42613	37056	43636	58085	06766	60227	96414
95457	30566	65482	25596	02678	54592	63607
95276	17894	63564	95958	39750	64379	46059
66954	52324	64776	92345	95110	59448	77249
17457	18481	14113	62462	02798	54977	48349
03704	36872	83214	59337	01695	60666	97410
21538	86497	33210	60337	27976	70661	08250
57178	67619	98310	70348	11317	71623	55510
31048	97558	94953	55866	96283	46620	52087
69799	55380	16498	80733	96422	58078	99643
90595	61867	59231	17772	67831	33317	00520
33570	04981	98939	78784	09977	29398	93896
15340	93460	57477	13898	48431	72936	78160
64079	42483	36512	56186	99098	48850	72527
63491	05546	67118	62063	74958	20946	28147
92003	63868	41034	28260	79708	00770	88643
52360	46658	66511	04172	73085	11795	52594
74622	12142	68355	65635	21828	39539	18988
04157	50079	61343	64315	70836	82857	35335
86003	60070	66241	32836	27573	11479	94114
41268	80187	20351	09636	84668	42486	71303

* Based on parts of *Table of 105,000 Random Decimal Digits*, Interstate Commerce Commission, Bureau of Transport Economics and Statistics, Washington, D.C.

Table IX

RANDOM NUMBERS (*Continued*)

48611	62866	33963	14045	79451	04934	45576
78812	03509	78673	73181	29973	18664	04555
19472	63971	37271	31445	49019	49405	46925
51266	11569	08697	91120	64156	40365	74297
55806	96275	26130	47949	14877	69594	83041
77527	81360	18180	97421	55541	90275	18213
77680	58788	33016	61173	93049	04694	43534
15404	96554	88265	34537	38526	67924	40474
14045	22917	60718	66487	46346	30949	03173
68376	43918	77653	04127	69930	43283	35766
93385	13421	67957	20384	58731	53396	59723
09858	52104	32014	53115	03727	98624	84616
93307	34116	49516	42148	57740	31198	70336
04794	01534	92058	03157	91758	80611	45357
86265	49096	97021	92582	61422	75890	86442
65943	79232	45702	67055	39024	57383	44424
90038	94209	04055	27393	61517	23002	96560
97283	95943	78363	36498	40662	94188	18202
21913	72958	75637	99936	58715	07943	23748
41161	37341	81838	19389	80336	46346	91895
23777	98392	31417	98547	92058	02277	50315
59973	08144	61070	73094	27059	69181	55623
82690	74099	77885	23813	10054	11900	44653
83854	24715	48866	65745	31131	47636	45137
61980	34997	41825	11623	07320	15003	56774
99915	45821	97702	87125	44488	77613	56823
48293	86847	43186	42951	37804	85129	28993
33225	31280	41232	34750	91097	60752	69783
06846	32828	24425	30249	78801	26977	92074
32671	45587	79620	84831	38156	74211	82752
82096	21913	75544	55228	89796	05694	91552
51666	10433	10945	55306	78562	89630	41230
54044	67942	24145	42294	27427	84875	37022
66738	60184	75679	38120	17640	36242	99357
55064	17427	89180	74018	44865	53197	74810
69599	60264	84549	78007	88450	06488	72274
64756	87759	92354	78694	63638	80939	98644
80817	74533	68407	55862	32476	19326	95558
39847	96884	84657	33697	39578	90197	80532
90401	41700	95510	61166	33757	23279	85523
78227	90110	81378	96659	37008	04050	04228
87240	52716	87697	79433	16336	52862	69149
08486	10951	26832	39763	02485	71688	90936
39338	32169	03713	93510	61244	73774	01245
21188	01850	69689	49426	49128	14660	14143
13287	82531	04388	64693	11934	35051	68576
53609	04001	19648	14053	49623	10840	31915
87900	36194	31567	53506	34304	39910	79630
81641	00496	36058	75899	46620	70024	88753
19512	50277	71508	20116	79520	06269	74173

Table IX

RANDOM NUMBERS (*Continued*)

24418	23508	91507	76455	54941	72711	39406
57404	73678	08272	62941	02349	71389	45605
77644	98489	86268	73652	98210	44546	27174
68366	65614	01443	07607	11826	91326	29664
64472	72294	95432	53555	96810	17100	35066
88205	37913	98633	81009	81060	33449	68055
98455	78685	71250	10329	56135	80647	51404
48977	36794	56054	59243	57361	65304	93258
93077	72941	92779	23581	24548	56415	61927
84533	26564	91583	83411	66504	02036	02922
11338	12903	14514	27585	45068	05520	56321
23853	68500	92274	87026	99717	01542	72990
94096	74920	25822	98026	05394	61840	83089
83160	82362	09350	98536	38155	42661	02363
97425	47335	69709	01386	74319	04318	99387
83951	11954	24317	20345	18134	90062	10761
93085	35203	05740	03206	92012	42710	34650
33762	83193	58045	89880	78101	44392	53767
49665	85397	85137	30496	23469	42846	94810
37541	82627	80051	72521	35342	56119	97190
22145	85304	35348	82854	55846	18076	12415
27153	08662	61078	52433	22184	33998	87436
00301	49425	66682	25442	83668	66236	79655
43815	43272	73778	63469	50083	70696	13558
14689	86482	74157	46012	97765	27552	49617
16680	55936	82453	19532	49988	13176	94219
86938	60429	01137	86168	78257	86249	46134
33944	29219	73161	46061	30946	22210	79302
16045	67736	18608	18198	19468	76358	69203
37044	52523	25627	63107	30806	80857	84383
61471	45322	35340	35132	42163	69332	98851
47422	21296	16785	66303	39249	51463	95963
24133	39719	14484	58613	88717	29289	77360
67253	67064	10748	16006	16767	57345	42285
62382	76941	01635	35829	77516	98468	51686
98011	16503	09201	03523	87192	66483	55649
37366	24386	20654	85117	74078	64120	04643
73587	83993	54176	05221	94119	20108	78101
33583	68291	50547	96085	62180	27453	18567
02878	33223	39109	49536	56199	05993	71201
91498	41673	17195	33175	04994	09879	70337
91127	19815	30219	55591	21725	43827	78862
12997	55013	18662	81724	24305	37661	18956
96098	13651	15393	69995	14762	69734	89150
97627	17837	10472	18983	28387	99781	52977
40064	47981	31484	76603	54088	91095	00010
16239	68743	71374	55863	22672	91609	51514
58354	24913	20435	30965	17453	65623	93058
52567	65085	60220	84641	18273	49604	47418
06236	29052	91392	07551	83532	68130	56970

Table IX

RANDOM NUMBERS (*Continued*)

94620	27963	96478	21559	19246	88097	44926
60947	60775	73181	43264	56895	04232	59604
27499	53523	63110	57106	20865	91683	80688
01603	23156	89223	43429	95353	44662	59433
00815	01552	06392	31437	70385	45863	75971
83844	90942	74857	52419	68723	47830	63010
06626	10042	93629	37609	57215	08409	81906
56760	63348	24949	11859	29793	37457	59377
64416	29934	00755	09418	14230	62887	92683
63569	17906	38076	32135	19096	96970	75917
22693	35089	72994	04252	23791	60249	83010
43413	59744	01275	71326	91382	45114	20245
09224	78530	50566	49965	04851	18280	14039
67625	34683	03142	74733	63558	09665	22610
86874	12549	98699	54952	91579	26023	81076
54548	49505	62515	63903	13193	33905	66936
73236	66167	49728	03581	40699	10396	81827
15220	66319	13543	14071	59148	95154	72852
16151	08029	36954	03891	38313	34016	18671
43635	84249	88984	80993	55431	90793	62603
30193	42776	85611	57635	51362	79907	77364
37430	45246	11400	20986	43996	73122	88474
88312	93047	12088	86937	70794	01041	74867
98995	58159	04700	90443	13168	31553	67891
51734	20849	70198	67906	00880	82899	66065
88698	41755	56216	66852	17748	04963	54859
51865	09836	73966	65711	41699	11732	17173
40300	08852	27528	84648	79589	95295	72895
02760	28625	70476	76410	32988	10194	94917
78450	26245	91763	73117	33047	03577	62599
50252	56911	62693	73817	98693	18728	94741
07929	66728	47761	81472	44806	15592	71357
09030	39605	87507	85446	51257	89555	75520
56670	88445	85799	76200	21795	38894	58070
48140	13583	94911	13318	64741	64336	95103
36764	86132	12463	28385	94242	32063	45233
14351	71381	28133	68269	65145	28152	39087
81276	00835	63835	87174	42446	08882	27067
55524	86088	00069	59254	24654	77371	26409
78852	65889	32719	13758	23937	90740	16866
11861	69032	51915	23510	32050	52052	24004
67699	01009	07050	73324	06732	27510	33761
50064	39500	17450	18030	63124	48061	59412
93126	17700	94400	76075	08317	27324	72723
01657	92602	41043	05686	15650	29970	95877
13800	76690	75133	60456	28491	03845	11507
98135	42870	48578	29036	69876	86563	61729
08313	99293	00990	13595	77457	79969	11339
90974	83965	62732	85161	54330	22406	86253
33273	61993	88407	69399	17301	70975	99129

Table X

LOGARITHMS

N	0	1	2	3	4	5	6	7	8	9
10	0000	0043	0086	0128	0170	0212	0253	0294	0334	0374
11	0414	0453	0492	0531	0569	0607	0645	0682	0719	0755
12	0792	0828	0864	0899	0934	0969	1004	1038	1072	1106
13	1139	1173	1206	1239	1271	1303	1335	1367	1399	1430
14	1461	1492	1523	1553	1584	1614	1644	1673	1703	1732
15	1761	1790	1818	1847	1875	1903	1931	1959	1987	2014
16	2041	2068	2095	2122	2148	2175	2201	2227	2253	2279
17	2304	2330	2355	2380	2405	2430	2455	2480	2504	2529
18	2553	2577	2601	2625	2648	2672	2695	2718	2742	2765
19	2788	2810	2833	2856	2878	2900	2923	2945	2967	2989
20	3010	3032	3054	3075	3096	3118	3139	3160	3181	3201
21	3222	3243	3263	3284	3304	3324	3345	3365	3385	3404
22	3424	3444	3464	3483	3502	3522	3541	3560	3579	3598
23	3617	3636	3655	3674	3692	3711	3729	3747	3766	3784
24	3802	3820	3838	3856	3874	3892	3909	3927	3945	3962
25	3979	3997	4014	4031	4048	4065	4082	4099	4116	4133
26	4150	4166	4183	4200	4216	4232	4249	4265	4281	4298
27	4314	4330	4346	4362	4378	4393	4409	4425	4440	4456
28	4472	4487	4502	4518	4533	4548	4564	4579	4594	4609
29	4624	4639	4654	4669	4683	4698	4713	4728	4742	4757
30	4771	4786	4800	4814	4829	4843	4857	4871	4886	4900
31	4914	4928	4942	4955	4969	4983	4997	5011	5024	5038
32	5051	5065	5079	5092	5105	5119	5132	5145	5159	5172
33	5185	5198	5211	5224	5237	5250	5263	5276	5289	5302
34	5315	5328	5340	5353	5366	5378	5391	5403	5416	5428
35	5441	5453	5465	5478	5490	5502	5514	5527	5539	5551
36	5563	5575	5587	5599	5611	5623	5635	5647	5658	5670
37	5682	5694	5705	5717	5729	5740	5752	5763	5775	5786
38	5798	5809	5821	5832	5843	5855	5866	5877	5888	5899
39	5911	5922	5933	5944	5955	5966	5977	5988	5999	6010
40	6021	6031	6042	6053	6064	6075	6085	6096	6107	6117
41	6128	6138	6149	6160	6170	6180	6191	6201	6212	6222
42	6232	6243	6253	6263	6274	6284	6294	6304	6314	6325
43	6335	6345	6355	6365	6375	6385	6395	6405	6415	6425
44	6435	6444	6454	6464	6474	6484	6493	6503	6513	6522
45	6532	6542	6551	6561	6571	6580	6590	6599	6609	6618
46	6628	6637	6646	6656	6665	6675	6684	6693	6702	6712
47	6721	6730	6739	6749	6758	6767	6776	6785	6794	6803
48	6812	6821	6830	6839	6848	6857	6866	6875	6884	6893
49	6902	6911	6920	6928	6937	6946	6955	6964	6972	6981
50	6990	6998	7007	7016	7024	7033	7042	7050	7059	7067
51	7076	7084	7093	7101	7110	7118	7126	7135	7143	7152
52	7160	7168	7177	7185	7193	7202	7210	7218	7226	7235
53	7243	7251	7259	7267	7275	7284	7292	7300	7308	7316
54	7324	7332	7340	7348	7356	7364	7372	7380	7388	7396

Table X

LOGARITHMS (*Continued*)

N	0	1	2	3	4	5	6	7	8	9
55	7404	7412	7419	7427	7435	7443	7451	7459	7466	7474
56	7482	7490	7497	7505	7513	7520	7528	7536	7543	7551
57	7559	7566	7574	7582	7589	7597	7604	7612	7619	7627
58	7634	7642	7649	7657	7664	7672	7679	7686	7694	7701
59	7709	7716	7723	7731	7738	7745	7752	7760	7767	7774
60	7782	7789	7796	7803	7810	7818	7825	7832	7839	7846
61	7853	7860	7868	7875	7882	7889	7896	7903	7910	7917
62	7924	7931	7938	7945	7952	7959	7966	7973	7980	7987
63	7993	8000	8007	8014	8021	8028	8035	8041	8048	8055
64	8062	8069	8075	8082	8089	8096	8102	8109	8116	8122
65	8129	8136	8142	8149	8156	8162	8169	8176	8182	8189
66	8195	8202	8209	8215	8222	8228	8235	8241	8248	8254
67	8261	8267	8274	8280	8287	8293	8299	8306	8312	8319
68	8325	8331	8338	8344	8351	8357	8363	8370	8376	8382
69	8388	8395	8401	8407	8414	8420	8426	8432	8439	8445
70	8451	8457	8463	8470	8476	8482	8488	8494	8500	8506
71	8513	8519	8525	8531	8537	8543	8549	8555	8561	8567
72	8573	8579	8585	8591	8597	8603	8609	8615	8621	8627
73	8633	8639	8645	8651	8657	8663	8669	8675	8681	8686
74	8692	8698	8704	8710	8716	8722	8727	8733	8739	8745
75	8751	8756	8762	8768	8774	8779	8785	8791	8797	8802
76	8808	8814	8820	8825	8831	8837	8842	8848	8854	8859
77	8865	8871	8876	8882	8887	8893	8899	8904	8910	8915
78	8921	8927	8932	8938	8943	8949	8954	8960	8965	8971
79	8976	8982	8987	8993	8998	9004	9009	9015	9020	9025
80	9031	9036	9042	9047	9053	9058	9063	9069	9074	9079
81	9085	9090	9096	9101	9106	9112	9117	9122	9128	9133
82	9138	9143	9149	9154	9159	9165	9170	9175	9180	9186
83	9191	9196	9201	9206	9212	9217	9222	9227	9232	9238
84	9243	9248	9253	9258	9263	9269	9274	9279	9284	9289
85	9294	9299	9304	9309	9315	9320	9325	9330	9335	9340
86	9345	9350	9355	9360	9365	9370	9375	9380	9385	9390
87	9395	9400	9405	9410	9415	9420	9425	9430	9435	9440
88	9445	9450	9455	9460	9465	9469	9474	9479	9484	9489
89	9494	9499	9504	9509	9513	9518	9523	9528	9533	9538
90	9542	9547	9552	9557	9562	9566	9571	9576	9581	9586
91	9590	9595	9600	9605	9609	9614	9619	9624	9628	9633
92	9638	9643	9647	9652	9657	9661	9666	9671	9675	9680
93	9685	9689	9694	9699	9703	9708	9713	9717	9722	9727
94	9731	9736	9741	9745	9750	9754	9759	9763	9768	9773
95	9777	9782	9786	9791	9795	9800	9805	9809	9814	9818
96	9823	9827	9832	9836	9841	9845	9850	9854	9859	9863
97	9868	9872	9877	9881	9886	9890	9894	9899	9903	9908
98	9912	9917	9921	9926	9930	9934	9939	9943	9948	9952
99	9956	9961	9965	9969	9974	9978	9983	9987	9991	9996

Table XI

VALUES OF e^{-x}

x	e^{-x}	x	e^{-x}	x	e^{-x}	x	e^{-x}
0.0	1.000	2.5	0.082	5.0	0.0067	7.5	0.00055
0.1	0.905	2.6	0.074	5.1	0.0061	7.6	0.00050
0.2	0.819	2.7	0.067	5.2	0.0055	7.7	0.00045
0.3	0.741	2.8	0.061	5.3	0.0050	7.8	0.00041
0.4	0.670	2.9	0.055	5.4	0.0045	7.9	0.00037
0.5	0.607	3.0	0.050	5.5	0.0041	8.0	0.00034
0.6	0.549	3.1	0.045	5.6	0.0037	8.1	0.00030
0.7	0.497	3.2	0.041	5.7	0.0033	8.2	0.00028
0.8	0.449	3.3	0.037	5.8	0.0030	8.3	0.00025
0.9	0.407	3.4	0.033	5.9	0.0027	8.4	0.00023
1.0	0.368	3.5	0.030	6.0	0.0025	8.5	0.00020
1.1	0.333	3.6	0.027	6.1	0.0022	8.6	0.00018
1.2	0.301	3.7	0.025	6.2	0.0020	8.7	0.00017
1.3	0.273	3.8	0.022	6.3	0.0018	8.8	0.00015
1.4	0.247	3.9	0.020	6.4	0.0017	8.9	0.00014
1.5	0.223	4.0	0.018	6.5	0.0015	9.0	0.00012
1.6	0.202	4.1	0.017	6.6	0.0014	9.1	0.00011
1.7	0.183	4.2	0.015	6.7	0.0012	9.2	0.00010
1.8	0.165	4.3	0.014	6.8	0.0011	9.3	0.00009
1.9	0.150	4.4	0.012	6.9	0.0010	9.4	0.00008
2.0	0.135	4.5	0.011	7.0	0.0009	9.5	0.00008
2.1	0.122	4.6	0.010	7.1	0.0008	9.6	0.00007
2.2	0.111	4.7	0.009	7.2	0.0007	9.7	0.00006
2.3	0.100	4.8	0.008	7.3	0.0007	9.8	0.00006
2.4	0.091	4.9	0.007	7.4	0.0006	9.9	0.00005

Although square root tables are relatively easy to use, most beginners seem to have some difficulty in choosing the right column and in placing the decimal point correctly in the answer. Table XII, in addition to containing the *squares* of the numbers from 1.00 to 9.99 spaced at intervals of 0.01, gives the *square roots* of these numbers rounded to 6 decimals. To find the square root of any positive number rounded to 3 significant digits, we have only to use the following rule in deciding whether to take the entry of the $\sqrt{n}$ or the $\sqrt{10n}$ column:

> *Move the decimal point an even number of places to the right or to the left until a number greater than or equal to 1 but less than 100 is reached. If the resulting number is less than 10 go to the $\sqrt{n}$ column; if it is 10 or more go to the $\sqrt{10n}$ column.*

Thus, to find the square root of 14,600, 459, or 0.0315 we go to the $\sqrt{n}$ column since the decimal point has to be moved, respectively, 4 places to the left, 2 places to the left, or 2 places to the right, to give 1.46, 4.59, or 3.15. Similarly, to find the square root of 2,163, 0.192, or 0.0000158 we go to the $\sqrt{10n}$ column since the decimal point has to be moved, respectively. 2 places to the left, 2 places to the right, or 6 places to the right, to give 21.63, 19.2, or 15.8.

Having found the entry in the appropriate column of Table XII, the only thing that remains to be done is to put the decimal point in the right position in the answer. Here it will help to use the following rule:

> *Having previously moved the decimal point an even number of places to the left or right to get a number greater than or equal to 1 but less than 100, the decimal point of the entry of the appropriate column in Table XII is moved half as many places in the opposite direction.*

For example, to determine the square root of 14,600 we first note that the decimal point has to be moved *four places to the left* to give 1.46. We thus take the entry of the $\sqrt{n}$ column corresponding to 1.46, move its decimal point *two places to the right*, and get $\sqrt{14,600}$ = 120.8305. Similarly, to find the square root of 0.0000158, we note that the decimal point has to be moved *six places to the right* to give 15.8. We thus take the entry of the $\sqrt{10n}$ column corresponding to 1.58, move the decimal point *three places to the left*, and get $\sqrt{0.0000158}$ = 0.003974921. In actual practice, if a number whose square root we want to find is *rounded*, the square root should be rounded to as many significant digits as the original number.

Table XII

SQUARES AND SQUARE ROOTS

n	n^2	$\sqrt{n}$	$\sqrt{10n}$	n	n^2	$\sqrt{n}$	$\sqrt{10n}$
1.00	1.0000	1.000000	3.162278	1.50	2.2500	1.224745	3.872983
1.01	1.0201	1.004988	3.178050	1.51	2.2801	1.228821	3.885872
1.02	1.0404	1.009950	3.193744	1.52	2.3104	1.232883	3.898718
1.03	1.0609	1.014889	3.209361	1.53	2.3409	1.236932	3.911521
1.04	1.0816	1.019804	3.224903	1.54	2.3716	1.240967	3.924283
1.05	1.1025	1.024695	3.240370	1.55	2.4025	1.244990	3.937004
1.06	1.1236	1.029563	3.255764	1.56	2.4336	1.249000	3.949684
1.07	1.1449	1.034408	3.271085	1.57	2.4649	1.252996	3.962323
1.08	1.1664	1.039230	3.286335	1.58	2.4964	1.256981	3.974921
1.09	1.1881	1.044031	3.301515	1.59	2.5281	1.260952	3.987480
1.10	1.2100	1.048809	3.316625	1.60	2.5600	1.264911	4.000000
1.11	1.2321	1.053565	3.331666	1.61	2.5921	1.268858	4.012481
1.12	1.2544	1.058301	3.346640	1.62	2.6244	1.272792	4.024922
1.13	1.2769	1.063015	3.361547	1.63	2.6569	1.276715	4.037326
1.14	1.2996	1.067708	3.376389	1.64	2.6896	1.280625	4.049691
1.15	1.3225	1.072381	3.391165	1.65	2.7225	1.284523	4.062019
1.16	1.3456	1.077033	3.405877	1.66	2.7556	1.288410	4.074310
1.17	1.3689	1.081665	3.420526	1.67	2.7889	1.292285	4.086563
1.18	1.3924	1.086278	3.435113	1.68	2.8224	1.296148	4.098780
1.19	1.4161	1.090871	3.449638	1.69	2.8561	1.300000	4.110961
1.20	1.4400	1.095445	3.464102	1.70	2.8900	1.303840	4.123106
1.21	1.4641	1.100000	3.478505	1.71	2.9241	1.307670	4.135215
1.22	1.4884	1.104536	3.492850	1.72	2.9584	1.311488	4.147288
1.23	1.5129	1.109054	3.507136	1.73	2.9929	1.315295	4.159327
1.24	1.5376	1.113553	3.521363	1.74	3.0276	1.319091	4.171331
1.25	1.5625	1.118034	3.535534	1.75	3.0625	1.322876	4.183300
1.26	1.5876	1.122497	3.549648	1.76	3.0976	1.326650	4.195235
1.27	1.6129	1.126943	3.563706	1.77	3.1329	1.330413	4.207137
1.28	1.6384	1.131371	3.577709	1.78	3.1684	1.334166	4.219005
1.29	1.6641	1.135782	3.591657	1.79	3.2041	1.337909	4.230839
1.30	1.6900	1.140175	3.605551	1.80	3.2400	1.341641	4.242641
1.31	1.7161	1.144552	3.619392	1.81	3.2761	1.345362	4.254409
1.32	1.7424	1.148913	3.633180	1.82	3.3124	1.349074	4.266146
1.33	1.7689	1.153256	3.646917	1.83	3.3489	1.352775	4.277850
1.34	1.7956	1.157584	3.660601	1.84	3.3856	1.356466	4.289522
1.35	1.8225	1.161895	3.674235	1.85	3.4225	1.360147	4.301163
1.36	1.8496	1.166190	3.687818	1.86	3.4596	1.363818	4.312772
1.37	1.8769	1.170470	3.701351	1.87	3.4969	1.367479	4.324350
1.38	1.9044	1.174734	3.714835	1.88	3.5344	1.371131	4.335897
1.39	1.9321	1.178983	3.728270	1.89	3.5721	1.374773	4.347413
1.40	1.9600	1.183216	3.741657	1.90	3.6100	1.378405	4.358899
1.41	1.9881	1.187434	3.754997	1.91	3.6481	1.382027	4.370355
1.42	2.0164	1.191638	3.768289	1.92	3.6864	1.385641	4.381780
1.43	2.0449	1.195826	3.781534	1.93	3.7249	1.389244	4.393177
1.44	2.0736	1.200000	3.794733	1.94	3.7636	1.392839	4.404543
1.45	2.1025	1.204159	3.807887	1.95	3.8025	1.396424	4.415880
1.46	2.1316	1.208305	3.820995	1.96	3.8416	1.400000	4.427189
1.47	2.1609	1.212436	3.834058	1.97	3.8809	1.403567	4.438468
1.48	2.1904	1.216553	3.847077	1.98	3.9204	1.407125	4.449719
1.49	2.2201	1.220656	3.860052	1.99	3.9601	1.410674	4.460942

Table XII

SQUARES AND SQUARE ROOTS (*Continued*)

n	n^2	$\sqrt{n}$	$\sqrt{10n}$	n	n^2	$\sqrt{n}$	$\sqrt{10n}$
2.00	4.0000	1.414214	4.472136	2.50	6.2500	1.581139	5.000000
2.01	4.0401	1.417745	4.483302	2.51	6.3001	1.584298	5.009990
2.02	4.0804	1.421267	4.494441	2.52	6.3504	1.587451	5.019960
2.03	4.1209	1.424781	4.505552	2.53	6.4009	1.590597	5.029911
2.04	4.1616	1.428286	4.516636	2.54	6.4516	1.593738	5.039841
2.05	4.2025	1.431782	4.527693	2.55	6.5025	1.596872	5.049752
2.06	4.2436	1.435270	4.538722	2.56	6.5536	1.600000	5.059644
2.07	4.2849	1.438749	4.549725	2.57	6.6049	1.603122	5.069517
2.08	4.3264	1.442221	4.560702	2.58	6.6564	1.606238	5.079370
2.09	4.3681	1.445683	4.571652	2.59	6.7081	1.609348	5.089204
2.10	4.4100	1.449138	4.582576	2.60	6.7600	1.612452	5.099020
2.11	4.4521	1.452584	4.593474	2.61	6.8121	1.615549	5.108816
2.12	4.4944	1.456022	4.604346	2.62	6.8644	1.618641	5.118594
2.13	4.5369	1.459452	4.615192	2.63	6.9169	1.621727	5.128353
2.14	4.5796	1.462874	4.626013	2.64	6.9696	1.624808	5.138093
2.15	4.6225	1.466288	4.636809	2.65	7.0225	1.627882	5.147815
2.16	4.6656	1.469694	4.647580	2.66	7.0756	1.630951	5.157519
2.17	4.7089	1.473092	4.658326	2.67	7.1289	1.634013	5.167204
2.18	4.7524	1.476482	4.669047	2.68	7.1824	1.637071	5.176872
2.19	4.7961	1.479865	4.679744	2.69	7.2361	1.640122	5.186521
2.20	4.8400	1.483240	4.690416	2.70	7.2900	1.643168	5.196152
2.21	4.8841	1.486607	4.701064	2.71	7.3441	1.646208	5.205766
2.22	4.9284	1.489966	4.711688	2.72	7.3984	1.649242	5.215362
2.23	4.9729	1.493318	4.722288	2.73	7.4529	1.652271	5.224940
2.24	5.0176	1.496663	4.732864	2.74	7.5076	1.655295	5.234501
2.25	5.0625	1.500000	4.743416	2.75	7.5625	1.658312	5.244044
2.26	5.1076	1.503330	4.753946	2.76	7.6176	1.661325	5.253570
2.27	5.1529	1.506652	4.764452	2.77	7.6729	1.664332	5.263079
2.28	5.1984	1.509967	4.774935	2.78	7.7284	1.667333	5.272571
2.29	5.2441	1.513275	4.785394	2.79	7.7841	1.670329	5.282045
2.30	5.2900	1.516575	4.795832	2.80	7.8400	1.673320	5.291503
2.31	5.3361	1.519868	4.806246	2.81	7.8961	1.676305	5.300943
2.32	5.3824	1.523155	4.816638	2.82	7.9524	1.679286	5.310367
2.33	5.4289	1.526434	4.827007	2.83	8.0089	1.682260	5.319774
2.34	5.4756	1.529706	4.837355	2.84	8.0656	1.685230	5.329165
2.35	5.5225	1.532971	4.847680	2.85	8.1225	1.688194	5.338539
2.36	5.5696	1.536229	4.857983	2.86	8.1796	1.691153	5.347897
2.37	5.6169	1.539480	4.868265	2.87	8.2369	1.694107	5.357238
2.38	5.6644	1.542725	4.878524	2.88	8.2944	1.697056	5.366563
2.39	5.7121	1.545962	4.888763	2.89	8.3521	1.700000	5.375872
2.40	5.7600	1.549193	4.898979	2.90	8.4100	1.702939	5.385165
2.41	5.8081	1.552417	4.909175	2.91	8.4681	1.705872	5.394442
2.42	5.8564	1.555635	4.919350	2.92	8.5264	1.708801	5.403702
2.43	5.9049	1.558846	4.929503	2.93	8.5849	1.711724	5.412947
2.44	5.9536	1.562050	4.939636	2.94	8.6436	1.714643	5.422177
2.45	6.0025	1.565248	4.949747	2.95	8.7025	1.717556	5.431390
2.46	6.0516	1.568439	4.959839	2.96	8.7616	1.720465	5.440588
2.47	6.1009	1.571623	4.969909	2.97	8.8209	1.723369	5.449771
2.48	6.1504	1.574802	4.979960	2.98	8.8804	1.726268	5.458938
2.49	6.2001	1.577973	4.989990	2.99	8.9401	1.729162	5.468089

Table XII

SQUARES AND SQUARE ROOTS (*Continued*)

n	n^2	$\sqrt{n}$	$\sqrt{10n}$	n	n^2	$\sqrt{n}$	$\sqrt{10n}$
3.00	9.0000	1.732051	5.477226	3.50	12.2500	1.870829	5.916080
3.01	9.0601	1.734935	5.486347	3.51	12.3201	1.873499	5.924525
3.02	9.1204	1.737815	5.495453	3.52	12.3904	1.876166	5.932959
3.03	9.1809	1.740690	5.504544	3.53	12.4609	1.878829	5.941380
3.04	9.2416	1.743560	5.513620	3.54	12.5316	1.881489	5.949790
3.05	9.3025	1.746425	5.522681	3.55	12.6025	1.884144	5.958188
3.06	9.3636	1.749286	5.531727	3.56	12.6736	1.886796	5.966574
3.07	9.4249	1.752142	5.540758	3.57	12.7449	1.889444	5.974948
3.08	9.4864	1.754993	5.549775	3.58	12.8164	1.892089	5.983310
3.09	9.5481	1.757840	5.558777	3.59	12.8881	1.894730	5.991661
3.10	9.6100	1.760682	5.567764	3.60	12.9600	1.897367	6.000000
3.11	9.6721	1.763519	5.576737	3.61	13.0321	1.900000	6.008328
3.12	9.7344	1.766352	5.585696	3.62	13.1044	1.902630	6.016644
3.13	9.7969	1.769181	5.594640	3.63	13.1769	1.905256	6.024948
3.14	9.8596	1.772005	5.603570	3.64	13.2496	1.907878	6.033241
3.15	9.9225	1.774824	5.612486	3.65	13.3225	1.910497	6.041523
3.16	9.9856	1.777639	5.621388	3.66	13.3956	1.913113	6.049793
3.17	10.0489	1.780449	5.630275	3.67	13.4689	1.915724	6.058052
3.18	10.1124	1.783255	5.639149	3.68	13.5424	1.918333	6.066300
3.19	10.1761	1.786057	5.648008	3.69	13.6161	1.920937	6.074537
3.20	10.2400	1.788854	5.656854	3.70	13.6900	1.923538	6.082763
3.21	10.3041	1.791647	5.665686	3.71	13.7641	1.926136	6.090977
3.22	10.3684	1.794436	5.674504	3.72	13.8384	1.928730	6.099180
3.23	10.4329	1.797220	5.683309	3.73	13.9129	1.931321	6.107373
3.24	10.4976	1.800000	5.692100	3.74	13.9876	1.933908	6.115554
3.25	10.5625	1.802776	5.700877	3.75	14.0625	1.936492	6.123724
3.26	10.6276	1.805547	5.709641	3.76	14.1376	1.939072	6.131884
3.27	10.6929	1.808314	5.718391	3.77	14.2129	1.941649	6.140033
3.28	10.7584	1.811077	5.727128	3.78	14.2884	1.944222	6.148170
3.29	10.8241	1.813836	5.735852	3.79	14.3641	1.946792	6.156298
3.30	10.8900	1.816590	5.744563	3.80	14.4400	1.949359	6.164414
3.31	10.9561	1.819341	5.753260	3.81	14.5161	1.951922	6.172520
3.32	11.0224	1.822087	5.761944	3.82	14.5924	1.954483	6.180615
3.33	11.0889	1.824829	5.770615	3.83	14.6689	1.957039	6.188699
3.34	11.1556	1.827567	5.779273	3.84	14.7456	1.959592	6.196773
3.35	11.2225	1.830301	5.787918	3.85	14.8225	1.962142	6.204837
3.36	11.2896	1.833030	5.796551	3.86	14.8996	1.964688	6.212890
3.37	11.3569	1.835756	5.805170	3.87	14.9769	1.967232	6.220932
3.38*	11.4244	1.838478	5.813777	3.88	15.0544	1.969772	6.228965
3.39	11.4921	1.841195	5.822371	3.89	15.1321	1.972308	6.236986
3.40	11.5600	1.843909	5.830952	3.90	15.2100	1.974842	6.244998
3.41	11.6281	1.846619	5.839521	3.91	15.2881	1.977372	6.252999
3.42	11.6964	1.849324	5.848077	3.92	15.3664	1.979899	6.260990
3.43	11.7649	1.852026	5.856620	3.93	15.4449	1.982423	6.268971
3.44	11.8336	1.854724	5.865151	3.94	15.5236	1.984943	6.276942
3.45	11.9025	1.857418	5.873670	3.95	15.6025	1.987461	6.284903
3.46	11.9716	1.860108	5.882176	3.96	15.6816	1.989975	6.292853
3.47	12.0409	1.862794	5.890671	3.97	15.7609	1.992486	6.300794
3.48	12.1104	1.865476	5.899152	3.98	15.8404	1.994994	6.308724
3.49	12.1801	1.868154	5.907622	3.99	15.9201	1.997498	6.316645

Table XII

SQUARES AND SQUARE ROOTS (*Continued*)

n	*n²*	$\sqrt{n}$	$\sqrt{10n}$	*n*	*n²*	$\sqrt{n}$	$\sqrt{10n}$
4.00	16.0000	2.000000	6.324555	4.50	20.2500	2.121320	6.708204
4.01	16.0801	2.002498	6.332456	4.51	20.3401	2.123676	6.715653
4.02	16.1604	2.004994	6.340347	4.52	20.4304	2.126029	6.723095
4.03	16.2409	2.007486	6.348228	4.53	20.5209	2.128380	6.730527
4.04	16.3216	2.009975	6.356099	4.54	20.6116	2.130728	6.737952
4.05	16.4025	2.012461	6.363961	4.55	20.7025	2.133073	6.745369
4.06	16.4836	2.014944	6.371813	4.56	20.7936	2.135416	6.752777
4.07	16.5649	2.017424	6.379655	4.57	20.8849	2.137756	6.760178
4.08	16.6464	2.019901	6.387488	4.58	20.9764	2.140093	6.767570
4.09	16.7281	2.022375	6.395311	4.59	21.0681	2.142429	6.774954
4.10	16.8100	2.024846	6.403124	4.60	21.1600	2.144761	6.782330
4.11	16.8921	20.27313	6.410928	4.61	21.2521	2.147091	6.789698
4.12	16.9744	2.029778	6.418723	4.62	21.3444	2.149419	6.797058
4.13	17.0569	2.032240	6.426508	4.63	21.4369	2.151743	6.804410
4.14	17.1396	2.034699	6.434283	4.64	21.5296	2.154066	6.811755
4.15	17.2225	2.037155	6.442049	4.65	21.6225	2.156386	6.819091
4.16	17.3056	2.039608	6.449806	4.66	21.7156	2.158703	6.826419
4.17	17.3889	2.042058	6.457554	4.67	21.8089	2.161018	6.833740
4.18	17.4724	2.044505	6.465292	4.68	21.9024	2.163331	6.841053
4.19	17.5561	2.046949	6.473021	4.69	21.9961	2.165641	6.848357
4.20	17.6400	2.049390	6.480741	4.70	22.0900	2.167948	6.855655
4.21	17.7241	2.051828	6.488451	4.71	22.1841	2.170253	6.862944
4.22	17.8084	2.054264	6.496153	4.72	22.2784	2.172556	6.870226
4.23	17.8929	2.056696	6.503845	4.73	22.3729	2.174856	6.877500
4.24	17.9776	2.059126	6.511528	4.74	22.4676	2.177154	6.884766
4.25	18.0625	2.061553	6.519202	4.75	22.5625	2.179449	6.892024
4.26	18.1476	2.063977	6.526868	4.76	22.6576	2.181742	6.899275
4.27	18.2329	2.066398	6.534524	4.77	22.7529	2.184033	6.906519
4.28	18.3184	2.068816	6.542171	4.78	22.8484	2.186321	6.913754
4.29	18.4041	2.071232	6.549809	4.79	22.9441	2.188607	6.920983
4.30	18.4900	2.073644	6.557439	4.80	23.0400	2.190890	6.928203
4.31	18.5761	2.076054	6.565059	4.81	23.1361	2.193171	6.935416
4.32	18.6624	2.078461	6.572671	4.82	23.2324	2.195450	6.942622
4.33	18.7489	2.080865	6.580274	4.83	23.3289	2.197726	6.949820
4.34	18.8356	2.083267	6.587868	4.84	23.4256	2.200000	6.957011
4.35	18.9225	2.085665	6.595453	4.85	23.5225	2.202272	6.964194
4.36	19.0096	2.088061	6.603030	4.86	23.6196	2.204541	6.971370
4.37	19.0969	2.090454	6.610598	4.87	23.7169	2.206808	6.978539
4.38	19.1844	2.092845	6.618157	4.88	23.8144	2.209072	6.985700
4.39	19.2721	2.095233	6.625708	4.89	23.9121	2.211334	6.992853
4.40	19.3600	2.097618	6.633250	4.90	24.0100	2.213594	7.000000
4.41	19.4481	2.100000	6.640783	4.91	24.1081	2.215852	7.007139
4.42	19.5364	2.102380	6.648308	4.92	24.2064	2.218107	7.014271
4.43	19.6249	2.104757	6.655825	4.93	24.3049	2.220360	7.021396
4.44	19.7136	2.107131	6.663332	4.94	24.4036	2.222611	7.028513
4.45	19.8025	2.109502	6.670832	4.95	24.5025	2.224860	7.035624
4.46	19.8916	2.111871	6.678323	4.96	24.6016	2.227106	7.042727
4.47	19.9809	2.114237	6.685806	4.97	24.7009	2.229350	7.049823
4.48	20.0704	2.116601	6.693280	4.98	24.8004	2.231591	7.056912
4.49	20.1601	2.118962	6.700746	4.99	24.9001	2.233831	7.063993

Table XII

SQUARES AND SQUARE ROOTS *(Continued)*

n	n^2	$\sqrt{n}$	$\sqrt{10n}$	n	n^2	$\sqrt{n}$	$\sqrt{10n}$
5.00	25.0000	2.236068	7.071068	5.50	30.2500	2.345208	7.416198
5.01	25.1001	2.238303	7.078135	5.51	30.3601	2.347339	7.422937
5.02	25.2004	2.240536	7.085196	5.52	30.4704	2.349468	7.429670
5.03	25.3009	2.242766	7.092249	5.53	30.5809	2.351595	7.436397
5.04	25.4016	2.244994	7.099296	5.54	30.6916	2.353720	7.443118
5.05	25.5025	2.247221	7.106335	5.55	30.8025	2.355844	7.449832
5 06	25.6036	2.249444	7.113368	5.56	30.9136	2.357965	7.456541
5.07	25.7049	2.251666	7.120393	5.57	31.0249	2.360085	7.463243
5.08	25.8064	2.253886	7.127412	5.58	31.1364	2.362202	7.469940
5.09	25.9081	2.256103	7.134424	5.59	31.2481	2.364318	7.476630
5.10	26.0100	2.258318	7.141428	5.60	31.3600	2.366432	7.483315
5.11	26.1121	2.260531	7.148426	5.61	31.4721	2.368544	7.489993
5.12	26.2144	2.262742	7.155418	5.62	31.5844	2.370654	7.496666
5.13	26.3169	2.264950	7.162402	5.63	31.6969	2.372762	7.503333
5.14	26.4196	2.267157	7.169379	5.64	31.8096	2.374868	7.509993
5.15	26.5225	2.269361	7.176350	5.65	31.9225	2.376973	7.516648
5.16	26.6256	2.271563	7.183314	5.66	32.0356	2.379075	7.523297
5.17	26.7289	2.273763	7.190271	5.67	32.1489	2.381176	7.529940
5.18	26.8324	2.275961	7.197222	5.68	32.2624	2.383275	7.536577
5.19	26.9361	2.278157	7.204165	5.69	32.3761	2.385372	7.543209
5.20	27.0400	2.280351	7.211103	5.70	32.4900	2.387467	7.549834
5.21	27.1441	2.282542	7.218033	5.71	32.6041	2.389561	7.556454
5.22	27.2484	2.284732	7.224957	5.72	32.7184	2.391652	7.563068
5.23	27.3529	2.286919	7.231874	5.73	32.8329	2.393742	7.569676
5.24	27.4576	2.289105	7.238784	5.74	32.9476	2.395830	7.576279
5.25	27.5625	2.291288	7.245688	5.75	33.0625	2.397916	7.582875
5.26	27.6676	2.293469	7.252586	5.76	33.1776	2.400000	7.589466
5.27	27.7729	2.295648	7.259477	5.77	33.2929	2.402082	7.596052
5.28	27.8784	2.297825	7.266361	5.78	33.4084	2.404163	7.602631
5.29	27.9841	2.300000	7.273239	5.79	33.5241	2.406242	7.609205
5.30	28.0900	2.302173	7.280110	5.80	33.6400	2.408319	7.615773
5.31	28.1961	2.304344	7.286975	5.81	33.7561	2.410394	7.622336
5.32	28.3024	2.306513	7.293833	5.82	33.8724	2.412468	7.628892
5.33	28.4089	2.308679	7.300685	5.83	33.9889	2.414539	7.635444
5.34	28.5156	2.310844	7.307530	5.84	34.1056	2.416609	7.641989
5.35	28.6225	2.313007	7.314369	5.85	34.2225	2.418677	7.648529
5.36	28.7296	2.315167	7.321202	5.86	34.3396	2.420744	7.655064
5.37	28.8369	2.317326	7.328028	5.87	34.4569	2.422808	7.661593
5.38	28.9444	2.319483	7.334848	5.88	34.5744	2.424871	7.668116
5.39	29.0521	2.321637	7.341662	5.89	34.6921	2.426932	7.674634
5.40	29.1600	2.323790	7.348469	5.90	34.8100	2.428992	7.681146
5.41	29.2681	2.325941	7.355270	5.91	34.9281	2.431049	7.687652
5.42	29.3764	2.328089	7.362065	5.92	35.0464	2.433105	7.694154
5.43	29.4849	2.330236	7.368853	5.93	35.1649	2.435159	7.700649
5.44	29.5936	2.332381	7.357636	5.94	35.2836	2.437212	7.707140
5.45	29.7025	2.334524	7.382412	5.95	35.4025	2.439262	7.713624
5.46	29.8116	2.336664	7.389181	5.96	35.5216	2.441311	7.720104
5.47	29.9209	2.338803	7.395945	5.97	35.6409	2.443358	7.726578
5.48	30.0304	2.340940	7.402702	5.98	35.7604	2.445404	7.733046
5.49	30.1401	2.343075	7.409453	5.99	35.8801	2.447448	7.739509

Table XII

SQUARES AND SQUARE ROOTS (*Continued*)

n	n^2	$\sqrt{n}$	$\sqrt{10n}$	n	n^2	$\sqrt{n}$	$\sqrt{10n}$
6.00	36.0000	2.449490	7.745967	6.50	42.2500	2.549510	8.062258
6.01	36.1201	2.451530	7.752419	6.51	42.3801	2.551470	8.068457
6.02	36.2404	2.453569	7.758866	6.52	42.5104	2.553429	8.074652
6.03	36.3609	2.455606	7.765307	6.53	42.6409	2.555386	8.080842
6.04	36.4816	2.457641	7.771744	6.54	42.7716	2.557342	8.087027
6.05	36.6025	2.459675	7.778175	6.55	42.9025	2.559297	8.093207
6.06	36.7236	2.461707	7.784600	6.56	43.0336	2.561250	8.099383
6.07	36.8449	2.463737	7.791020	6.57	43.1649	2.563201	8.105554
6.08	36.9664	2.465766	7.797435	6.58	43.2964	2.565151	8.111720
6.09	37.0881	2.467793	7.803845	6.59	43.4281	2.567100	8.117881
6.10	37.2100	2.469818	7.810250	6.60	43.5600	2.569047	8.124038
6.11	37.3321	2.471841	7.816649	6.61	43.6921	2.570992	8.130191
6.12	37.4544	2.473863	7.823043	6.62	43.8244	2.572936	8.136338
6.13	37.5769	2.475884	7.829432	6.63	43.9569	2.574879	8.142481
6.14	37.6996	2.477902	7.835815	6.64	44.0896	2.576820	8.148620
6.15	37.8225	2.479919	7.842194	6.65	44.2225	2.578759	8.154753
6.16	37.9456	2.481935	7.848567	6.66	44.3556	2.580698	8.160882
6.17	38.0689	2.483948	7.854935	6.67	44.4889	2.582634	8.167007
6.18	38.1924	2.485961	7.861298	6.68	44.6224	2.584570	8.173127
6.19	38.3161	2.487971	7.867655	6.69	44.7561	2.586503	8.179242
6.20	38.4400	2.489980	7.874008	6.70	44.8900	2.588436	8.185353
6.21	38.5641	2.491987	7.880355	6.71	45.0241	2.590367	8.191459
6.22	38.6884	2.493993	7.886698	6.72	45.1584	2.592296	8.197561
6.23	38.8129	2.495997	7.893035	6.73	45.2929	2.594224	8.203658
6.24	38.9376	2.497999	7.899367	6.74	45.4276	2.596151	8.209750
6.25	39.0625	2.500000	7.905694	6.75	45.5625	2.598076	8.215838
6.26	39.1876	2.501999	7.912016	6.76	45.6976	2.600000	8.221922
6.27	39.3129	2.503997	7.918333	6.77	45.8329	2.601922	8.228001
6.28	39.4384	2.505993	7.924645	6.78	45.9684	2.603843	8.234076
6.29	39.5641	2.507987	7.930952	6.79	46.1041	2.605763	8.240146
6.30	39.6900	2.509980	7.937254	6.80	46.2400	2.607681	8.246211
6.31	39.8161	2.511971	7.943551	6.81	46.3761	2.609598	8.242272
6.32	39.9424	2.513961	7.949843	6.82	46.5124	2.611513	8.258329
6.33	40.0689	2.515949	7.956130	6.83	46.6489	2.613427	8.264381
6.34	40.1956	2.517936	7.962412	6.84	46.7856	2.615339	8.270429
6.35	40.3225	2.519921	7.968689	6.85	46.9225	2.617250	8.276473
6.36	40.4496	2.521904	7.974961	6.86	47.0596	2 619160	8.282512
6.37	40.5769	2.523886	7.981228	6.87	47.1969	2.621068	8.288546
6.38	40.7044	2.525866	7.987490	6.88	47.3344	2.622975	8.294577
6.39	40.8321	2.527845	7.993748	6.89	47.4721	2.624881	8.300602
6.40	40.9600	2.529822	8.000000	6.90	47.6100	2.626785	8.306624
6.41	41.0881	2.531798	8.006248	6.91	47.7481	2.628688	8.312641
6.42	41.2164	2.533772	8.012490	6.92	47.8864	2.630589	8.318654
6.43	41.3449	2.535744	8.018728	6.93	48.0249	2.632489	8.324662
6.44	41.4736	2.537716	8.024961	6.94	48.1636	2.634388	8.330666
6.45	41.6025	2.539685	8.031189	6.95	48.3025	2.636285	8.336666
6.46	41.7316	2.541653	8.037413	6.96	48.4416	2.638181	8.342661
6.47	41.8609	2.543619	8.043631	6.97	48.5809	2.640076	8.348653
6.48	41.9904	2.545584	8.049845	6.98	48.7204	2.641969	8.354639
6.49	42.1201	2.547548	8.056054	6.99	48.8601	2.643861	8.360622

Table XII

SQUARES AND SQUARE ROOTS *(Continued)*

n	n^2	$\sqrt{n}$	$\sqrt{10n}$	n	n^2	$\sqrt{n}$	$\sqrt{10n}$
7.00	49.0000	2.645751	8.366600	7.50	56.2500	2.738613	8.660254
7.01	49.1401	2.647640	8.372574	7.51	56.4001	2.740438	8.660026
7.02	49.2804	2.649528	8.378544	7.52	56.5504	2.742262	8.671793
7.03	49.4209	2.651415	8.384510	7.53	56.7009	2.744085	8.677557
7.04	49.5616	2.653300	8.390471	7.54	56.8516	2.745906	8.683317
7.05	49.7025	2.655184	8.396428	7.55	57.0025	2.747726	8.689074
7.06	49.8436	2.657066	8.402381	7.56	57.1536	2.749545	8.694826
7.07	49.9849	2.658947	8.408329	7.57	57.3049	2.751363	8.700575
7.08	50.1264	2.660827	8.414274	7.58	57.4564	2.753180	8.706320
7.09	50.2681	2.662705	8.420214	7.59	57.6081	2.754995	8.712061
7.10	50.4100	2.664583	8.426150	7.60	57.7600	2.756810	8.717798
7.11	50.5521	2.666458	8.432082	7.61	57.9121	2.758623	8.723531
7.12	50.6944	2.668333	8.438009	7.62	58.0644	2.760435	8.729261
7.13	50.8369	2.670206	8.443933	7.63	58.2169	2.762245	8.734987
7.14	50.9796	2.672078	8.449852	7.64	58.3696	2.764055	8.740709
7.15	51.1225	2.673948	8.455767	7.65	58.5225	2.765863	8.746428
7.16	51.2656	2.675818	8.461678	7.66	58.6756	2.767671	8.752143
7.17	51.4089	2.677686	8.467585	7.67	58.8289	2.769476	8.757854
7.18	51.5524	2.679552	8.473488	7.68	58.9824	2.771281	8.763561
7.19	51.6961	2.681418	8.479387	7.69	59.1361	2.773085	8.769265
7.20	51.8400	2.683282	8.485281	7.70	59.2900	2.774887	8.774964
7.21	51.9841	2.685144	8.491172	7.71	59.4441	2.776689	8.780661
7.22	52.1284	2.687006	8.497058	7.72	59.5984	2.778489	8.786353
7.23	52.2729	2.688866	8.502941	7.73	59.7529	2.780288	8.792042
7.24	52.4176	2.690725	8.508819	7.74	59.9076	2.782086	8.797727
7.25	52.5625	2.692582	8.514693	7.75	60.0625	2.783882	8.803408
7.26	52.7076	2.694439	8.520563	7.76	60.2176	2.785678	8.809086
7.27	52.8529	2.696294	8.526429	7.77	60.3729	2.787472	8.814760
7.28	52.9984	2.698148	8.532292	7.78	60.5284	2.789265	8.820431
7.29	53.1441	2.700000	8.538150	7.79	60.6841	2.791057	8.826098
7.30	53.2900	2.701851	8.544004	7.80	60.8400	2.792848	8.831761
7.31	53.4361	2.703701	8.549854	7.81	60.9961	2.794638	8.837420
7.32	53.5824	2.705550	8.555700	7.82	61.1524	2.796426	8.843076
7.33	53.7289	2.707397	8.561542	7.83	61.3089	2.798214	8.848729
7.34	53.8756	2.709243	8.567380	7.84	61.4656	2.800000	8.854377
7.35	54.0225	2.711088	8.573214	7.85	61.6225	2.801785	8.860023
7.36	54.1696	2.712932	8.579044	7.86	61.7796	2.803569	8.865664
7.37	54.3169	2.714774	8.584870	7.87	61.9369	2.805352	8.871302
7.38	54.4644	2.716616	8.590693	7.88	62.0944	2.807134	8.876936
7.39	54.6121	2.718455	8.596511	7.89	62.2521	2.808914	8.882567
7.40	54.7600	2.720294	8.602325	7.90	62.4100	2.810694	8.888194
7.41	54.9081	2.722132	8.608136	7.91	62.5681	2.812472	8.893818
7.42	55.0564	2.723968	8.613942	7.92	62.7264	2.814249	8.899438
7.43	55.2049	2.725803	8.619745	7.93	62.8849	2.816026	8.905055
7.44	55.3536	2.727636	8.625543	7.94	63.0436	2.817801	8.910668
7.45	55.5025	2.729469	8.631338	7.95	63.2025	2.819574	8.916277
7.46	55.6516	2.731300	8.637129	7.96	63.3616	2.821347	8.921883
7.47	55.8009	2.733130	8.642916	7.97	63.5209	2.823119	8.927486
7.48	55.9504	2.734959	8.648699	7.98	63.6804	2.824889	8.933085
7.49	56.1001	2.736786	8.654479	7.99	63.8401	2.826659	8.938680

Table XII

SQUARES AND SQUARE ROOTS (*Continued*)

n	n^2	$\sqrt{n}$	$\sqrt{10n}$	n	n^2	$\sqrt{n}$	$\sqrt{10n}$
8.00	64.0000	2.828427	8.944272	8.50	72.2500	2.915476	9.219544
8.01	64.1601	2.830194	8.949860	8.51	72.4201	2.917190	9.224966
8.02	64.3204	2.831960	8.955445	8.52	72.5904	2.918904	9.230385
8.03	64.4809	2.833725	8.961027	8.53	72.7609	2.920616	9.235800
8.04	64.6416	2.835489	8.966605	8.54	72.9316	2.922328	9.241212
8.05	64.8025	2.837252	8.972179	8.55	73.1025	2.924038	9.246621
8.06	64.9636	2.839014	8.977750	8.56	73.2736	2.925748	9.252027
8.07	65.1249	2.840775	8.983318	8.57	73.4449	2.927456	9.257429
8.08	65.2864	2.842534	8.988882	8.58	73.6164	2.929164	9.262829
8.09	65.4481	2.844293	8.994443	8.59	73.7881	2.930870	9.268225
8.10	65.6100	2.846050	9.000000	8.60	73.9600	2.932576	9.273618
8.11	65.7721	2.847806	9.005554	8.61	74.1321	2.934280	9.279009
8.12	65.9344	2.849561	9.011104	8.62	74.3044	2.935984	9.284396
8.13	66.0969	2.851315	9.016651	8.63	74.4769	2.937686	9.289779
8.14	66.2596	2.853069	9.022195	8.64	74.6496	2.939388	9.295160
8.15	66.4225	2.854820	9.027735	8.65	74.8225	2.941088	9.300538
8.16	66.5856	2.856571	9.033272	8.66	74.9956	2.942788	9.305912
8.17	66.7489	2.858321	9.038805	8.67	75.1689	2.944486	9.311283
8.18	66.9124	2.860070	9.044335	8.68	75.3424	2.946184	9.316652
8.19	67.0761	2.861818	9.049862	8.69	75.5161	2.947881	9.322017
8.20	67.2400	2.863564	9.055385	8.70	75.6900	2.949576	9.327379
8.21	67.4041	2.865310	9.060905	8.71	75.8641	2.951271	9.332738
8.22	67.5684	2.867054	9.066422	8.72	76.0384	2.952965	9.338094
8.23	67.7329	2.868798	9.071935	8.73	76.2129	2.954657	9.343447
8.24	67.8976	2.870540	9.077445	8.74	76.3876	2.956349	9.348797
8.25	68.0625	2.872281	9.082951	8.75	76.5625	2.958040	9.354143
8.26	68.2276	2.874022	9.088454	8.76	76.7376	2.959730	9.359487
8.27	68.3929	2.875761	9.093954	8.77	76.9129	2.961419	9.364828
8.28	68.5584	2.877499	9.099451	8.78	77.0884	2.963106	9.370165
8.29	68.7241	2.879236	9.104944	8.79	77.2641	2.964793	9.375500
8.30	68.8900	2.880972	9.110434	8.80	77.4400	2.966479	9.380832
8.31	69.0561	2.882707	9.115920	8.81	77.6161	2.968164	9.386160
8.32	69.2224	2.884441	9.121403	8.82	77.7924	2.969848	9.391486
8.33	69.3889	2.886174	9.126883	8.83	77.9689	2.971532	9.396808
8.34	69.5556	2.887906	9.132360	8.84	78.1456	2.973214	9.402127
8.35	69.7225	2.889637	9.137833	8.85	78.3225	2.974895	9.407444
8.36	69.8896	2.891366	9.143304	8.86	78.4996	2.976575	9.412757
8.37	7.00569	2.893095	9.148770	8.87	78.6769	2.978255	9.418068
8.38	70.2244	2.894823	9.154234	8.88	78.8544	2.979933	9.423375
8.39	70.3921	2.896550	9.159694	8.89	79.0321	2.981610	9.428680
8.40	70.5600	2.898275	9.165151	8.90	79.2100	2.983287	9.433981
8.41	70.7281	2.900000	9.170605	8.91	79.3881	2.984962	9.439280
8.42	70.8964	2.901724	9.176056	8.92	79.5664	2.986637	9.444575
8.43	71.0649	2.903446	9.181503	8.93	79.7449	2.988311	9.449868
8.44	71.2336	2.905168	9.186947	8.94	79.9236	2.989983	9.455157
8.45	71.4025	2.906888	9.192388	8.95	80.1025	2.991655	9.460444
8.46	71.5716	2.908608	9.197826	8.96	80.2816	2.993326	9.465728
8.47	71.7409	2.910326	9.203260	8.97	80.4609	2.994996	9.471008
8.48	71.9104	2.912044	9.208692	8.98	80.6404	2.996665	9.476286
8.49	72.0801	2.913760	9.214120	8.99	80.8201	2.998333	9.481561

Table XII

SQUARES AND SQUARE ROOTS *(Continued)*

n	n^2	$\sqrt{n}$	$\sqrt{10n}$	n	n^2	$\sqrt{n}$	$\sqrt{10n}$
9.00	81.0000	3.000000	9.486833	9.50	90.2500	3.082207	9.746794
9.01	81.1801	3.001666	9.492102	9.51	90.4401	3.083829	9.751923
9.02	81.3604	3.003331	9.497368	9.52	90.6304	3.085450	9.757049
9.03	81.5409	3.004996	9.502631	9.53	90.8209	3.087070	9.762172
9.04	81.7216	3.006659	9.507891	9.54	91.0116	3.088689	9.767292
9.05	81.9025	3.008322	9.513149	9.55	91.2025	3.090307	9.772410
9.06	82.0836	3.009983	9.518403	9.56	91.3936	3.091925	9.777525
9.07	82.2649	3.011644	9.523655	9.57	91.5849	3.093542	9.782638
9.08	82.4464	3.013304	9.528903	9.58	91.7764	3.095158	9.787747
9.09	82.6281	3.014963	9.534149	9.59	91.9681	3.096773	9.792855
9.10	82.8100	3.016621	9.539392	9.60	92.1600	3.098387	9.797959
9.11	82.9921	3.018278	9.544632	9.61	92.3521	3.100000	9.803061
9.12	83.1744	3.019934	9.549869	9.62	92.5444	3.101612	9.808160
9.13	83.3569	3.021589	9.555103	9.63	92.7369	3.103224	9.813256
9.14	83.5396	3.023243	9.560335	9.64	92.9296	3.104835	9.818350
9.15	83.7225	3.024897	9.565563	9.65	93.1225	3.106445	9.823441
9.16	83.9056	3.026549	9.570789	9.66	93.3156	3.108054	9.828530
9.17	84.0889	3.028201	9.576012	9.67	93.5089	3.109662	9.833616
9.18	84.2724	3.029851	9.581232	9.68	93.7024	3.111270	9.838699
9.19	84.4561	3.031501	9.586449	9.69	93.8961	3.112876	9.843780
9.20	84.6400	3.033150	9.591663	9.70	94.0900	3.114482	9.848858
9.21	84.8241	3.034798	9.596874	9.71	94.2841	3.116087	9.853933
9.22	85.0084	3.036445	9.602083	9.72	94.4784	3.117691	9.859006
9.23	85.1929	3.038092	9.607289	9.73	94.6729	3.119295	9.864076
9.24	85.3776	3.039737	9.612492	9.74	94.8676	3.120897	9.869144
9.25	85.5625	3.041381	9.617692	9.75	95.0625	3.122499	9.874209
9.26	85.7476	3.043025	9.622889	9.76	95.2576	3.124100	9.879271
9.27	85.9329	3.044667	9.628084	9.77	95.4529	3.125700	9.884331
9.28	86.1184	3.046309	9.633276	9.78	95.6484	3.127299	9.889388
9.29	86.3041	3.047950	9.638465	9.79	95.8441	3.128898	9.894443
9.30	86.4900	3.049590	9.643651	9.80	96.0400	3.130495	9.899495
9.31	86.6761	3.051229	9.648834	9.81	96.2361	3.132092	9.904544
9.32	86.8624	3.052868	9.654015	9.82	96.4324	3.133688	9.909591
9.33	87.0489	3.054505	9.659193	9.83	96.6289	3.135283	9.914636
9.34	87.2356	3.056141	9.664368	9.84	96.8256	3.136877	9.919677
9.35	87.4225	3.057777	9.669540	9.85	97.0225	3.138471	9.924717
9.36	87.6096	3.059412	9.674709	9.86	97.2196	3.140064	9.929753
9.37	87.7969	3.061046	9.679876	9.87	97.4169	3.141656	9.934787
9.38	87.9844	3.062679	9.685040	9.88	97.6144	3.143247	9.939819
9.39	88.1721	3.064311	9.690201	9.89	97.8121	3.144837	9.944848
9.40	88.3600	3.065942	9.695360	9.90	98.0100	3.146427	9.949874
9.41	88.5481	3.067572	9.700515	9.91	98.2081	3.148015	9.954898
9.42	88.7364	3.069202	9.705668	9.92	98.4064	3.149603	9.959920
9.43	88.9249	3.070831	9.710819	9.93	98.6049	3.151190	9.964939
9.44	89.1136	3.072458	9.715966	9.94	98.8036	3.152777	9.969955
9.45	89.3025	3.074085	9.721111	9.95	99.0025	3.154362	9.974969
9.46	89.4916	3.075711	9.726253	9.96	99.2016	3.155947	9.979980
9.47	89.6809	3.077337	9.731393	9.97	99.4009	3.157531	9.984989
9.48	89.8704	3.078961	9.736529	9.98	99.6004	3.159114	9.989995
9.49	90.0601	3.080584	9.741663	9.99	99.8001	3.160696	9.994999

Answers to Odd-Numbered Exercises

In exercises involving extensive calculations, the reader may well get answers differing somewhat from those given here due to rounding at various intermediate stages.

Page 5

1. (a) Description; (b) inference; (c) inference; (d) description; (e) description; (f) inference; (g) description; (h) inference.

3. (a) The results will apply only to persons having telephones; (b) due to a defect in the machine every 5th, 10th, 12th, ..., can might be improperly sealed; (c) persons with low incomes are less likely to return the questionnaires; (d) it is very difficult to get honest answers about personal habits; (e) "Xerox" is used by many as a generic term for copying machines; (f) if there is no provision for return visits, women who hold jobs are less likely to be included in the survey.

Page 23

1. (a) 103; (b) cannot be determined; (c) cannot be determined; (d) 138; (e) cannot be determined; (f) 105.

3. (a) 405; (b) cannot be determined; (c) cannot be determined; (d) 317; (e) cannot be determined; (f) cannot be determined.

5. (a) The lower class limits are 15.5, 15.7, 15.9, 16.1, 16.3, 16.5, and the upper class limits are 15.6, 15.8, 16.0, 16.2, 16.4, 16.6; (b) the class marks are 15.55, 15.75, 15.95, 16.15, 16.35, 16.55; (c) the class boundaries are 15.45, 15.65, 15.85, 16.05, 16.25, 16.45, 16.65; (d) the class interval is 0.20.

7. The class boundaries are 105.5, 120.5, 135.5, 150.5, 165.5, 180.5, and the class limits are 106–120, 121–135, 136–150, 151–165, 166–180.

9. (a) the class frequencies are 9, 21, 30, 16, 3, 1; (b) the cumulative frequencies are 9, 30, 60, 76, 79, 80.

11. (a) the class frequencies are 1, 4, 14, 28, 10, 2, 1; (c) the cumulative frequencies are 60, 59, 55, 41, 13, 3, 1, 0.

13. (a) the class frequencies are 3, 15, 24, 12, 6; (c) the cumulative percentages are 100, 95, 70, 30, 10, 0.

21. The pictogram gives the impression that the 1970 figure is nine times the 1960 figure, since the area is nine times as large; to avoid this, make each side of the 1970 certificate $\sqrt{3} = 1.73$ times as large as the corresponding side of the 1960 certificate.

Page 41

1. (a) For instance, if we are interested only in determining the average high temperature in Atlanta, Georgia, for 1972; (b) for instance, if we want to infer something about daily high temperatures in Atlanta, Georgia, in other years.

3. (a) For instance, if he wants to predict how many A's, B's, C's, ..., his faculty will give in some other year; (b) for instance, if he wants to compare the grades given that year by the instructors of various departments.

5. $17.15.

7. (a) 11.996 ounces; (b) 11.996 ounces.

9. 0° and 360° are the same direction, yet their average is 180°; to get a meaningful average, it would be appropriate to replace angles exceeding 180 degrees by the corresponding *negative* angles, namely, replace 349 by -11 and 350 by -10. This will give a mean of 0°.

11. 16.0.

13. (a) 58.308; (b) 58.292; (c) 58.292.

15. (a) 5.54; (b) 5.57.

17. 60.0.

19. 114.3.

21. 121.6, which differs considerably from the value of 107.3 obtained in (a) of Exercise 20.

23. (a) 117; (b) 115.

25. 73.

27. $98.

29. $6,887.

31. (a) 116.9; (b) 117.5; (c) 97.3.

Page 55

1. The mean is 57, the median is 57, and the mid-range is 59.

3. 66.

5. 106.

7. The median is 94 and the mean is 114 (to the nearest per cent); the median is better because the mean is strongly affected by the one very large value.

11. (a) 0.11; (b) 0.112.

13. (a) \$923.50; (b) \$919.14.

15. 3.45, 4.17, 4.73, 5.18, 5.57, 5.95, 6.45, 6.95, 7.66.

17. (a) 23.95 and 31.20; (b) 21.95 and 33.70.

19. So far as the mode is concerned, 51, 54, 55, 61, and 69 each occur six times; the mid-range is 53.5.

21. The mode is 67 and the mid-range is 65.

23. Monkeys.

Page 60

1. (a) $x_1 + x_2 + x_3 + x_4 + x_5 + x_6 + x_7 + x_8$; (b) $y_1 + y_2 + y_3 + y_4 + y_5 + y_6$; (c) $x_1y_1 + x_2y_2 + x_3y_3 + x_4y_4$; (d) $x_1f_1 + x_2f_2 + x_3f_3 + x_4f_4$; (e) $x_2^2 + x_3^2 + x_4^2 + x_5^2 + x_6^2$; (f) $(x_1 - y_1) + (x_2 - y_2) + (x_3 - y_3) - (x_4 - y_4) + (x_5 - y_5)$.

3. (a) 7; (b) 0; (c) 35.

5. (a) 7; (b) 5; (c) -4; (d) -5; (e) 12; (f) 7.

9. (a) 19; (b) 19.

11. No.

13. 93.

Page 75

1. 18.

3. 0.26 parts per million.

5. $\dfrac{\Sigma|x - \bar{x}|}{n}$; 3.

7. (a) 30.1; (b) 30.1; (c) 30.1; (d) 22.8, so that the difference is 7.3.

9. 0.10.

11. 5.3.

13. (a) 0.23; (b) 0.23.

15. (a) 38.4; (b) 38.4; (c) 5.2.

17. 6.8.

19. 2.33.

21. At least 8/9.

23. 27.2%.

25. 0.83%.

27. 29%.

29. (a) 20.4 and 18.0%; (b) 4.35 and 7.4%; (c) 3.6 and 13.1%.

31. The chicken dinner is *relatively* most overpriced.

35. 5.1.

Page 82

1. 0.069.

3. 0.

5. (a) 3.2; (b) 2.6.

Page 92

1. (b) 6; (c) 2.

3. 9.

9. (a) 5; (b) 25; (c) 20.

11. (a) 300; (b) 120; (c) 15.

13. 36.

15. 65,536.

17. $2^6 - 1 = 63$.

Page 98

1. 4,896.

3. 13,800.

5. (a) 30,240; (b) 252.

7. (a) 720; (b) 72; (c) 6.

11. 78.

13. 5,005.

15. 700.

17. (a) 165; (b) 55.

Page 107

1. (a) 1/6; (b) 1/2; (c) 2/3.

3. 1/4, 1/2, and 1/4.

5. (a) 17/50; (b) 3/5; (c) 3/50; (d) 37/50.

7. (a) 6/11; (b) 1/22; (c) 9/22.

9. (a) 0.40; (b) 26/165.

11. 0.31.

13. (a) 1/5; (b) 4 to 1; (c) the other person would be favored.

15. (a) 2/3; (b) 2 to 1; (c) the bet is fair, that is, neither party would be favored.

17. 0.45.

19. 0.80.

21. The probability is at least 5/6 but less than 7/8.

Page 120

1. (a) The second person is undecided; (b) the first person is in favor of the legislation; (c) the second person is either for the legislation or undecided; (d) either both persons are undecided or one is in favor of the legislation while the other is against it.

3. $T = \{(1,1), (1,2), (1,3)\}$, $U = \{(1,1), (2,1), (3,1)\}$, and $V = \{(1,1), (1,2), (2,1), (2,2)\}$.

5. (a) Not mutually exclusive; (b) not mutually exclusive; (c) not mutually exclusive; (d) mutually exclusive.

7. (b) E is the event that the number of bedrooms and baths add up to 5, F is the event that there are at least 2 fewer baths than bedrooms, and G is the event that there will be 2 or 3 bedrooms; (c) F' is the event that there are as many bedrooms as baths or one more bedroom than baths, G' is the event that there will be 4 bedrooms, $E \cap F$ is the event that there will be 4 bedrooms and one bath, $F \cup G$ is the event that there will be 2 or 3 bedrooms or 4 bedrooms with 1 or 2 baths, $E \cap G$ is the event that there will be 3 bedrooms and 2 baths, and $E \cup G'$ is the event that there will be 4 bedrooms or 3 bedrooms and 2 baths; (d) not mutually exclusive, mutually exclusive, not mutually exclusive, not mutually exclusive.

9. (a) The 27 points of the sample space have the coordinates (1,1,1), (1,1,2), (1,1,3), (1,2,1), (1,2,2), (1,2,3), (1,3,1), (1,3,2), (1,3,3), (2,1,1), (2,1,2), (2,1,3), (2,2,1), (2,2,2), (2,2,3), (2,3,1), (2,3,2), (2,3,3), (3,1,1), (3,1,2), (3,1,3), (3,2,1), (3,2,2), (3,2,3), (3,3,1), (3,3,2), and (3,3,3); (b) J is the event that the first two persons favor the legislation, K is the event that all three persons respond the same way, L is the event that the second person is undecided while the first and third are not undecided, and M is the event that all three respond differently; (c) $J \cap K$ is the event that all three are for the legislation, $K \cup M$ is the event that either all three respond in the same way or all three respond differently, $L \cap M$ is the event that the second person is undecided while one of the other two is for the legislation and the other is against it, and $J \cap M'$ is the event that the first two persons favor the legislation; (d) not mutually exclusive, mutually exclusive, mutually exclusive, and not mutually exclusive.

11. (a) Visit Rome, have a good time, and run out of cash; (b) not visit Rome, have a good time, and not run out of cash; (c) visit Rome and have a good time; (d) have a good time but not visit Rome; (e) have a good time or run out of cash, but not visit Rome; (f) visit Rome.

Page 131

1. (a) The probability that one of these delinquents has not dropped out of school; (b) the probability that one of these delinquents' parents are not on welfare; (c) the probability that one of these delinquents' parents are on welfare and he has dropped out of school; (d) the probability that either one of these delinquents' parents are on welfare or he has not dropped out of school; (e) the probability that one of these delinquents' parents are not on welfare and he has not dropped out of school; (f) the probability that either one of these delinquents' parents are not on welfare or he has dropped out of school.

3. The claim of the first college president is impossible because the sum of the probabilities is less than 1; the claim of the second college president is possible; the claim of the third college president is impossible because the sum of the probabilities exceeds 1.

5. (a) 0.79; (b) 0.67; (c) 0.54; (d) 0.21; (e) 0.33; (f) 1.

7. The corresponding probabilities are $3/5$, $1/5$, and $1/10$, and since their sum is not 1, they are inconsistent.

9. The corresponding probabilities are $2/3$ and $3/4$, and since their sum exceeds 1, they cannot be right.

11. (a) 0.172; (b) 0.971; (c) 0.442.

13. (a) $3/10$, $1/5$, $3/10$, and $1/5$; (b) $1/2$, $1/20$, $1/2$, and $1/20$.

15. (a) 0.24, 0.35, 0.12, and 0.24; (b) 0.38, 0.33, 0.17, 0.08, and 0.04; (c) 0.38, 0.33, 0.17, 0.08, and 0.04; (d) 0.14, 0.24, 0.20, 0.18, and 0.24.

17. (a) 0.34; (b) 0.66.

19. (a) 0.37; (b) 0.28; (c) 0.63.

Page 144

1. (a) The probability that a student who is a good athlete will get high grades; (b) the probability that a student who gets high grades is also a good athlete; (c) the probability that a student who is not a good athlete will get high grades; (d) the probability that a student who is a good athlete will not get high grades; (e) the probability that a student who is not a good athlete will not get high grades; (f) the probability that a student who does not get high grades is also not a good athlete.

3. (a) The probability that one of the cars with bucket seats will also have air conditioning; (b) the probability that one of the cars without power steering will have air conditioning; (c) the probability that one of the cars without air conditioning will have bucket seats; (d) the probability that one of the cars with air conditioning will also have bucket seats as well as power steering; (e) the probability that one of the cars without bucket seats will not have power steering; (f) the probability that one of the cars with air conditioning and power steering will also have bucket seats.

5. (a) 43/60; (b) 17/60; (c) 2/3; (d) 1/3; (e) 8/15; (f) 9/60; (g) 4/5; (h) 32/43; (i) 9/20; (j) 9/17.

7. (a) 3/4; (b) 1/4; (c) 4/5; (d) 1/5; (e) 16/25; (f) 9/100; (g) 4/5; (h) 64/75; (i) 9/20; (j) 9/25.

9. (a) 15/26; (b) 15/19.

11. (a) 9/16; (b) 1/4.

13. Events Q and R are not independent.

15. (a) 1/256; (b) 1/256; (c) 1/81; (d) 1/64; (e) 0.59049.

17. 10/13.

19. (a) 0.29; (b) 0.39.

21. 5/19.

23. 3/4.

Page 154

1. 20 cents.

3. (a) $19.50; (b) $2.50.

5. 60 cents.

7. $5\frac{13}{16}$ games.

9. $430.

11. 1.47.

13. The probability is less than 0.40.

15. The probability is greater than 0.75.

17. (a) The probability of the client winning his case is less than $1/3$; (b) the probability of the client winning is greater than $1/3$; (c) the probability of the client winning equals $1/3$.

19. The utility is at least 4.

Page 160

1. (a) the expected gains are $-\$420,000$ and $-\$54,000$, and it would be preferable to discontinue the project; (b) the expected gains are $-\$300,000$ and $-\$71,250$, and it would be preferable to discontinue the project; (c) the expected gains are $-\$100,000$ and $-\$100,000$, so that it does not matter whether they continue or discontinue the project.

3. (b) The expected profits are $\$175,000$ and $\$200,000$, and it would be preferable to use the college gymnasium; (c) the expected profits are $\$260,000$ and $\$230,000$, and it would be preferable to build the new arena; (d) he would vote against it; (e) he would vote for it.

5. (b) The expected expenses are $\$27.20$ and $\$25.60$, and he should make his reservation at Hotel II; (c) the expected expenses are $\$26.20$ and $\$26.60$, and he should make his reservation at Hotel I; (d) the expected expenses are $\$26.40$ and $\$26.40$, and it does not matter where he makes his reservation; (e) Hotel II; (f) Hotel II.

7. He should continue the project.

9. He should make his reservation at Hotel II.

11. (a) $\$585,000$; (b) $-\$442,500$.

13. In that case, the worst that can happen is less.

15. Yes.

17. (a) The first job; (b) the second job.

Page 168

1. To minimize the expected losses he should stock none of the items. Of course, he will then have no chance of making a profit.

3. 19, which is the mid-range.

5. (a) 18; (b) 19; (c) 20; (d) 21.

Page 182

3. (a) No, the sum of the probabilities is less than 1; (b) yes; (c) yes; (d) no, the sum of the probabilities is less than 1.

5. (a) $\dfrac{1,792}{6,561}$; (b) $\dfrac{256}{6,561}$; (c) $\dfrac{129}{6,561}$.

7. (a) 0.2936 (approximately); (b) 0.294.

9. (a) 0.282; (b) 0.889; (c) 0.111.

11. (a) 0.207; (b) 0.280; (c) 0.515; (d) 0.811.

15. 25/66.

17. (a) 5/28; (b) 1/56.

19. (a) 0.1167; (b) 0.1148.

21. 0.26.

23. 0.18.

25. (a) 0.819; (b) 0.164; (c) 0.999.

27. (a) 0.741; (b) 0.222; (c) 0.033; (d) 0.003.

Page 192

1. $\mu = 1.00$ and $\sigma^2 = 0.74$.

3. (a) 0.74; (b) 1.34.

5. (a) 2/3; (b) 2/3; (c) 0.745; (d) 0.745.

9. (b) $\mu = 2.5$ and $\sigma^2 = 0.918$; (c) $\mu = 2.5$.

11. $\mu = 0.444$ and $\sigma^2 = 0.326$; $\mu = 4/9$.

13. (a) $\mu = 3.997$ and $\sigma^2 = 4.003$; these values are very close to $\mu = 4$ and $\sigma^2 = 4$.

15. (a) The probability that anywhere between 137 and 171 marriage licenses will be issued is at least 0.75; (b) the probability that at most 128 or at least 180 marriage licenses will be issued is at most 1/9.

Page 206

1. (a) 1/3; (b) 4/15; (c) 7/30; (d) 3/5.

3. (a) 0.330, 0.549, and 0.148; (b) 0.393 and 0.085; (c) 0.082.

5. (a) 2.37; (b) 1.23; (c) −0.31; (d) −2.02; (e) 0.88; (f) 1.81.

7. (a) 1.28; (b) 1.64; (c) 1.96; (d) 2.05; (e) 2.33; (f) 2.58.

9. 15.48.

13. The shape of the distribution is quite close to that of a normal distribution.

Page 216

1. (a) 0.3707; (b) 0.0013; (c) 0.5705; (d) 0.0618.

3. (a) 27.76%; (b) 37.45%; (c) 34.94%; (d) 17.01 ounces; (e) 20.77 ounces.

5. 0.0594 and 0.0116.

7. (a) 10.4%; (b) 13.8%; (c) 85; (d) 56.

9. (a) 0.0985; (b) 0.1335; (c) 0.2981; (d) 0.2389.

11. (a) 0.1290; (b) 0.0637; (c) 0.1587; (d) 0.3707.

13. 0.0823.

15. 0.1587.

Page 223

1. (a) 15; (b) 66; (c) 435.

3. (a) 1/210; (b) 1/12,650.

5. (a) 3/5; (b) 3/10.

9. 44, 326, 558, 353, 577, 305, 24, 189, 285, 442, 569, and 555.

13. Each subset of a given size does not have the same chance of being selected.

Page 235

1. (b) 4 and 5, 4 and 6, 4 and 7, 4 and 8, 4 and 9, 5 and 6, 5 and 7, 5 and 8, 5 and 9, 6 and 7, 6 and 8, 6 and 9, 7 and 8, 7 and 9, 8 and 9; (c) the values 4.5, 5.0, 8.0, and 8.5 have probabilities of 1/15, the values 5.5, 6.0, 7.0, and 7.5 have probabilities of 2/15, and 6.5 has the probability 3/15; (d) $\mu = 6.5$ and $\sigma = 1.08$.

3. (a) 4 and 4, 4 and 5, 4 and 6, 4 and 7, 4 and 8, 4 and 9, 5 and 4, 5 and 5, 5 and 6, 5 and 7, 5 and 8, 5 and 9, 6 and 4, 6 and 5, 6 and 6, 6 and 7, 6 and 8, 6 and 9, 7 and 4, 7 and 5, 7 and 6, 7 and 7, 7 and 8, 7 and 9, 8 and 4, 8 and 5, 8 and 6, 8 and 7, 8 and 8, 8 and 9, 9 and 4, 9 and 5, 9 and 6, 9 and 7, 9 and 8, 9 and 9; (b) the values 4 and 9 have probabilities of 1/36, the values 4.5 and 8.5 have probabilities of 2/36, the values 5 and 8 have probabilities of 3/36, the values 5.5 and 7.5 have probabilities of 4/36, the values 6 and 7 have probabilities of 5/36, and 6.5 has the probability 6/36; (c) $\mu = 6.5$ and $\sigma = 1.21$.

5. The frequencies corresponding to 11, 12, ..., and 21 are, respectively, 1, 1, 4, 4, 12, 12, 5, 8, 1, 1, and 1; the standard deviation is 1.96.

7. (a) 3.64 and 1.41; (b) 1.27.

9. (a) 0.990; (b) 0.988; (c) 0.990.

11. 2.13; there is a fifty-fifty chance that the mean of a random sample of size 60 (from the given population) will differ from the population mean by less than 2.13.

13. (a) The probability is at least 0.84; (b) 0.9876.

Page 241

7. The numbers assigned to 0, 1, 2, 3, 4, and 5 are, respectively, 000–009, 010–086, 087–316, 317–662, 663–921, 922–999.

Page 258

1. We can say with a probability of 0.95 that the error is less than 0.76 minutes.

3. $41.4 < \mu < 44.2$.

5. (a) $\$417.80 < \mu < \$447.32$; (b) we can assert with a probability of 0.99 that the error is less than $19.42.

7. 0.92.

9. $n = 711$.

11. $n = 87$.

13. We can assert with a probability of 0.95 that the error is less than 2.06.

15. 0.96, compared to the theoretical value of 0.95.

17. $1.43 < \mu < 2.77$.

19. (a) $231.48 < \mu < 233.06$; (b) we can assert with a probability of 0.98 that the error is less than 1.07 degrees.

21. (a) $4.89 < \mu < 8.11$; (b) we can assert with a probability of 0.99 that the error is less than 1.85.

23. 70.2.

Page 273

1. Type II; Type I.

3. Type I; Type II.

5. (a) The hypothesis that the technique is not effective; (b) the hypothesis that the technique is effective.

7. (a) 0.026; (b) 0.008, 0.068, 0.288, 0.644, 0.903, 0.903, 0.644, 0.288, 0.068, 0.008.

11. (a) $\mu_2 > \mu_1$, and switch to the radial tire only if the hypothesis $\mu_1 = \mu_2$ can be rejected; (b) $\mu_2 < \mu_1$, and switch to the radial tires unless the hypothesis $\mu_1 = \mu_2$ can be rejected; (c) $\mu_2 \neq \mu_1$.

Page 283

1. $z = -1.97$, and the hypothesis $\mu = 19.4$ cannot be rejected against the alternative $\mu < 19.4$.

3. $z = 3.04$, and we can conclude that the course is effective.

5. $z = 2.85$, the hypothesis $\mu = 3.0$ must be rejected, and the process will have to be adjusted.

7. (a) $z = 1.80$, reject the hypothesis; (b) $z = 1.80$, cannot reject the hypothesis.

9. $t = 3.10$, the hypothesis can be rejected against the alternative that the weight loss exceeds 10.5 pounds.

11. $t = -3.85$, and we conclude that the new clubs have improved his game.

13. $t = -3.54$, so that the underfill is significant.

15. $z = 1.44$, and we cannot conclude that the claim is justified.

17. $z = 1.22$, and the null hypothesis $\mu_2 - \mu_1 = 0.5$ cannot be rejected against the alternative $\mu_2 - \mu_1 > 0.5$.

19. $t = 1.32$, which is not significant, and hence does not support the claim.

21. $t = 3.66$, so that the difference is significant.

23. $t = 4.06$, which is significant and supports the claim that the program of exercise is effective.

Page 291

1. $0.36 < \sigma < 1.10$.

3. $1.84 < \sigma < 4.42$.

5. $2.01 < \sigma < 3.14$.

7. $12.7 < \sigma < 17.8$.

9. (a) 16.7; (b) 7.3.

Page 297

1. $\chi^2 = 14.7$, and the null hypothesis cannot be rejected.

3. $\chi^2 = 10.5$, and the null hypothesis cannot be rejected.

5. (a) $z = -1.98$, and the null hypothesis $\sigma = 2.0$ will have to be rejected; (b) $z = 2.25$, and the null hypothesis $\sigma = 2,400$ cannot be rejected.

7. $F = 2.09$, and the null hypothesis cannot be rejected.

9. $F = 1.07$, and the null hypothesis cannot be rejected.

Page 308

1. 0.41–0.51.

3. 0.60–0.72.

5. (a) 0.16–0.32; (b) 0.153–0.307.

7. (a) 0.30–0.46; (b) 0.30–0.46.

9. (a) 0.176–0.264; (b) we can assert with a probability of 0.98 that the error is less than 0.039.

11. (a) 0.163–0.315; (b) 0.101–0.244; (c) 0.771–0.86υ.

13. $n = 385$.

15. (a) $n = 267$; (b) $n = 201$.

17. (a) 3/4 and 1/4; (b) 0.40 and 0.60; $\mu = 0.72$.

19. (a) 1/3, 1/3, and 1/3; (b) 0.63, 0.31, and 0.06.

Page 316

1. $x = 4$ or more; $\alpha = 0.012$.

3. $x = 6$ or less; $\alpha = 0.007$.

5. $z = -1.19$, and the claim cannot be rejected.

7. $z = -1.90$, so that the claim can be rejected.

9. $z = -2.66$, which makes the data inconsistent with the claim.

11. $z = 2.20$; (a) reject the hypothesis; (b) cannot reject the hypothesis.

Page 323

1. $z = 1.87$, and the difference is not significant.

3. $z = -1.20$, and the difference is not significant.

5. $z = 2.31$, and he can conclude that the second machine is better.

7. $\chi^2 = 3.5$, and $(1.87)^2 = 3.50$.

9. $\chi^2 = 4.2$, which is not significant.

11. $\chi^2 = 19.65$, so that there are significant differences.

Page 331

1. $\chi^2 = 11.8$, so that there are significant differences in attitude toward the legislation.

3. $\chi^2 = 1.3$, and the differences in quality are not significant.

5. $\chi^2 = 0.2$, and there is no significant relationship between handicap and performance.

7. (a) 0.204; (b) 0.175.

9. (a) 15, 45, 45, and 15; (b) $\chi^2 = 1.78$, and there is a good fit.

11. (a) 48, 126, 156, 126, 78, 42, 18, and 6; (b) $\chi^2 = 7.79$, and the given Poisson distribution provides a good fit.

13. (a) 0.0078, 0.0306, 0.0930, 0.1878, 0.2522, 0.2253, 0.1339, 0.0528, and 0.0166; (b) 1, 5, 14, 28, 38, 34, 20, 8, and 2; (c) $\chi^2 = 3.6$, and there is a good fit.

Page 345

3. (a) $F = 3.08$, which is not significant; (b) the analysis of variance table is

Source of variation	Degrees of freedom	Sum of squares	Mean square	F
Treatments	4	64	16	3.08
Error	15	78	78/15	
Total	19	142		

5. $F = 0.4$, which is not significant.

7. $F = 0.85$, which is not significant.

9. $F = 3.19$, which is not significant.

Page 356

1. For temperatures $F = 0.6$, which is not significant, and for detergents $F = 5.75$, which is also not significant.

3. For packaging $F = 0.92$, which is not significant, and for markets $F = 1.40$, which is also not significant.

5. For threads $F = 8.3$, which is significant, and for instruments $F = 0.06$, which is not significant.

7. For temperatures $F = 1.3$, which is not significant, for detergents $F = 16.1$, which is significant, and for interactions $F = 2.7$, which is not significant.

9. For mixes $F = 1.63$, which is not significant, for recipes $F = 0.09$, which is not significant, and for interactions $F = 2.67$, which is not significant.

Page 370

1. $z = 2.67$ (2.41 with continuity correction), so that the null hypothesis can be rejected.

3. Reject the null hypothesis.

5. (a) Reject the null hypothesis; (b) $z = 2.32$ (2.06 with continuity correction), so that the null hypothesis can be rejected.

7. $z = 2.06$ (1.83 with continuity correction), and the null hypothesis cannot be rejected.

9. The program is effective.

11. $z = 3.4$, so that the null hypothesis can be rejected.

17. (b) $\chi^2 = 6.86$, so that the null hypothesis can be rejected.

19. $\chi^2 = 3.86$, and the null hypothesis cannot be rejected.

Page 378

1. $z = 0.98$, and the null hypothesis cannot be rejected.

3. $z = 0.15$, and the null hypothesis cannot be rejected.

5. $z = -1.34$, and the null hypothesis cannot be rejected.

7. $z = 0.098$, and the null hypothesis cannot be rejected.

9. $H = 1.5$, and the null hypothesis cannot be rejected.

11. $H = 12.2$, and the null hypothesis can be rejected.

Page 383

5. $z = 0.24$, and the hypothesis of randomness cannot be rejected.

7. $z = -0.51$, and the hypothesis of randomness cannot be rejected.

9. $z = -3.63$; there is a significant trend.

Page 395

1. (a) $a = 28.5$ and $b = 10.9$; (b) $a = 28.5$ and $b = 10.9$; (d) 72.1.

3. (a) $y = -1.77 + 4.42x$; (b) 38.0.

5. $y = -1.05 + 1.68x$; 14.1.

7. 11.8 inches.

9. $y = 4.83 + 0.84x$; 14.1.

11. 433, 460, 508, 553, 577, 581, 591, 593, 556, 544, 544, 511, 482, 473, 463, 456, 462, 452, 456, 446, 427, 413, 422, 437, 457, 483, 509, 526, and 540.

Page 408

1. $y = (0.88)(1.50)^x$; 32.9.

3. $y = 1,420(1.29)^x$; 5,000 thousands.

5. $y = 6,170 \cdot x^{-1.5}$; 49.4 psi.

7. (a) $y = 1,354 + 397.6x + 64.86x^2$; 4,964 thousands; (b) $y = 5,487 + 1,137x + 346.5x^2$; 43,800 thousands, which is a questionable extrapolation.

9. $y = -33.1 + 3.92x_1 + 0.585x_2$; 75.4.

Page 416

1. $s_e = 5.4$.

3. $17.36 < \alpha < 43.30$.

5. (a) The hypothesis that *on the average* a person who has not studied French will get a grade of 30 on the test; $t = -0.3$, and the hypothesis cannot be rejected; (b) the hypothesis that *on the average* when no money is spent on advertising, the net operating profit will be 0.5 per cent of total sales; $t = 2.11$, and the hypothesis can be rejected.

7. (a) $12.8 < \alpha < 44.2$; (b) $0.38 < \alpha < 1.08$.

9. (a) $57.4 - 86.8$; (b) $48.9 - 80.2$; (c) $4.49 - 6.85$.

Page 430

1. (a) 0.98; (b) 0.98; $t = 15.5$, so that the null hypothesis of no correlation can be rejected.

3. $r = 0.995$; significant.

5. $r = 0.91$; significant.

7. $r = 0.98$; significant.

9. (a) $t = 1.82$, which is not significant; (b) $z = 1.68$, which is not significant.

11. $z = 1.23$, and the null hypothesis cannot be rejected.

13. $z = 0.63$, and the null hypothesis cannot be rejected.

15. (a) $0.39 < \rho < 0.91$; (b) $-0.77 < \rho < -0.28$; (c) $0.21 < \rho < 0.55$; (d) $-0.25 < \rho < 0.52$.

17. (a) 0.197; (b) 0.139; (c) 0.088; (d) 0.062.

Page 437

1. $r' = -1$; significant.

3. $r' = 0.98$; significant.

5. $r' = 0.64$.

7. $r' = 0.88$.

9. 0.999.

11. (a) 0.98; (b) 0.99.

Page 444

1. $r = 0.58$; significant.

3. $r = 0.76$; significant.

Page 454

1. (a) 7, 17, 7, 26, and 11; 16, 10, 18, 5, and 114; 6, 23, 81, 7, and 13; 21, 8, 9, 9, and 32; (b) the means of the four samples are 13.6, 32.6, 26, and 15.8, and the over-all mean is 22.

3. $\sigma = 8.6$.

5. (a) $n_1 = 10$, $n_2 = 24$, $n_3 = 4$ and $n_4 = 2$; (b) $n_1 = 80$, $n_2 = 60$, $n_3 = 20$, $n_4 = 32$, and $n_5 = 8$.

9. (a) $n_1 = 25$ and $n_2 = 75$; (b) $n_1 = 20$ and $n_2 = 80$.

Page 466

1. 180.

3. $A_L B_L C_L D_L$, $A_L B_L C_L D_H$, $A_L B_L C_H D_L$, $A_L B_L C_H D_H$, $A_L B_H C_L D_L$, $A_L B_H C_L D_H$, $A_L B_H C_H D_L$, $A_L B_H C_H D_H$,. $A_H B_L C_L D_L$, $A_H B_L C_L D_H$, $A_H B_L C_H D_L$, $A_H B_L C_H D_H$, $A_H B_H C_L D_L$, $A_H B_H C_L D_H$, $A_H B_H C_H D_L$, $A_H B_H C_H D_H$.

5.

A	D	B	C
C	B	D	A
D	A	C	B
B	C	A	D

7. (a) $F = 2.65$ which is not significant; (b) $F = 27.8$, which is significant; (c) $F = 97.0$, which is significant.

9. B, C, and D on March 2, C, G, and H on March 5, D, E, and I on March 9, C, F, and I on March 11, B, G, and I on March 13, and C, E, and A on March 16.

11. 4, 3, and 7, on Tuesday; 2, 5, and 7 on Thursday; 5, 3, and 6 on Saturday.

Index

Absolute value, 66, 75
Addition rule, 127, 130
Aggregative index, 44, 47
α (alpha), probability of Type I error, 272
α (alpha), regression coefficient, 412
α_4, alpha-four, 83
α_3, alpha-three, 82
Alternative hypothesis, 271, 276
  one-sided and two-sided, 271
Analysis of variance:
  Latin square, 460
  one-way, 341
    unequal sample sizes, 344
  table, 342, 350, 353, 360, 461
  two-way with interaction, 353
  two-way without interaction, 349, 350
  with replication, 359
ANOVA (*see* Analysis of variance)
Area sampling, 453
Arithmetic mean (*see also* Mean), 33
Arithmetic probability paper, 204
Average (*see* Mean)
Axioms, 124

Balanced design, 448
Balanced incomplete block design, 464
Bar chart, 17
Base year, 43
Bayes' rule, 141, 142
Bayesian analysis, 159
Bayesian estimate, 256, 306
Bayesian inference, 249
Bayesian statistician, 257
Bayesian statistics, 258
Bell-shaped distribution, 79
β (beta), probability of Type II error, 273
β (beta), regression coefficient, 412

Bias in sampling, 5
Binomial coefficients, 97, 101
  table, 488
Binomial distribution, 174
  and hypergeometric distribution, 180
  and normal distribution, 213
  and Poisson distribution, 180
  mean, 188
  standard deviation, 190
  table, 483
    use of, 175, 218
  variance, 190
Block, 348
Block effect, 349
Block sum of squares:
  Latin square, 461
  two-way analysis with interaction, 353
  two-way analysis without interaction, 350
Boundary, class, 14

Categorical distribution, 10, 16
Causation and correlation, 426
Central limit theorem, 232, 233
Certainty, 124
Chance variation, 64, 225
Change of scale (*see* Coding)
Chebyshev's theorem, 72, 77, 191, 232
Chi-square distribution, 289, 322, 328
  degrees of freedom, 289, 322, 328, 331
  table, 477
Chi-square statistic, 289, 294, 322, 328, 330
Class:
  boundary, 14
  frequency, 13
  interval, 11, 15

Class: *cont.*
  limit, 13
  mark, 14
  modal, 53
  open, 12, 39
  "real" limits, 14
  unequal intervals, 18, 39
Classical probability concept, 112
Cluster sampling, 453
Coding, 38, 61, 71, 398, 410
Coefficient of correlation (*see* Correlation
    coefficient)
Coefficient of determination, 422
Coefficient of quartile variation, 75, 78
Coefficient of rank correlation, 433
Coefficient of skewness, Pearsonian, 81
Coefficient of variation, 74
Coefficients, binomial, 97, 101
  table, 488
Coefficients, regression (*see* Regression
    coefficients)
Collateral information, 258
Column sum of squares, 461
Combinations, 97
Complement, 118
  probability of, 126
Completely balanced design, 458
Condition probability, 134, 136
Confidence, degree of, 253
Confidence interval:
  correlation coefficient, 429
  mean, 253, 255
  mean of y for given x, 417
  proportion, 300, 301
    table, 481, 482
  regression coefficients, 415
  standard deviation, 290, 291
Confidence limits (*see* Confidence
    interval)
Consistency criterion, 126, 132
Contingency coefficient, 333
Contingency table, 326
  chi-square statistic, 328
  degrees of freedom, 328
  expected cell frequencies, 326
  hypothesis of independence, 326
Continuity correction, 212, 214, 314,
    364
Continuous distribution (*see also* Prob-
    ability density), 196, 197
  mean and standard deviation, 198
Continuous random variable, 196
Correction, continuity, 212, 214, 314, 364
Correction factor, finite population, 231,
    237, 260, 310
Correlation:
  and causation, 426
  linear, 426
  multiple, 435
  negative, 424
  partial, 436
  positive, 424
  rank, 433
Correlation analysis, normal, 427

Correlation coefficient:
  abuse of, 426
  and covariance, 422
  computing formula, 422
  confidence interval, 429
  definition, 422
  grouped data, 443
  interpretation, 425
  multiple, 435
  partial, 436
  population, 427
  rank, 433
  significance test, 427, 433, 434
Count data, 299, 325
Countably infinite sample space, 180
Covariance, 422
Cross-stratification, 453
Cumulative frequency distribution, 15
Curve fitting:
  exponential, 400
  linear, 389
  parabola, 403
  polynomial, 400
  power function, 403

Data:
  collateral, 258
  count, 299, 325
  paired, 392
  raw, 10
Deciles, 52
Decision making, 54
  based on optimism, 159
  based on pessimism, 159
  Bayesian, 159
  minimax, 159
Decision theory, 247
Degree of confidence, 253
Degrees of freedom (*see also* individual
    tests):
  chi-square distribution, 289
  F distribution, 295
  t distribution, 254, 279
Density, probability (*see also* Probability
    density), 196
Dependent events, 138
Descriptive statistics, 3
Design, sample, 448
Design of survey, 4
Design of experiment, 4, 457
  completely balanced, 458
  balanced incomplete blocks, 464
Determinants, 394
Deviation from mean, 66
Difference between means:
  large-sample test, 280
  sampling distribution, 280
  small-sample test, 281
  standard error, 280
Difference between proportions:
  significance test, 319
  standard error, 318
Difference between standard deviations,
    294

Discrete random variable, 195
Distribution (*see also* Frequency distri-
        bution, Probability function,
        Probability density, *and* Sampling
        distribution):
    bell-shaped, 79
    continuous, 196
    posterior, 307
    prior, 257, 306
    probability, 172
    symmetrical, 79
Distribution-free methods, 362
Double subscripts, 59, 341
Double summations, 59, 341

Effects:
    block, 349
    treatment, 338
Empty set, 118
    probability of, 126
Equiprobable events, special rule for,
        102, 129
Error:
    experimental, 341
    probable, 237
    standard, 230
    Type I and Type II, 266
Error sum of squares:
    Latin square, 461
    one-way analysis, 341
        unequal sample sizes, 344
    two-way analysis, 350, 353
        with replication, 359
Estimate:
    Bayesian, 257
    interval, 254
    point, 249
Estimation of mean:
    Bayesian, 257
    confidence interval, 253, 255
    finite population, 260
    maximum error, 251, 256
    sample size, 252
Estimation of proportion:
    Bayesian, 306
    confidence interval, 300, 303
    finite population, 310
    maximum error, 304
    pooled, 319
    sample size, 305
Events, 116
    dependent, 138
    independent, 138
    mutually exclusive, 117
Expectation, mathematical, 151, 152
Expected frequencies, 321
Expected profit, 153
Expected value of perfect information,
        163
Experiment, 113
    outcomes of, 113
Experimental error, 341, 459
Experimental sampling distribution, 225
Experiments, design of, 4, 457

Exponential curve, 400
    fitting of, 401
Exponential distribution, 207
Exponential function, table, 495
Extrapolation, 397
Extreme value, 65

F distribution, 294, 340
    degrees of freedom, 295, 340
    table, 479, 480
F statistic, 296, 340, 343
Factorial notation, 96
Finite population, 219
    correction factor, 231, 237, 260, 310
    random sample from, 219
Finite sample space, 116, 124
Fisher Z-transformation, 428
    table, 487
Fractiles, 51
    grouped data, 52
    ungrouped data, 58
Frequency:
    class, 13
    expected, 321
    observed, 321
    relative, 299
Frequency distribution, 10
    bell-shaped, 79
    categorical, 10, 16
    class boundaries, 14
    class frequency, 13
    class interval, 11, 15
    class limits, 13
    class mark, 14
    classes, number of, 11
    cumulative, 15
    deciles, 52
    histogram, 17
    J-shaped, 81
    mean, 37
    median, 51
    numerical, 10
    ogive, 22
    open classes, 12, 39
    peakedness, 83
    percentage, 16
    percentiles, 52
    qualitative, 10, 16
    quantitative, 10
    quartiles, 52
    "real" class limits, 14
    skewness, 79
    standard deviation, 71
    symmetrical, 79
    two-way, 441
    unequal classes, 18, 39
    U-shaped, 81
Frequency interpretation of probability,
        103, 112
Frequency polygon, 19

General addition rule, 130
General multiplication rule, 138

Generalized addition rule, 127
Geometric distribution, 184
Geometric mean, 33, 45
    weighted, 54
Given year, 43
Goodness of fit, 329
    least-squares line, 420
    normal distribution, 336
Grand mean, 338, 341, 349
Graphical presentation:
    bar chart, 17
    frequency polygon, 19
    histogram, 19
        three-dimensional, 441
    ogive, 22
    pictogram, 29
    pie chart, 28
Grouping error, 72

H test, 376
Harmonic mean, 33, 46
Histogram, 17
    three-dimensional, 441
    with unequal classes, 18
Hypergeometric distribution, 178, 179
    binomial approximation, 180
    mean, 189
Hypothesis:
    alternative, 271
        one-sided, 276
        two-sided, 276
    null, 269
Hypothesis testing (see Tests of
    hypotheses)

Incomplete block design, 461
Independent events, 138
Independent samples, 280
Index numbers, 43
    aggregative, 44, 47
    base year, 43
    given year, 43
    mean of price relatives, 44
    quantity, 44
    units test, 45
Inductive statistics, 4
Inference, statistical, 4
Infinite population, 219
    random sample from, 223
Infinite sample space, 116
Interaction, 349
    effects, 354
    sum of squares, 353
Interquartile range, 78
Intersection, 118
Interval, class, 11, 15
Interval estimate, 254

J-shaped distribution, 81
Judgment sample, 454

Kruskal-Wallis test, 377
Kurtosis, 83

Large numbers, law of, 104, 241, 242
Large-sample confidence interval:
    mean, 253
    proportion, 303
    standard deviation, 291
Large-sample test:
    difference between means, 281
    difference between proportions, 319
    mean, 278
    proportion, 314
    standard deviation, 297
Latin square, 460
    analysis of, 461, 462
Law of large numbers, 104, 241, 242
Least squares, method of, 391
Least squares line, 392
Level of significance, 272
Limits, class, 13
Limits of prediction, 418
Linear equation:
    k unknowns, 406
    two unknowns, 387
Linear regression analysis, 413
Location, measures of, 31
Logarithms, table, 493
Log-log graph paper, 402
Lower class limit, 13

Mann-Whitney test, 373, 375
Mark, class, 14
Mathematical expectation, 151, 152
Maximum error:
    estimation of means, 251, 256
    estimation of proportions, 304
Mean:
    and median, 49
    arithmetic, 33
    binomial distribution, 188
    combined data, 41
    confidence interval for, 253, 255
    definition, 34
    deviation from, 66
    distribution of random variable, 187
    geometric, 33, 45
    grand, 338, 341, 349
    grouped data, 37
    hypergeometric distribution, 189
    in decision making, 167
    Poisson distribution, 189, 194
    population, 34
    probability density, 198
    probability distribution, 187
    probable error of, 237
    properties, 35
    reliability of, 36
    sample, 34
    sampling distribution, 225, 226, 227
    standard error of, 230
    tests, 263, 278, 279
    weighted, 40
Mean deviation, 66, 75, 76
Mean squares, 343
Measures of central tendencies, 32
Measures of location, 31

Measures of position, 32
Measures of relative variation, 74
Measures of variation, 31, 63
Median:
  and mean, 49
  definition, 48
  grouped data, 51
  in decision making, 166
  population, 49, 365
  properties, 49
  sampling distribution, 236
  standard error of, 236
Median test:
  two-sample, 367
  k-sample, 369
Method of elimination, 394
Method of least squares, 391
Mid-quartile, 53, 78
Mid-range, 53
Minimax criterion, 159
Modal class, 53
Mode, 53
  in decision making, 166
Moment about mean, 82
Monte Carlo methods, 238
Moving average, 389, 399
μ (mu), mean of population or prob-
    ability distribution, 35, 187
Multiple correlation coefficient, 435
Multiple regression, 406
Multiplication of choices, 92
Multiplication rule:
  general, 138, 147
  special, 139
Mutually exclusive events, 117

Negative correlation, 424
Negative skewness, 79
Nonparametric methods, 362
  Kruskal-Wallis test, 377
  Mann-Whitney test, 375
  median tests, 367, 369
  run tests, 380, 382
  sign tests, 363, 365
Normal correlation analysis, 427
Normal distribution, 198, 199
  and binomial distribution, 213
  areas under, 200
  fitting of, 336
  standard form, 200
  symmetry, 201
  table, 473
Normal equations, 392, 401, 403, 407
Normal regression analysis, 413
Null hypothesis, 269
Numbers, random, 220
  table, 489
Numerical distribution, 10

Observed frequencies, 10
OC-curve, 267
Odds and probabilities, 106, 107
Ogive, 22

One-sample sign test, 363
One-sided alternative, 271, 276
One-sided test, 272
One-tail test, 272
One-way analysis of variance, 341
Open class, 12, 39
Operating characteristic curve, 267
Opportunity loss, 163
Optimum allocation, 452
Outcome of an experiment, 113

Paired data, 392
Parabola, 388
  fitting of, 400, 403
  maximum (minimum) of, 405
Parameter, 35
Partial correlation coefficient, 436
Pascal's triangle, 101
Peakedness, 83
Pearsonian coefficient of skewness, 81
Percentage distribution, 16
Percentiles, 52
Perfect information, expected value of,
    163
Permutations, 95
  indistinguishable objects, 99
Personal probability, 106, 112
Pictogram, 29
Pie chart, 28
Point estimate, 249
Poisson distribution, 180
  and binomial distribution, 180
  mean, 189, 194
  parameter, 181
  standard deviation, 194
Polygon, frequency, 19
Polynomial curve fitting, 400
Pooled estimate, 319
Population:
  definition, 32
  finite, 219
  infinite, 219
  mean, 34
  median, 49, 365
  size, 32, 35
  standard deviation, 67, 193
  variability of, 64
  variance, 67
Position, measures of, 32
Positive correlation, 424
Positive skewness, 79
Posterior distribution, 307
Postulate, 124
Postulates of probability, 124
  countably infinite sample space, 180
Power function, 268, 403
  fitting of, 403
Prediction, limits of, 418
Price relative, 44
Prior distribution, 257, 306
Probability:
  and area under curve, 197
  and odds, 106, 107
  classical concept, 112

Probability: *cont.*
 conditional, 134, 136
 consistency criterion, 126, 132
 equiprobable events, 102, 112
 frequency interpretation, 103, 112
 general addition rule, 130
 general multiplication rule, 138
 generalized addition rule, 127
 independence, 138
 of empty set, 126
 of complement, 126
 of nonrepeatable event, 105
 personal, 106, 112, 154
 postulates of, 124
 Rule of Bayes, 141, 142
 special multiplication rule, 139
 special rule for equiprobable events,
  102, 129
 subjective, 106, 112, 154
Probability density, 196, 197
 chi-square, 289
 exponential, 207
 F, 294
 mean and standard deviation, 198
 normal, 198, 199, 200
 triangular, 206
 uniform, 206
Probability distribution (*see* Probability
  density *and* Probability function)
Probability function, 172
 binomial, 174
 geometric, 184
 hypergeometric, 178, 179
 mean of, 187
 Poisson, 180
 standard deviation, 189
 variance, 189
Probability paper, arithmetic, 204
Probable error, 237
Proportions:
 Bayesian estimate, 306
 confidence interval, 300, 303
  table, 481, 482
 difference between, 318
 maximum error of estimate, 304
 standard error, 303
 tests for, 313, 314
Proportional allocation, 451

Qualitative distribution, 10, 16
Quality control, 65
Quantitative distribution, 10
Quantity index, 44
Quartile variation, coefficient of, 78
Quartiles, 52
Quota sampling, 454

r (*see* Correlation coefficient)
Random numbers, 221
 table, 489
Random sample:
 finite population, 220
 infinite population, 223
 simple, 200

Random variable, 170
 continuous, 196
 discrete, 195
Randomization, 458
Randomness, tests of, 380, 382
Range, 65
Range, interquartile, 78
Range, semi-interquartile, 78
Rank correlation coefficient, 433
 standard error, 434
Rank-sum tests, 373
Raw data, 10
Real class limits, 14
Regression, 386
 analysis, 413
 coefficients, 412
  confidence intervals for, 415
  tests, 415
 line, 412
  slope of, 414
  true and estimated, 412
  y-intercept, 415
 multiple, 406
 non-linear, 400
Regression analysis, normal, 413
Regret, 163
Relative frequency, 299
Relative variation, measures of, 74
Reliability, 36, 49, 65
Replication, 352, 459
 sum of squares, 359
ρ (rho), population correlation coefficient,
  427
Root-mean-square deviation, 66
Rule of Bayes, 141, 142
Runs, 380
 above and below median, 382

Sample:
 covariance, 422
 definition, 32
 mean, 34
 proportion, 299
 random, 220, 223
 range, 65
 size, 32, 34, 35
  estimation of mean, 252
  estimation of proportion, 305
 standard deviation, 67
 variance, 67
Samples:
 independent, 280
 paired, 286
Sample space, 144
 countably infinite, 180
 finite, 116
 infinite, 116
Sampling:
 area, 453
 bias in, 5
 cluster, 453
 judgment, 454
 quota, 454
 random, 220, 223

Sampling: *cont.*
  stratified, 449
    optimum allocation, 452
    proportional allocation, 451
  systematic, 449
  with replacement, 223
Sampling distribution, 225
  difference between means, 280
  difference between proportions, 318
  experimental, 225
  mean, 225
  median, 236
  standard deviation, 236
  theoretical, 225
Scores, standard, 74
Semi-interquartile range, 78
Semi-log graph paper, 400
Set:
  complement, 118
  distributive laws, 119
  empty, 111
  intersection, 118
  union, 117
Sheppard's correction, 72, 84
σ (sigma), standard deviation of population or probability distribution, 67, 189
Σ (sigma), summation sign, 34, 59
Sign tests:
  one-sample, 363
  two-sample, 365
Significance, level of, 272
Significance test (*see also* Tests of hypotheses), 270
Simple aggregative index, 44
Simple random sample, 220
Simulation, 238
Size:
  population, 32
  sample, 32, 34
Skewness:
  α_3 (alpha-three), 82
  Pearsonian coefficient, 81
  positive and negative, 79
Slope of regression line, 414
Small-sample confidence interval:
  mean, 255
  standard deviation, 290
Small-sample test:
  difference between means, 281
  mean, 279
  proportion, 313
  standard deviation, 294
Spearman's rank correlation coefficient, 433
Special multiplication rule, 139
  generalization, 139
Square root table, 497
  use of, 496
Standard deviation:
  binomial distribution, 190
  confidence interval for, 290, 291
  definition, 66

Standard deviation: *cont.*
  estimation of, 288
  grouped data, 71
  Poisson distribution, 194
  population, 67
  probability distribution, 189
  probability density, 198
  sample, 67
  sampling distribution of, 236
  short-cut estimates, 292, 293
  significance tests, 294, 297
  standard error, 236, 237, 291
Standard deviations, equality of, 294
Standard error:
  difference between means, 280
  difference between proportions, 318
  mean, 230
  median, 236
  proportion, 303
  standard deviation, 236, 237, 291
Standard error of estimate, 414
Standard normal distribution, 200
  table, 473
Standard scores, 74
Standard units, 74
State of Nature, 159
Statistic, 35
Statistics:
  descriptive, 3
  inductive, 4
Statistical inference, 4
Stratified sampling, 450
  cross stratification, 453
  optimum allocation, 452
  proportional allocation, 451
Student-*t* distribution (*see t* distribution)
Subjective prior distribution, 257
Subjective probability, 106, 112
  consistency criterion, 126
  determination of, 154
Subscripts, 58
  double, 59
Subset, 116
Sum of squares:
  block, 350, 353, 461
  column, 461
  error, 341, 344, 350, 353, 359, 461
  interaction, 353
  replication, 359
  row, 461
  total, 341, 344, 353, 461
  treatment, 341, 344, 353, 461
Summation, 34, 59
  double, 59
Survey, design of, 4
Symmetrical distribution, 79
Systematic sampling, 449

t distribution, 254, 279
  degrees of freedom, 254, 279
  reference to theory, 287
  table, 475
t statistic, 254, 279, 281, 413, 427

Table, analysis of variance:
  Latin square, 462
  one-way, 342
  two-way with interaction, 353
  two-way without interaction, 350
  with replication, 360
Tests of hypotheses:
  analysis of variance, 343
  correlation coefficient, 427
    rank, 434
  differences between means, 280, 281
  differences between proportions, 319, 320
  equality of variances, 272
  goodness of fit, 329
  Kruskal-Wallis test, 377
  level of significance, 272
  Mann-Whitney test, 375
  means, 263, 278, 279
  median tests, 367, 369
  nonparametric, 362
  null hypothesis, 269
  one-sided, 272
  one-tail, 272
  operating characteristic curve, 267
  proportions, 313, 314
  randomness, 380
  rank correlation coefficient, 434
  runs, 380
  sign tests, 363, 365
  significance tests, 270
  standard deviations, 294
  two-sided, 272
  two-tail, 272
  Type I error, 266
  Type II error, 266
Theoretical sampling distribution, 225
Three-dimensional sample space, 115
Ties in rank, 374
Total sum of squares:
  Latin square, 461
  one-way analysis, 341
    unequal sample sizes, 344
  two-way analysis with interaction, 353
Treatments, 341
Treatment effects, 338, 349
Treatment sum of squares:
  Latin square, 461
  one-way analysis, 341
    unequal sample sizes, 344
  two-way analysis with interaction, 353
Tree diagram, 88
Trend, 382
Triangular density, 206

Two-dimensional sample space, 115
Two-sample median test, 367
Two-sample sign test, 365
Two-sided alternative, 271, 276
Two-sided test, 272
Two-tail test, 272
Two-way analysis of variance, 348
  with interaction, 353
  without interaction 349, 350
Two-way frequency distribution, 441
Type I error, 266
  probability of, 272
Type II error, 266
  probability of, 273

u, total number of runs, 380
U statistic, 374
  sampling distribution, 375
U test, 373
U-shaped distribution, 81
Unbiased estimate, 67
Unequal classes, 18, 39
Uniform density, 206
Union, 117
Units, standard, 74
Units test, 45
Universe, 33
Upper class limits, 14

Variable, random, 170, 195, 196
Variance:
  binomial distribution, 190
  Poisson distribution, 194
  population, 67
  probability distribution, 189
  sample, 67
Variance, analysis of (see Analysis of variance)
Variance ratio, 294, 340
Variation:
  chance, 64
  coefficient of, 74
  measures of, 31, 63
Venn diagram, 119

Weight, 40
Weighted aggregative index, 47
Weighted mean, 40

z-scores, 74
Z-transformation, 428
  table, 487